AGROFORESTRY AND FORESTRY

AGROFORESTRY AND FORESTRY

Editor

Marius Portner

scitus
academics

Agroforestry and Forestry
Edited by **Marius Portner**

This Book is Printed on Acid-Free Paper. ♾

ISBN: 9781681179346
Library of Congress Control Number: 2017949072

616, Corporate Way, Suite 2, 4766,
Valley Cottage, NY 10989

www.scitusacademics.com

Notice

Printed in United States of America.

Preface

Agroforestry (AF) is an ecofriendly and sustainable modern farming land use practice that maintains or increases total yields by combining food crops (annuals) with tree crops (perennials) and/or livestock on the same unit of land, either alternately or at the same time, using management practices that suit the social and cultural characteristics of the local people and the economic and eco-logical conditions of the area. Scientifically speaking, agroforestry is derived from ecology and is one of the three principal land-use sciences, the other two being agriculture and forestry. In this system, agricultural crops are intercropped with tree crops in the interspace between the trees. Under this system agricultural crops can be grown upto two years under protective irrigated condition and under rainfed farming upto four years. The crops can be grown profitably upto the above said period beyond which it is uneconomical to grow grain crops. However fodder crops, shade loving crops and shallow rooted crops can be grown economically. Over the past two decades, a number of studies have been carried out analysing the viability of agroforestry. The combined research has highlighted that agroforestry can reap substantial benefits both economically and environmentally, producing more output and proving to be more sustainable than forestry or agricultural monocultures. Agroforestry systems have already been adopted in many parts of the world. However, forest ecosystems have come to be seen as the most important component of the biosphere, and forestry has emerged as a vital applied science, craft, and technology. Forestry is an important economic segment in various industrial countries.

This volume 'Agroforestry and Forestry' demonstrate the role of agroforestry in providing commodity as well non-commodity benefits such as ecosystem services and forest ecological aspects and conservation and looks into new techniques for conserving the forests. This book will bridge the gaps in the knowledge about some new emerging issues on forest ecology and conservation. It covers the management of forest resources, within sites and across landscapes, to provide sustainable, desirable habitat conditions for all forest-dependent fauna while concurrently yielding economically viable, quality timber products. This volume will be an interesting and helpful resource to all those in the field of forestry working for its sustainable use and conservation.

Table of Contents

CHAPTER 1

Factors Influencing Biomass and Carbon Storage Potential of Different Land Use Systems along an Elevational Gradient in Temperate Northwestern Himalaya

Bhalendra Singh Rajput, D. R. Bhardwaj, Nazir A. Pala

Department of Agronomy (Agroforestry), BHU, Rajiv Gandhi South Campus, Barkachha, Mirzapur, U.P., India

ABSTRACT

We observed the influence of five different altitudes and prevailing agro ecosystems on biomass and carbon sequestration potential in Kullu district of Himachal Pradesh, India. The study area had five prevailing land uses viz., agriculture, agro-horticulture, horticulture, silvi-pasture, and forest at four elevations representing about 1 °C temperature change. The results showed that maximum total biomass of 404.35 Mg C ha^{-1} was accumulated by forest landuse and followed a decreasing trend in the order as forest > silvi-pasture > agro-horticulture > horticulture > agriculture.
Similar trends were also seen with respect to biomass carbon (C) density and C-sequestration potential of different land uses. Biomass and carbon density potential enhanced with the increase in the altitudinal ranges from

1100–1400 to 2000–2300 m a.s.l. But, the rate of C-sequestration potential enhanced from 1100 to 2000 m and declined at 2000–2300 m a.s.l. Maximum carbon density (393.29 Mg C ha^{-1}) of both plant as well as soil was displayed by the forest-based land use systems situated at an altitudinal gradient of 2000–2300 m a.s.l. The rate of C-sequestration was maximum (2.17 Mg ha^{-1}) in the agro-horticulture at 2000–2300 m a.s.l. This study brings out the potential of different land use systems influenced by varying factors on their C-sequestration potential in western Himalayan elevation gradient, thereby providing useful information for effective management in a climate change mitigation and carbon budget.

Keywords: Biomass, Carbon density, Land use, Climate change, Himalaya

INTRODUCTION

Many environmental factors viz., temperature, precipitation, atmospheric pressure, solar radiation, and wind velocity change systematically with altitude. Therefore, altitudinal gradients are among the most powerful 'natural experiments' for testing ecological and evolutionary responses of biota to environmental changes (Korner 2007). Although changes in species composition and distribution, biodiversity, and community structure along altitudinal gradients have been well documented, the altitudinal patterns of carbon (C) storage in different land uses remain poorly studied (Zhu et al. 2010). A recent global assessment of the impacts of climate change suggests that during twenty-first century, mountain ecosystems experiences two to three times' greater rates of warming than during the twentieth century (Nogues-Bravo and Araiijo 2006).

Vegetation biomass is a crucial variable for understanding the potential future changes of the climate system. Depending on the quantity of biomass, vegetation cover can have a direct influence on local, regional, and even global climate, particularly on air temperature and humidity (Bombelli et al. 2009). Biomass and carbon storage in forest ecosystems play an important role in the global carbon cycle (Li et al. 2013; Zhao et al. 2014). Soil carbon, whereas, depends on the aboveground input received from plant litter and on the decomposition of fine roots belowground (Rasse et al. 2006). The aboveground tree biomass and belowground root biomass both need to be assessed to enable better estimations of total carbon (Hamburg 2000). Whereas, soil organic C-stock (SOC) exhibits considerable spatial variability both horizontally as well as vertically. The SOC diminishes with depth regardless of vegetation type and soil texture (Trujilo et al. 1997).

Major policy initiatives, including the National Forest Policy 1988, the National Agriculture Policy 2000, Planning Commission Task Force on Greening India 2001, National Bamboo Mission 2002, National Policy on Farmers, 2007 and Green India Mission 2010, emphasize the role of agroforestry for efficient nutrient cycling, organic matter addition for sustainable agriculture, and for improving vegetation cover (Rajput et al. 2015). As mountain regions cover about 27.2 % of the global land area and there have been rapid climate changes in mountain regions during the past few decades (IPCC 2007) understanding the shifts in forest C-storage and allocation along altitudinal gradients in mountain region will help us better predict the response of regional and global C balance to future climate change. Information on variation in biomass and C-stocks along the altitudinal gradients in different land use types of the temperate region is still lacking. Keeping the above facts in view, the study was undertaken in Kullu district of Himachal Pradesh, which falls in temperate region of northwestern Himalaya with the objective of studying the biomass, C-stock, and C-sequestration potential of different land uses along an altitudinal gradient influenced by varying locality factors.

MATERIALS AND METHODS

Study area

The study was carried out in the mountainous region of Kullu district of Himachal Pradesh, India located between 31°58′00″N latitude and 77°06′04″E longitude. The district Kullu forms a transitional zone between the lesser and the greater Himalaya and presents a typical rugged mountainous terrain with moderate to high relief (Kumar 2010; Rajput et al. 2015). Climate varies from hot and dry at (1100–1800 m), moderate between 1800 and 2500 m, and intensely cold between 2500 and 4300 m elevations. The monthly maximum and minimum temperature ranged between 8.5–31.8 and 10–21.5 °C, respectively. The district receives moderate rainfall and bulk of it is received during the months of July, August, December, and January. Soil pH of the area ranges from 6.3 to 6.6 with entisols, inceptisols, and mollisol as the dominant soil types (Kumar 2010; Rajput et al. 2015). The forest in the study area comprised low Himalayan temperate forest, oak forest, upper west Himalayan temperate forest, low-level blue pine forest, west Himalayan high-level dry blue pine forest etc. (Champion and Seth 1968).

Experimental methodology

Four mountainous ranges having altitudinal ranges of 1100–2300 m a.s.l. were selected as replicates. Each range was then further divided into four altitudinal gradients viz., 1100–1400, 1400–1700, 1700–2000, and 2000–2300 m a.s.l. In each altitudinal range, five land use systems viz., agriculture, agro-horticulture, horticulture, silvi-pasture, and forests were selected. This experiment was laid out as randomized block design (factorial experiment), comprising 20 treatment combinations [5 (land use systems) × 4 (altitudinal ranges)] having specific tree-crop combinations.

Estimation of vegetation biomass and soil carbon content

The entire trees falling in the plot (50 × 10 m^2) were enumerated. DBH (diameter at breast height) and height was measured with caliper and Ravi's multimeter, respectively. Local volume equation for specific tree species was used for calculating the volume. Wherever volume equation was not available for the species, form factor was calculated using Pressler (1865) and Bitterlich (1984). Specific gravity values were used to determine the biomass and stem cores were taken to find out specific gravity using maximum moisture method (Smith 1954). Total numbers of branches, irrespective of size were counted on each of the sample tree, and categorized on the basis of basal diameter into three groups, viz., <6, 6–10, and >10 cm. Branch biomass and leaf biomass of forest tree species was measured by methods given by Chidumaya (1990) and Jenkins et al. (2003), respectively. Leaf carbon content was estimated by multiplying with a factor of 0.5 (IPCC default value). The total tree biomass was calculated as the sum of stem biomass, branch biomass, and leaf biomass. The tree biomass was converted into its carbon content by multiplying a factor of 0.5 (IPCC default value). Fruit tree root biomass was determined using the root–shoot (apple = 0.33; plum = 0.35) developed by Rajput et al. (2015).

$$\text{Apple: } Y=1.052X^{1.076}(R^2=0.823)$$

$$\text{Plum: } Y=0.008X^2+1.86X-0.754(R^2=0.894)$$

Fallen leaves and pruned wood under each tree was collected, weighed, subsampled, and oven dried at 65 ± 5 °C to a constant weight. Five plots of 1 × 1 m^2 were used for estimation of crop biomass. All the crop biomass occurring within the borders of the quadrates were cut at ground level and collected samples were weighed, subsampled, and oven dried at 65 ± 5 °C to a constant weight. Belowground biomass of crops and grasses was calculated by multiplying aboveground biomass of crops/grasses with a factor of root:shoot ratio of particular crop/grass (Rajput et al. 2015). C-stock was obtained by multiplying the biomass with the IPCC default value (0.5) and C-sequestration was calculated deducting the C-loss from the system through removal of biomass, fruit yield, or through pruned wood with the total C captured by plants.

Soil samples were collected, air dried in shade, grinded with wooden pestle, passed through 2 mm sieve, and stored in cloth bags for further laboratory analysis. The bulk density (g cm^{-3}) and organic carbon content were estimated by the specific gravity method (Singh 1980) and Walkley and Black (1934) method, respectively. The soil organic C-stock for a specific depth was computed using the formulae given by Nelson and Sommers (1996). The bulk density and carbon concentration data were used to compute amounts of carbon per unit area of land use:

$$C(\text{Mg ha}^{-1})=[\text{soil bulk density (g cm}-3)\times\text{soil depth (cm)}\times C(\%)]\times 100,$$

where C expressed in decimal fraction.

The data obtained were subjected to statistical analysis as per the procedure suggested by Gomez and Gomez (1984).

RESULTS AND DISCUSSION

Biomass production (t ha^{-1})

Total highest biomass (404.35 Mg ha^{-1}) was reported in forest landuse, which is 3.5–4 times more than total biomass accumulation in other perennial component-based land use systems viz., agro-horticulture, horticulture, silvi-pasture; and about 20 times more than annual cropping system i.e., agriculture (Table 1). Highest biomass stock (Mg ha^{-1}) in the forest system may be because of age and tree density that ranged between 90–180 years and 190–650 trees per hectare, respectively. The average total biomass of forest ecosystem (404.35 Mg ha^{-1}) in the present study is on the higher side than other temperate and boreal forest ecosystems (326.0 Mg ha^{-1}) of the world (Anonymous 1999). Higher biomass of our temperate forest ecosystem can be ascribed to low biotic interference and inaccessibility of these landscapes in the past. Whittaker and Marks (1975) cited biomass data for a number of temperate forests that indicated a range of aboveground biomass from 113 to 340 t ha^{-1} for mature pine-oak woodland and a mature spruce-fir forest, respectively. Total biomass production (Table 1) increased with altitudinal ranges from 1100–1400 m to 2000–2300 m a.s.l. The physiographic factors are widely known to show a major impact on plant microhabitat especially in hill slope form (Sharma et al. 2010). Biomass increased with increasing altitude in the present study is also supported by the reported values of Zhu et al. (2010) and Gairola et al. (2011). In our study area dominance of mature large conifers at higher altitude as compared to lower can explain the cause. Different treatment combinations of forest land use system and altitudinal gradient (T_5A) exhibited significantly higher values of biomass density in comparison to all other systems and the values of silvi-pasture and forest land use systems enhanced with the increasing altitudinal ranges (Tables 1).

Table 1 Biomass production, C-density, and C-sequestration of different land use systems

Land use systems	Altitudinal ranges (A) Biomass production (Mg C ha^{-1})				Mean	Altitudinal ranges (A) Carbon density (Mg C ha^{-1} year^{-1})				Mean	Altitudinal ranges (A) C-sequestration (Mg C ha^{-1})				Mean
	A_1 (1100–1400 m)	A_2 (1400–1700 m)	A_3 (1700–2000 m)	A_4 (2000–2300 m)		A_1 (1100–1400 m)	A_2 (1400–1700 m)	A_3 (1700–2000 m)	A_4 (2000–2300 m)		A_1 (1100–1400 m)	A_2 (1400–1700 m)	A_3 (1700–2000 m)	A_4 (2000–2300 m)	
T_1 (agriculture)	29.00	28.45	15.33	8.38	20.28	14.50	13.61	7.63	4.31	10.01	0.39	0.52	0.51	0.56	0.50
T_2 (agro-horticulture)	93.84	103.2	103.3	91.60	97.51	46.92	51.58	51.65	46.04	49.05	1.93	2.11	2.12	2.07	2.08
T_3 (horticulture)	87.49	94.69	95.33	89.92	91.32	44.51	47.41	47.66	44.96	46.13	1.62	1.55	1.57	0.97	1.43
T_4 (silvi-pasture)	102.10	116.15	131.50	143.29	123.31	51.16	58.09	65.77	71.61	61.66	1.26	1.21	1.37	0.93	1.19
T_5 (forest)	322.50	353.03	367.37	574.52	404.35	161.30	176.5	183.6	287.3	202.2	1.78	1.77	1.64	1.60	1.69
Mean	126.60	139.09	142.54	181.11		63.67	69.44	71.27	90.57		1.39	1.42	1.44	1.25	
		SE±		$CD_{0.05}$			SE±		$CD_{0.05}$			SE±			$CD_{0.05}$
T		4.44		8.99			2.20		4.45			0.05			0.10
A		3.97		8.04			1.96		3.98			0.05			0.09
T × A		8.88		17.94			4.40		8.89			0.11			0.22

Biomass carbon density of different land use systems

Maximum biomass carbon density in the forest land use system differed significantly ($P < 0.05$) from other land use systems and followed the trend: forest > silvi-pasture > agro-horticulture > horticulture > agriculture (Table 1). Higher carbon density of the perennial component-based land use systems can be attributed to continuous accumulation of carbon in the woody component. The biomass carbon storage capacity (49.05 Mg C ha^{-1}) as calculated in our fruit-based temperate agroforestry system is similar to the value reported by Sanneh (2007) for fruit-based agroforestry systems (51.85 Mg ha^{-1}) of wet temperate northwestern Himalaya. The total biomass carbon storage potential (202.2 Mg ha^{-1}) recorded in the present study is almost same (i.e. 190 Mg C ha^{-1}) as given by Singh et al. (1994) for Himalayan forest and by Sanneh (2007) for wet temperate Himalayan forest (185.0 Mg C ha^{-1}). But, the value is on higher side in comparison to the average value of 160.0 Mg C ha^{-1} given by Houghton (1995) for world's temperate forest ecosystems. This variation can mainly be due to difference in the nature of the temperate forest ecosystems as average annual temperature of Himalayan forest (warm temperate) is higher than the temperate forest found in other parts of the world. The biomass carbon density increased with increasing altitudinal ranges from 1100–1400 to 2000–2300 m a.s.l. The altitudinal range of 2000–2300 m a.s.l. displayed maximum biomass density (90.57 Mg C ha^{-1}) and is significantly higher than other altitudinal ranges. In the interaction effect, maximum carbon density (287.3 Mg C ha^{-1}) is found on elevation range of 2000–2300 m a.s.l. (Table 1).

Rate of C-sequestration

Maximum rate of C-sequestration potential (2.08 Mg C ha^{-1} $year^{-1}$) has been found in agro-horticulture system followed by forest (1.69 Mg ha^{-1} $year^{-1}$), horticulture (1.43 Mg C ha^{-1} $year^{-1}$), silvi-pasture (1.19 Mg C ha^{-1} $year^{-1}$), and agriculture (0.50 Mg C ha^{-1} $year^{-1}$), respectively (Table 1). The estimate of C-sequestration potential in agroforestry systems are highly variable, ranging from 0.29 to 15.21 Mg C ha^{-1} $year^{-1}$ (Nair et al. 2009), depending on the site characteristics, land use types, species involved, stand age, and management practices. Maximum C-sequestration potential is displayed by agro-horticulture land use system situated at the altitudinal range of 2000–2300 m a.s.l and minimum by agriculture land use system at the altitudinal range of 1100–1400 m a.s.l. The rate of C-sequestration potential in fruit-based agro-horticulture land use is higher than all combinations of forests, silvi-pasture systems. This is because of the fact that agriculture crops are put to intensive management practices resulting into their higher biomass production but whatever produced is removed annually from the system leading to their lower sequestration potential. Whereas, in fruit-based agroforestry systems, which are again put to intensive management, the biomass keep on piling year after year in fruit trees, and only pruned wood and fruits are removed annually resulting into their higher C-sequestration potential.

Soil organic carbon (%)

Maximum organic carbon (2.5 %) is found in the forest system followed by agri-horticulture (0.89 %), horticulture (0.87 %), silvi-pasture (0.81 %), and agriculture (0.68 %) in the descending order (Fig. 1). Leaf litter and root litter input play a major role in forest soil, while agricultural practices such as tillage, FYM, fertilizer inputs, and the return of crop residues determine the SOC dynamics in cultivated soils. Significant increase in organic carbon content in soils under tree-based land use systems may be ascribed to more leaf litter deposition followed by decomposition and root turnover from trees (Rhodes 1995). The maximum accumulation of soil organic carbon (1.32 %) is in the surface layer (0–20 cm), and decreased to 0.98 % at 20–40 cm depth. The greater accumulation of soil organic carbon on the surface is due to the greater incorporation of leaf litter on it. The findings are supported by the results of Minhas et al. (1997) and Shah et al. (2013) for Himalayan region. The organic matter has a significant positive correlation ($r^2 = 0.77$) with altitude (Banerjee et al. 1998). In our study (Fig. 2), we also found that the organic carbon increased with increasing altitudinal ranges, which can be owed to continuous accumulation of leaf litter and slower decomposition rate at higher altitude than at lower ones. The increase in organic matter with altitude has also been reported by Minhas and Bora (1982) in soil profiles of Himachal Pradesh.

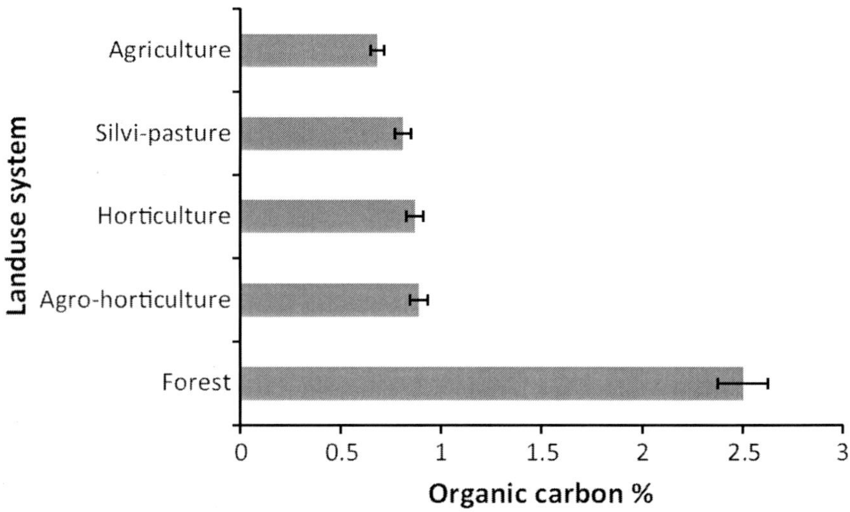

Figure. 1 Organic carbon (%) in different land use systems across the study area

Soil organic carbon pool inventory (Mg ha^{-1})

Maximum soil organic pool in 0–40 cm layer has been recorded in forest system (98.08 Mg ha^{-1}), followed by agro-horticulture (41.05 Mg ha^{-1}), horticulture (39.16 Mg ha^{-1}), silvi-pasture (35.79 Mg ha^{-1}), and agriculture (33.88 Mg ha^{-1}), respectively in the descending order. In general, soil organic C-stock in the 0–40 cm depth showed an increasing trend with increasing altitudinal ranges, except a slight dip at 1700–2000 m elevation range. Perennial plant-based land use systems viz., agri-horticulture, silvi-pasture, horticulture, and forest use systems have displayed higher C-stock than in soils (Table 2). Whereas, in the annual or agriculture-based land use system, the C-stock is higher in soil than in the plant pool. These findings agree with the findings of Houghton (1995) and Sanneh (2007) implying that these ecosystems need to be protected and conserved.

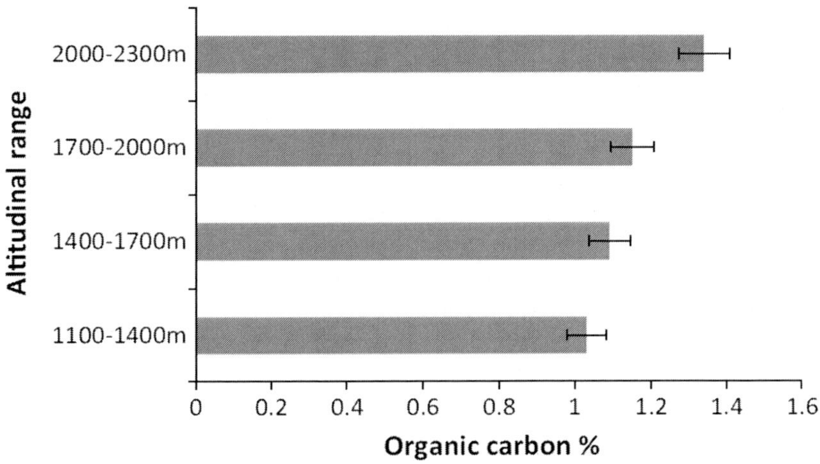

Figure. 2 Altitude-wise status of organic carbon (%) in the study area

Table 2 Soil (+leaf litter) and biomass carbon (Mg C ha^{-1}) inventory under different land use systems

Land use systems	Altitudinal ranges																Mean Soil:plant ratio	Mean Total
	A_1 (1100–1400)				A_2 (1400–1700)				A_3 (1700–2000)				A_4 (2000–2300)					
	Plant	Soil	Total	Soil:plant ratio	Plant	Soil	Total	Soil:plant ratio	Plant	Soil	Total	Soil:plant ratio	Plant	Soil	Total	Soil:plant ratio		
T_1—agriculture	14.50	32.53	47.03	2.24	13.61	33.34	46.95	2.45	7.64	33.88	41.52	4.43	10.01	39.14	49.15	3.19	2.61	36.93
T_2—agro-horticulture	45.89	40.49	86.38	0.88	51.58	41.19	92.77	0.80	51.65	45.02	96.67	0.87	44.96	43.33	88.29	0.96	0.70	72.82
T_3—horticulture	43.83	37.53	81.36	0.81	47.41	38.86	86.27	0.82	47.66	42.09	89.75	0.88	44.38	41.23	85.61	0.93	0.71	68.60
T_4—silvi-pastural	51.06	36.25	87.31	0.71	58.09	42.19	100.28	0.73	65.77	37.41	103.18	0.57	71.61	38.32	109.93	0.54	0.51	80.14
T_5—forest	161.30	94.15	255.45	0.58	176.50	104.14	280.64	0.59	183.60	98.98	282.58	0.54	287.30	105.99	393.29	0.37	0.42	303.39
Mean	63.31	48.19	111.51	1.06	69.44	51.94	121.38	1.08	71.27	51.48	122.74	1.46	63.31	53.60	145.25	1.34		

Total C-stock (Mg ha^{-1})

Mean total C-stock (303.39 Mg C ha^{-1}) is maximum in the forest system, which is around 2.5–4 times higher than perennial plant-based systems (Table 2). The average total C-stock of the forest in the mountainous ecosystem is 303.39 Mg C ha^{-1}, which is slightly higher than reported by Houghton (1995) for temperate evergreen forest (294.1 Mg C ha^{-1}). The C-sequestration potential of our horticulture-based land use systems ranged between 72.8 and 80.1 Mg C ha^{-1}. The C-sequestration potential of tropical agroforestry was estimated between 12 and 228 Mg C ha^{-1} with a medium value of 95 Mg C ha^{-1} by Albrecht and Kandji (2003).

The data presented in the Table 2 reveal that as we moved from lower elevation range (1100–1400 m) to higher elevation range (1700–2000 m), the ratio between soil and plant kept on increasing and further slightly decreased at (2000–2300 m) elevation range. Increasing soil–plant ratio indicates greater vulnerability of the soil pool to CO_2 emission. If in future temperature increases then there will be net emission of CO_2 from soil into the atmosphere, because of greater mineralization from the soil and this emission rate will be higher between 1400–1700 and 1700–2000 m a.s.l. altitudinal ranges. This shift between one altitudinal range to another altitudinal ranges signifies the difference of approximately I °C. In the mountainous ecosystem, elevation range of 2000–2300 m a.s.l. shows a higher total carbon density level in all the land use systems. Thus, this elevation range also needs greater attention from environmental point of view.

CONCLUSION

The land use systems viz., forest and silvi-pasture particularly at higher elevations is store house of C-stocks in both plant as well as soil, which needs to be conserved for environmental protection. From carbon dioxide mitigation point of view, agro-horticulture land use systems are found better than agriculture, horticulture, silvi-pasture, and forest land use systems at all the altitudinal gradients. It shows that fruit-based systems that are a common land use in many parts of the temperate ecosystem including Himalayan region with a significant economic, nutritional, and food security role can also provide an important environmental service through mitigation of atmospheric CO_2. In a broader context, these results have implications on the role of tree-based systems in greenhouse mitigation through soil and plant carbon storage. The results of the present study will be helpful for understanding the pattern of C-storage and C-sequestration potential in various land uses of temperate regions in other parts of Himalayan region having similarity in locality factors.

REFERENCES

1. Albrecht A, Kandji ST (2003) Carbon sequestration in tropical agroforestry systems. Agric Ecosyst Environ 99:15–27
2. Anonymous (1999) State of forest report—Govt. of India: forest survey of India. Ministry of Environment and Forests, Dehradun, pp 49–52
3. Banerjee SK, Gupta RD, Jha MN, Das TH (1998) Status, nature and characteristics of soil organic matter in Himalayan and sub-Himalayan region. Bull Indian Soc Soil Sci 19:14–30
4. Bitterlich W (1984) Relaskop idea slough: commonwealth Agricultural Bureaux. Farnham Royal, Slough
5. Bombelli A, Avitabile V, Balzter H (2009) Biomass. In: di Caracalla VDT (ed) Assessment of the status of the development of the standards for the terrestrial essential climate variables (T12). Global Terrestrial Observing System, Rome
6. Champion HG, Seth SK (1968) A revised survey of the forest types of India. Government of India Press, New Delhi
7. Chidumaya EN (1990) Above ground woody biomass structure and productivity in Zambezian Woodland. For Ecol Manage 36:33–46
8. Gairola S, Sharma CM, Ghildiyal SK, Suyal S (2011) Live tree biomass and carbon variation along an altitudinal gradient in moist temperate valley slopes of the Garhwal Himalaya (India). Curr Sci 100:1862–1870
9. Gomez KA, Gomez AA (1984) Statistical procedure for agriculture research, 2nd edn. Wiley, New York
10. Hamburg SP (2000) Simple rules for measuring changes in ecosystem carbon in forestry-offset projects. Mitig Adapt Strat Glob Change 5:25–37
11. Houghton RA (1995) Changes in the storage of terrestrial carbon since 1850. CRC/Kewis Publishers, Boca Raton
12. IPCC (2007) Climate change 2007: the physical science basis. Cambridge University Press, Cambridge
13. Jenkins JC, Chojnacky DC, Heath LS, Bridsey RA (2003) National scale biomass estimation for United States tree species. For Sci 49:12–35
14. Korner C (2007) The use of altitude in ecological research. Trends Ecol Evol 22:569–574PubMed
15. Kumar A (2010) Hydrological conditions of River Beas and its fish fauna in Kullu Valley, Himachal Pradesh, India. Environ Conserv J 11(3):7–10
16. Li C, Zha T, Liu J, Jia X (2013) Carbon and nitrogen distribution across a chronosequence of secondary lacebark pine in China. For Chron 89(2):192–198
17. Minhas RS, Bora NC (1982) Distribution of organic C and the forms of nitrogen in topographic sequence of soils. J Indian Soc Soil Sci 30:135–139
18. Minhas RS, Minhas H, Verma SD (1997) Soil characterization in relation to forest vegetation in the wet temperate zones of Himachal Pradesh. J Indian Soc Soil Sci 45(1):146–151
19. Nair PKR, Kumar BM, Nair VD (2009) Agroforestry as a strategy for C-sequestration. J Soil Sci Plant Nutr 172:10–23
20. Nelson DW, Sommers LE (1996) Total carbon, organic matter. In: Sparks DL (ed) Method of soil analysis—part 3: chemical methods. Soil Science Society of America Inc., Madison, pp 961–1010
21. Nogues-Bravo D, Araiijo M (2006) Exposure of global mountain systems to climate warming during the 21st century. Glob Environ Change Human Policy Dimens 1:420–428
22. Pressler M (1865) Das Gesetz der Stambildung Leipzig, pp 153
23. Rajput BS, Bhardwaj DR, Pala NA (2015) Carbon dioxide mitigation potential and carbon density of different land use systems along an altitudinal gradient in north-western Himalayas. Agroforest Syst. doi:10.1007/s10457-015-9788-8

24. Rasse DP, Mulder J, Moni C, Chenu C (2006) Carbon turnover kinetics with depth in a French loamy soil. Soil Sci Soc Am J 70(6):2097–2105
25. Rhodes C (1995) Seasonal pattern of nitrogen mineralization and soil moisture beneath *Faidherbia albida* (syn. *Acacia albida*) in Central Malawi. Agroforest Syst 29:133–145
26. Sanneh A (2007) Status of C-stockunder different landuse systems in wet temperate north western Himalaya. M.Sc. Thesis, Dr Y.S. Parmar University of Horticulture and Forestry, Nauni, Solan
27. Shah S, Sharma DP, Pala NA, Tripathi P, Dar MA (2013) Carbon stock and density of soils under chir pine (*Pinus roxburghii* Sargent) forests of Solan forest division, Himachal Pradesh. Ind J Soil Conserv 41(3):279–286
28. Sharma CM, Gairola S, Ghildiyal S (2010) Physical properties of soils in relation to forest composition in moist temperate valley slopes of the Central Western Himalaya. J For Sci 26:117–129
29. Singh RA (1980) Soil physical analysis. Kalyani Publishers, New Delhi, pp 61–62
30. Singh SP, Adhikari BS, Zobel DB (1994) Biomass, productivity, leaf longevity and forest structure in the central Himalaya. Ecol Monogr 64:401–421
31. Smith DM (1954) Maximum moisture content method for determining specific gravity of small wood samples. Forest Products Laboratory, Madison
32. Trujilo W, Amezquita E, Fisher MJ, Lal R (1997) Soil organic carbon dynamics and land use in the Colombian savannas I. Aggregate size distribution. In: Lal R, Kimble JM, Follett RF, Stewart BA (eds) Soil processes and the carbon cycle. CRC Press, Boca Raton, pp 267–280
33. Walkley AJ, Black A (1934) Estimation of soil organic carbon by chromic acid titration method. Soil Sci 37:29–38
34. Whittaker RH, Marks PL (1975) Methods of assessing terrestrial productivity. In: Lieth H, Whittaker RH (eds) Primary productivity of the biosphere. Springer-Verlag, Berlin, pp 56–118
35. Zhao J, Kang F, Wang L, Yu X, Zhao W, Song X, Zhang Y, Chen F, Sun Y, He T, Han H (2014) Patterns of biomass and carbon distribution across a chronosequence of Chinese pine (*Pinus tabulaeformis*) forests. PLoS One 9(4):e94966. doi:10.1371/journal.pone.0094966PubMedPubMedCentral
36. Zhu B, Wang X, Fang J (2010) Altitudinal changes in carbon storage of temperate forests on Mt Changbai, Northeast China. J Plant Res 123:439–452

CHAPTER 2

Poplar Growth and Wood Production on a Grassland Irrigated for Decades with Potato Starch Wastewater

Gabriela Lorenc-Plucin´ska, Marta Walentynowicz, Andrzej Lewandowski

Institute of Dendrology, Polish Academy of Sciences, Parkowa 5, 62-035 Ko´rnik, Poland

ABSTRACT

Irrigation of grasslands with potato starch wastewater causes changes in soil quality parameters, often resulting in decreased crop yields and sometimes causing animal diseases. Reduced agricultural income leads to designation of such grasslands for afforestation aimed at production of bioenergy crops and improvement of soil quality. In this study, *Populus alba* L. and seven different poplar clones were planted in 2009 on the grassland irrigated in 1973–2008 with potato starch wastewater and in non-irrigated, experimental forest. The survival, growth and wood production potential of planted poplars after 2–3 growing seasons were analyzed. Morphophysiological parameters of leaves, nutrients in leaves and fine roots, sugars in fine roots, leaf rust resistance, frost hardiness and their effects on growth were also determined. Stem diameter and stem volume index of poplars growing on the grassland were higher than in the experimental forest. Only *Populus deltoides* Bartr. × *P. maximowiczii* Henry clone 'Eridano' (in short ERI) had a survival rate of 100 % and the highest values of stem diameter (77 mm) and stem volume index (0.034 m^3) in comparison with other poplars in both sites after 3 growing seasons. Clone ERI was also characterized by a high leaf rust resistance and frost hardiness. The results suggest that clone ERI is suitable as a bioenergy crop on grasslands irrigated with potato starch wastewater. Afforestation of the grassland improved the fertility of the soil by increasing concentrations of soil organic matter and availability of mineral nutrients (N, P, K, S and Fe).

Keywords: *Populus,* spp., Wastewater, Marginal and degraded land, Land-use change

INTRODUCTION

Fruit and vegetable processing industry, wineries, distilleries as well as olive oil and starch production use huge volumes of water and generate large quantities of wastewater (Hamilton et al. 2011; Kretschmer et al. 2002). Globally, production of olive mill wastewater itself attains about 7–30 million × 10^3 L per year (Mateo and Maicas 2015). The potato processing industry produces 8–28 × 10^3 L of wastewater per every 1000 kg of raw potatoes (Charmley et al. 2006). Wastewater reuse in agriculture irrigation is a common practice (Aryal and Reinhold 2015; Buelow et al. 2015). About 20 million ha of land worldwide are irrigated with wastewater (Hamilton et al. 2011).

Annual production of starch amounts to 72 million tones in a global scale and approximately 10 million tones in the European Union (EU) (The European Starch Industry Association 2014). In the EU, starch potatoes make 31 % of agricultural raw materials and the share of potato starch in the total starch production is about 20 % (The European Starch Industry Association 2014). In industrial starch production from potato (*Solanum tuberosum* L.), huge amounts of potato starch wastewater (PSW) are derived from potato washing, rinsing and starch processing. PSW is rich in organic substances, inorganic suspensions and macronutrients (Arienzo et al. 2009). Because of the high concentrations of organic substances and nutrients, attempts were made to use PSW as substrate for production of microorganisms (Chang et al. 2008; Vu et al. 2010) and electricity (Lu et al. 2009). However, most often PSW is used to irrigate and fertilize fields (Rosenwinkel et al. 2005).

Unfortunately, PSW application for many years leads to unfavorable changes in nutrient content of the soil. The wastewater usually includes high concentrations of N (1200 mg L^{-1}) and K (1800 mg L^{-1}) (Arienzo et al. 2009). Potassium from PSW may be immediately available for plants, so irrigation with PSW may also lead to increasing the total and bioavailable concentrations of K in soil (Arienzo et al. 2009). However, this surplus K concentration may induce Mg deficiency in soil (Römheld and Kirkby 2010) and a decrease in Mg, Ca, Zn, Al and Mo levels in plants (Öborn et al. 2005).

Potatoes, legumes and grasses are K accumulators, but grasses accumulate K mainly in leaves (Arienzo et al. 2009). Fodder production and grazing from grassland after long-term irrigation with PSW may lead to various animal diseases and lower growth rate of livestock (Öborn et al. 2005). At the time of increasing competition, the suboptimal productivity and insufficient finances for balanced fertilization of arable land and grassland lead to their exclusion from agricultural use and designation for afforestation aimed at biomass production for energy (Broeckx et al. 2012; Dillen et al. 2013).

Poplars are some of the most promising woody species that can be grown in environments of varying fertility, with relative tolerance to biotic and abiotic stress conditions and high biomass and stem-wood production. In Europe and North America, high yields of biomass from rural areas were obtained for poplar clones *Populus* × *interamericana* Brockh clones 'Hoogvorst' (in short HOO), 'Beaupré' (in short BEA) and 'Boelare' (in short BOE) (Dillen et al. 2013; Johansson and Karačić 2011; Truax et al. 2012), *Populus* × *euramericana* (Dode) Guinier clone 'Robusta' (in short ROB) (Broeckx et al. 2012; Kačík et al. 2012; Verlinden et al. 2013), *Populus maximowiczii* × *P. trichocarpa* Torr. et Gray clone 'Hybrid 275' (also known as 'NE 42', in short NE) (Baum et al. 2012; Gruenewald et al. 2007), *Populus alba* clone 'Villafranca' (in short VIL) (Kačík et al. 2012; Rosso et al. 2013), and *Populus deltoides* Bartr. × *P. maximowiczii*Henry clone 'Eridano' (in short ERI) (Coyle et al. 2006a, b; Kaczmarek et al. 2013). Therefore our working hypothesis was that these poplar clones can be also suitable for growth on grassland irrigated for a long time with PSW.

Biomass production depends on poplar clones' ability to cope with climatic conditions and diseases (Coyle et al. 2006b; Zalesny et al. 2009). For example, frost, drought, or leaf rust infection can distinctly lower yields (Aylott et al. 2008; Coyle et al. 2006b. Poplar clones HOO, BEA, BOE, ROB, NE, VIL and ERI grow in cutting orchards and in an old poplar plantation of our Institute of Dendrology (ID PAS). Therefore, their susceptibility to leaf rust and extreme weather conditions sometimes occurring in Poland was known to us.

The first objective of the present study was to determine which poplar clones are the most productive on the grassland irrigated with PSW. Therefore we evaluated the survival rate of selected poplar clones, their height, stem diameter and volume. Secondly, we investigated the effect of frost hardiness, leaf rust resistance as well as leaf and root morphophysiological characteristics on poplar growth and biomass production. We planted rooted cuttings, as differences in rooting ability of poplars may also affect their growth potential (Zalesny Jr and Bauer 2007). Our third objective was to investigate the effect of poplar planting on soil quality related to soil mineral concentrations and enzymatic activities (phosphomonoesterases) important in the metabolic capacity of the soil and for nutrient cycling (Nannipieri et al. 2012). Afforestation of agricultural land leads to an increase, decrease or no change in soil organic matter, carbon and nitrogen content in the first few years, depending on soil quality, soil management and species of tree planted (Laganière et al. 2010). Poplars are fast-growing trees and their extensive root system as well as abundant leaf fall improve the soil structure and increase nutrient cycling (Truax et al. 2012). Finally, the field experiment with different poplar clones for 3 growing seasons (GS1, GS2, and GS3) on the grassland irrigated for decades with PSW is compared with a parallel experiment in a non-irrigated, experimental forest (where other poplars were grown previously). The two sites differ in soil texture, pH and concentrations of essential mineral elements. We discuss the suitability of selected poplar clones for survival, biotic and abiotic tolerance, growth and wood production potential in both environments as well as for revitalization of the grassland soil irrigated with PSW.

MATERIALS AND METHODS

Site description

The potato processing plant in Trzemeszno (52°33′N, 17°49′E) was established about 130 years ago and since 1973 the industrial wastewater after treatment has been sprayed onto 170 ha of grassland and about 500 ha of agricultural land. In the last 3 years before our study, according to data from the processing plant, it generated ca. 3400 m^3 of PSW per day, with pH varying from 4.71 to 5.86, biological oxygen demand (BOD_5) ranging from 2015 to 3777 mg O_2 dm^{-3}, relatively high concentrations (mg dm^{-3}) of N (515–751), P (128–361), K (1613–2222) and antifoaming agents in excess. Irrigation with PSW was repeated many times (about 90 days per year), usually from September to December.

Approximately 0.55 ha of grassland irrigated for decades with PSW (hereinafter referred to as grassland) was allotted for the study and the last sprinkling irrigation of this area with PSW was 4 months before its planting with poplars. The grassland was dominated by perennial ryegrass (*Lolium perenne* L.), white clover (*Trifolium repens* L.), cocksfoot (*Dactylis glomerata* L.) and in some places by couch grass (*Elymus repens* (L.)

Gould). Prior to poplar planting, the grassland was disked. In adjacent grassland, there were no areas non-irrigated with PSW that could be used as reference fields. Therefore a similar plot to the grassland was established in the experimental forest (hereinafter referred to as forest) of ID PAS in Kórnik (52°14′N, 17°05′E). The forest site was formerly used for cultivation of poplar trees of other varieties than those used in the present study, for about 20 years. Those trees were harvested for lumber, plywood, pallets and fuel in 2006. The site was then ploughed and disked, and white mustard was grown there. After 2 years, the mustard biomass was ploughed into the soil as green manure.

Climatic data were obtained from weather stations nearest to the study sites, i.e. in Kórnik (52°14′N 17°05′E) for forest and in Gniezno (52°31′N 17°36′E) for grassland. During the study period (2009–2011), mean annual temperatures ranged from 7.9 to 9.7 °C in Kórnik and from 6.9 to 8.8 °C in Gniezno (Fig. 1 in supplementary material). Annual precipitation was 577 mm in 2009, 751 mm in 2010 and 451 mm in 2011 in Kórnik, compared to 603 mm in 2009, 709 mm in 2010 and 427 mm in 2011 in Gniezno.

Soil analysis

Soil samples were collected at 16 points on each site from the depth of 0–25 cm (without undecomposed plant and animal parts) immediately before poplar planting in March 2009. Soil pH and concentrations of N_{Total} (organic and $N-NH_4$), C_{Total}, S_{Total}, P_{Total}, bioavailable forms of N ($N-NH_4$ and $N-NO_3$), S ($S-SO_4$), P ($P-PO_4$) as well as total and bioavailable (EDTA) forms of metals (K, Mg, Ca, Fe, Cu, Cr, Zn, Cd and Pb) were determined exactly as previously described (Lorenc-Plucińska et al. 2013). Each soil sample was analyzed separately (with 3 replications). Pooled samples for each site (6 replications) were used to determine soil texture and total organic matter (OM) as in our earlier study (Lorenc-Plucińska et al. 2013), while bulk density, electrical conductivity and cation exchange capacity as reported by Kalra and Maynard (1991).

The forest site was postagricultural land with sandy loam texture, alkaline pH and a low bulk density (Table 1). The concentrations of OM were high, like N_{Total} and C_{Total}, which resulted in a low C:N ratio. Concentrations of P_{Total}, K_{Total} and S_{Total} were in the normal range for forest soil. However, plant-available phosphorus ($P-PO_4$) was scanty, indicating P deficiency. There was a very high concentration of Ca_{Total} (4.1 %), and thereby of Mg_{Total}. High concentrations of Ca_{Total} resulted in high cation exchange capacity and electrical conductivity, dependent on cation exchange capacity. In a short distance from the experimental area, the soil contains marl and probably calcium carbonate was spread there.

Table 1 Physicochemical characteristics of the forest and grassland soil before poplar planting (March 2009) and after the last growing season (GS3, November 2011)

Characteristic	March 2009		November 2011		p values[a]	
	Forest	Grassland	Forest	Grassland	Forest	Grassland
Texture (% sand:silt:clay)	75:25:0	84:16:0				
pH (H_2O)	7.74 (0.18)	6.2 (0.4)**	7.59 (0.02)	5.4 (0.2)**	0.007	<0.0001
pH (1 n KCl)	7.37 (0.04)	4.9 (0.2)**	7.27 (0.02)	3.9 (0.4)**	<0.0001	<0.0001
Specific gravity (kg dm^3)	1.13 (0.04	1.38 (0.06)**				
EC(μS cm^{-1})[b]	112 (10)	56 (8)**				
CEC (cmol kg^{-1})[c]	16.4 (0.4)	2.3 (0.4)**				
Organic matter (%)	5.2 (0.8)	1.84 (0.32)**	7.3 (1)	2.4 (0.2)**	<0.0001	0.0005
C_{Total} (%)	3.3 (0.6)	0.9 (0.2)**	4.4 (0.4)	1.3 (0.06)**	0.001	<0.0001
C:N	12.3 (1.6)	7.8 (1)**	14.1 (0.6)	10.1 (0.6)**	0.012	0.0002
N_{Total} (%)	0.27 (0.04)	0.12 (0.04)**	0.31 (0.02)	0.13 (0.02)**	0.023	ns
N–NH_4 (mg kg^{-1})	1.9 (0.6)	1.5 (0.6)	1.9 (0.2)	2.4 (0.2)**	ns	ns
N–NO_3 (mg kg^{-1})	9 (2)	5 (1)*	12.4 (2)	33 (2)**	ns	<0.0001
P_{Total} (mg kg^{-1})	539 (60)	635 (96)	511 (78)	1152 (144)**	ns	<0.0001
P–PO_4 (mg kg^{-1})	0.9 (0.6)	11 (4)**	1.9 (0.2)	25.9 (1.4)**	0.003	<0.0001
K_{Total} (mg kg^{-1})	546 (86)	1771 (84)**	657 (36)	1696 (92)**	0.041	ns
K–EDTA (mg kg^{-1})	28 (16)	286 (20)**	44 (6)	386 (20)**	0.031	<0.0001
Mg_{Total} (mg kg^{-1})	1134 (162)	742 (90)**	1095 (46)	810 (72)**	ns	ns
Mg–EDTA (mg kg^{-1})	68 (10)	34 (6)**	64 (4)	36.5 (1.6)**	ns	ns
Ca_{Total} (mg kg^{-1})	41,026 (7436)	1077 (190)**	29,657 (2418)	582 (34)**	0.001	<0.001
Ca–EDTA (mg kg^{-1})	6626 (678)	952 (130)**	7084 (1375)	511 (54)**	ns	<0.0001
S_{Total} (mg kg^{-1})	375 (80)	170 (26)*	217 (20)	102 (30)**	ns	0.021
S–SO_4 (mg kg^{-1})	2.1 (0.8)	1.8 (0.4)	17.9 (1.8)	11.1 (1)**	< 0.0001	<0.0001
Fe_{Total} (mg kg^{-1})	5910 (1390)	3894 (220)*	6348 (852)	4229 (306)**	ns	ns
Fe–EDTA (mg kg^{-1})	34 (6)	64 (16)**	41.9 (1.6)	124 (10)**	ns	<0.0001
Cu_{Total} (mg kg^{-1}	8.1 (0.8)	6.7 (1)				
Cu–EDTA (mg kg^{-1}	0.78 (0.12)	0.99 (0.04)				
Cr_{Total} (mg kg^{-1})	20 (4)	19 (2)				
Cr–EDTA (mg kg^{-1})	0.68 (0.04)	0.15 (0.04)**				
Zn_{Total} (mg kg^{-1})	30 (4)	22 (2)*				
Zn–EDTA (mg kg^{-1})	3.3 (1.0)	3.9 (0.8)				
Cd_{Total} (mg kg^{-1})	0.26 (0.04)	0.32 (0.06)				
Cd–EDTA (mg kg^{-1})	0.108 (0.02)	0.06 (0.04)*				
Pb_{Total} (mg kg^{-1})	15 (2)	5 (2)**				
Pb–EDTA (mg kg^{-1})	1.2 (0.4)	0.91 (0.18)				

Mean and 2 SE values of 16 independent soil samples collected in 2009 and 45 and 41 independent soil samples collected in 2011 from grassland and forest, respectively. Significance of differences between samples from forest and grassland: * $p < 0.05$; ** $p < 0.01$

[a] Significance of differences between samples from 2009 and 2011: p values, ns = not significant

[b] Electrical conductivity

[c] Cation exchange capacity

The grassland soil had mainly sandy loam and sandy textures, irregularly arranged (in 70 and 30 % of the tested samples, respectively), slightly acidic, and had a low cation exchange capacity (Table 1). The concentrations of OM were low, which could be a result of the application of the PWS, which probably also contributed to the relatively high specific gravity of soil. The N_{Total} content was quite low, proportional to the amount of C_{Total}, which gave a narrow range of C:N ratio. The concentration of $N-NH_4$ was normal but $N-NO_3$ level was low. Total phosphorus concentrations were within the normal range, but the concentration of $P-PO_4$ was rather low. Total and bioavailable forms of Ca and Mg_{Total} concentrations were at the lower limit of standards for agricultural soils (Mocek and Drzymała 2010). In contrast, concentrations of both forms of potassium were higher than usual in fields and the K ions determined the cation exchange capacity, although it usually depends on Ca ions. In both sites, soil concentrations of Cu, Cr, Zn, Cd and Pb were within the normal range, but we detected an Fe deficit on the grassland (Kabata-Pendias and Pendias 2001).

Soil analysis was also performed after poplar planting. The fraction of soil adhering to roots of each poplar (rhizosphere) in both sites was collected in November of GS3, simultaneously with the harvest of fine roots. The fresh soil samples were split into two parts: one air-dried at room temperature for elemental analysis, and the other kept moist at −4 °C in a climate chamber for enzymatic analysis. The first part was used to determine soil pH and concentrations of C and C:N ratio, as well as total and bioavailable N, P, S, K, Mg, Ca and Fe. Each soil sample was analyzed in triplicate for element measurements and with 4 replications for the activities of alkaline phosphomonoesterase (EC 3.1.3.1, Alkali–Pase) and acid phosphomonoesterase (EC 3.1.3.2, Acid–Pase). Element concentrations and the Acid–Pase and Alkali–Pase activities were analyzed exactly as earlier described by Lorenc-Plucińska et al. (2013).

Plant material

The native European white poplar (*Populus alba* L., and poplar clones (HOO, BEA BOE, ROB, NE, VIL and ERI) were planted in the forest and grassland in March 2009. The plants derived from dormant hardwood cuttings (20–30 cm long) collected from one-year-old stems on stool beds established at cutting orchards of the ID PAS. The cuttings were rooted in the forest from March 2008 to February 2009. All new shoots except one were removed as they appeared. In early March 2009, the rooted cuttings with single stems of similar height were transplanted onto the grassland or to a new place in the forest. They were planted manually with a shovel at the depth of 35–45 cm. Within each site, each poplar clone was represented by 15 trees. Randomized block design was used, with 3 blocks of 8 poplar species or clones and 5 trees of each clone or species per plot, in 4 single rows (tree spacing: 4 m × 7 m, i.e. 358 trees ha^{-1}). Both sites were fenced to protect the young trees against deer browsing. During the study, poplars in both sites were not irrigated with PSW or water, and no supplemental nutrients were added to the soil. In May 2009 and 2011, glyphosate (a systemic herbicide) was applied around each planted poplar, protected by a covering during the treatment. Once a year the understory between the rows was mown to suppress plant competition. In winter of each year, tree branches were pruned in order to shape the main stem and the tree top, as the crop was to be the single-stem system in 10–15 years. Trees that died in April of the first growing season (GS1) were replaced with new planting, but the replacement trees were not included in the study.

Leaves: chlorophyll, specific leaf area (SLA) and nutrient concentrations

Ten leaves from different crown positions of trees were used for non-destructive chlorophyll measurements with SPAD–502 Minolta (Japan) chlorophyll meter in mid-June of GS3. Three SPAD readings were taken on each leaf and then averaged per leaf. After the readings, 5 leaves were randomly selected and served as one pooled sample for chlorophyll extraction. The other 5 leaves were used for nutrient analysis. Chlorophyll $a + b$ concentrations were analyzed according to Lichtenthaler and Wellburn (1983). In order to convert SPAD values into absolute units of chlorophyll $a + b$ concentrations, calibration curves were derived for each clone and site. Concentrations of C, N, P, K, Ca, Mg and Fe in leaves were determined exactly as in Lorenc-Plucińska et al. (2013). Each leaf sample (pooled from 5 leaves) was analyzed in triplicate.

Right after chlorophyll estimation, additional 10 leaves from different crown positions of trees were collected for morphometric measurements. Projected leaf area (PLA) of scanned leaves (Epson Perfection V700 Photo, Model: J221A, Seiko Epson Corp., Japan) was measured with the software WinFOLIA Pro (Regent Instruments Inc. Canada) and then they were oven-dried at 70 °C for 72 h. The SLA was calculated by dividing the PLA by dry mass of the leaf.

Sunlit, healthy and non-senescent 10 leaves from different crown positions of poplars were also taken for nutrient analysis in mid-August of GS3. Leaves were randomly grouped into two pooled samples (each composed of 5 leaves) and analyzed in triplicate. Elemental analyses were carried out exactly as for the samples harvested in mid-June.

Leaves for chlorophyll, SLA and nutrient concentrations analysis were taken from randomly selected 9 trees for ERI, VIL, ROB and NE (i.e. 3 trees from each block × 3 blocks) on each site and 9 and 5 trees for HOO from grassland and forest, respectively. In the forest site, only 5 trees of HOO survived to GS3.

Fine roots: carbohydrate and nutrient concentrations

The fine root fraction (≤2 mm in diameter; in short: fine roots) from growing poplars was picked out by hand in November of GS3. Fine roots were taken from the same trees that were selected for leaf analysis. After collection, fine roots were manually separated from the soil and quickly and carefully washed in clean water. Living fine roots were separated from dead roots based on their lighter colour and greater resilience. Mycorrhizae associated with roots were regarded as part of the roots. Collected fine roots were analyzed for carbohydrates with 4 replicates, while for element concentrations in triplicate. Soluble carbohydrates and starch were extracted and determined using a standard colorimetric assay exactly as described in Szuba et al. (2013) and expressed as percentage of dry mass. Elemental analyses were carried out exactly as for leaf samples harvested in mid-June.

Analysis of survival, resistance to frost damage and leaf rust, tree growth and stem volume

Survival rate (% of living trees out of the initially planted cuttings) was recorded in April of GS1, GS2 and GS3. Resistance of poplars to frost damage was assessed in March of 2011, because of the very hard winter of 2010/2011, and to late spring frost in May of GS3, because of unusually low temperature on 4 and 5 May 2011 (−4.5 and −2.7 °C, respectively) (Fig. 1). In both cases, damage was rated on a scale of 0–3, where 0 = no damage of the apical buds by frost and young leaves by spring frost, and 3 = severe (>75 %) terminal mortality by frost and blackened leaves by spring frost.

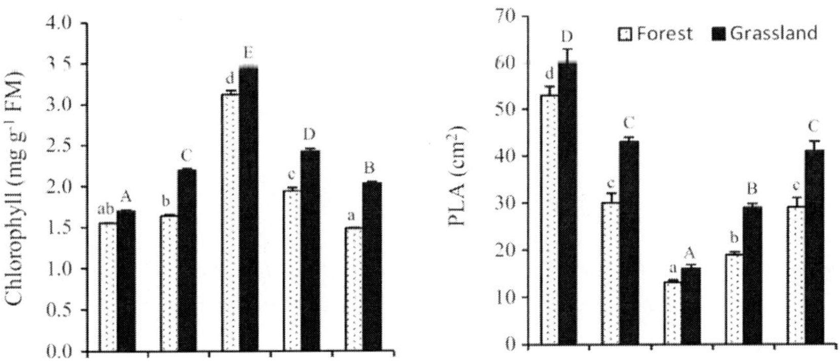

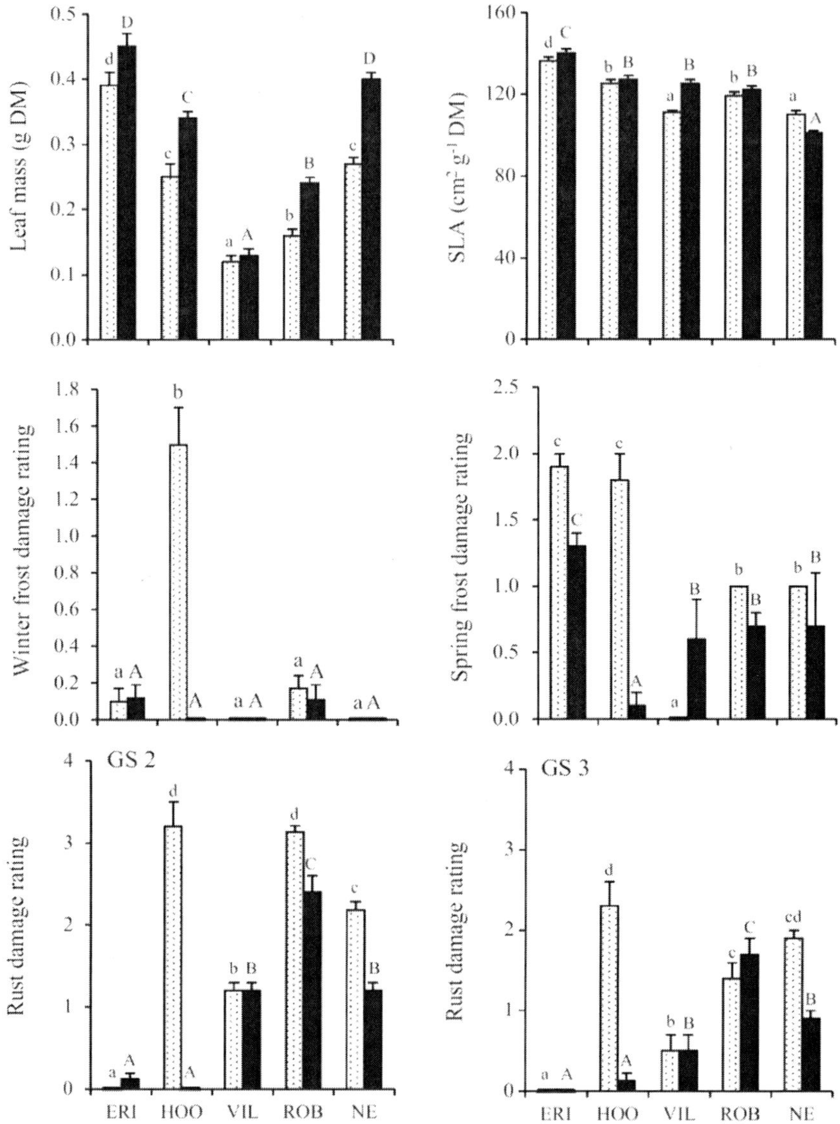

Figure. 1 Total chlorophyll concentration in leaves, projected leaf area (PLA), leaf dry mass, specific leaf area (SLA), as well as winter and spring frost and leaf rust damage in poplars from the forest and grassland. Measurements of chlorophyll, PLA, leaf dry mass, and SLA were carried out in June of the third growing season (GS3). 10 leaves were randomly sampled, from 9 trees of ERI, VIL, ROB and NE clones in both grassland and forest, and for HOO from 9 and 5 trees respectively in the grassland and

forest. Winter and spring frost damage was assessed in March and May of GS3, for 15 trees of ERI and VIL, 5 of HOO, 12 of ROB, and 11 of NE clone from the forest, as well as for 15 trees of ERI and HOO, 11 of VIL and ROB, and 10 of NE clone from the grassland. Leaf rust damage of poplars was assessed in September of the second growing season (GS2) and GS3. In GS2, 30 leaves were randomly sampled, from 15 trees of ERI, HOO, VIL, and ROB and 14 of NE clone in the forest, as well as from 15 trees of ERI, HOO and VIL, 11 of ROB and 14 of NE clone in the grassland. In GS3, numbers of trees were the same as in the assessment of winter and spring frost damage. The results are presented as means with one standard error. Significant differences between data are indicated with different letters (LSD test, $p < 0.05$). FM and DM: fresh and dry mass, respectively

Resistance of poplars to leaf rust (*Melampsora* spp.) was assessed in September of GS2 and GS3. The incidence of rust was recorded for 30 leaves from each crown position by visual assessment using a scale of 0–4, where: 0 = no incidence; 1 = up to 25 % of leaf area infected; 2 = 26–50 % of leaf area infected; 3 = 51–75 % of leaf area infected; and 4 = 76–100 % of leaf area infected. Assessment of winter and spring damages and rust incidence was done for all growing poplars in both sites.

Height and stem diameter at breast height (1.3 m, in short DBH) of all growing trees were measured in December of GS2 and GS3. Tree height was measured with a pole, while diameter, with a digital caliper. To assess wood productivity, potential stem volume index (hereinafter referred to as stem volume) was calculated as tree height × DBH^2 (m^3).

Statistical analysis

Data were analyzed using STATISTICA 10 (StatSoft Inc., USA). All results are presented as mean values with standard error (±SE). Means were compared by using Fisher's least significant difference (LSD) test. Differences were considered significant at $p < 0.05$. Two-way analysis of variance (ANOVA) was used to examine the significance of factors: poplar clones and sites and their interaction. For each site, linear correlations between leaf characteristics were estimated with Pearson's correlation coefficient on a clone mean basis.

RESULTS

Leaves: chlorophyll, SLA and nutrient concentrations

SPAD readings were closely correlated with chlorophyll concentration in the leaves of trees growing in the forest ($r = 0.96$; $p < 0.0001$) and grassland ($r = 0.78$; $p < 0.0001$). Chlorophyll concentrations in the leaves from the grassland were higher than in those from the forest ($p < 0.0001$). On both sites, clone VIL had the highest chlorophyll concentration in leaves and clone ERI had the lowest (Fig. 1).

The leaves of poplar clones growing in the two sites differed in PLA and dry mass (grassland > forest, $p < 0.0001$), while SLA was similar (grassland ≈ forest, $p > 0.05$). Leaves of ERI and NE had the highest and lowest SLA, respectively (Fig. 1). Leaves of ERI were also the heaviest, while VIL leaves were the lightest.

Younger leaves of all poplar clones from the grassland had higher concentrations of N, K and Fe, and lower values of Ca, Mg (except VIL) and P (except ERI and HOO) and C:N ratio, as compared with those from the forest (Table 1 in supplementary material). Older leaves of poplar from grassland were also characterized by higher concentrations of N, K, Mg and P and lower values of Ca content and C:N ratio (Table 1 in supplementary material). Concentrations of C and Fe were similar in leaves from both sites.

Fine roots: carbohydrate and nutrient concentrations

On the grassland, about 90 % of fine root biomass of planted poplar clones was distributed almost directly under the forest litter. This value was particularly high in ERI. In the forest site, nearly 80 % of fine roots were found at 10–15 cm soil depth. Concentrations of soluble carbohydrates (SC), starch and total non-structural carbohydrates (i.e. the sum of SC and starch; TNC) were lower in fine roots of poplar clones from the grassland than in those from forest ($p < 0.0001$ for SC; $p = 0.004$ for starch and $p < 0.0001$ for TNC) (Fig. 2 in supplementary material). In fine roots of poplar clones from grassland, concentrations of C, N, P, K, and Mg were higher, while Ca and Fe content and values of C:N ratio were lower than in those from the forest ($p < 0.0001$ for all elements) (Table 2 in supplementary material).

Survival, resistance to frost damage and leaf rust, tree growth and stem volume

Survival rate of planted poplar clones (i.e. ERI, HOO, VIL, ROB and NE) was variable (Fig. 2) but the site effect on clone survival after GS1, GS2 and GS3 was not significant (forest ≈ grassland, $p \geq 0.05$). All planted cuttings of ERI survived in both sites. HOO showed 100 % survival rate on the grassland but very low (33 %) in the forest. All specimens of VIL survived in the forest, while on the grassland site strong wind broke trunks of 4 trees weakened by small poplar borer (*Saperda populnea* L.), resulting in decreased survival rate (73 %) after GS3. Clones ROB and NE had survival rates of 80 and 73 % in the forest and 73 and 67 % on the grassland, respectively (Fig. 2). During the planting, *P. alba*, BOE and BEA had no signs of diseases or bud frost damages. Nevertheless, all specimens of BOE died already in April of GS1, whereas BEA and *P. alba* had very low (13–20 %) survival rates (data not shown). Due to the low survival rates these poplars were not included in the subsequent measurements.

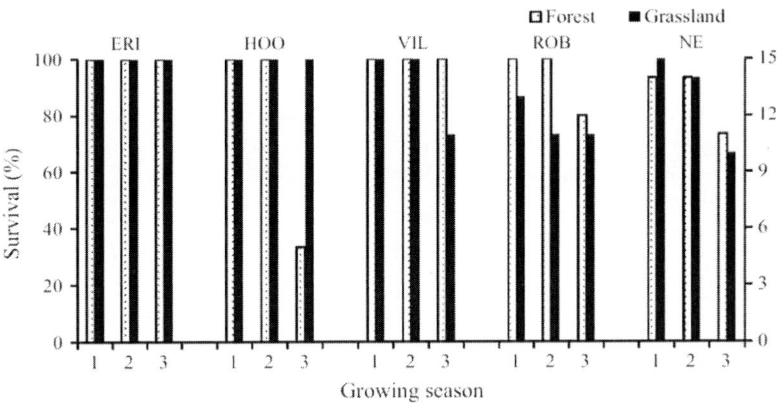

Figure. 2 Survival rate among poplar clones (ERI, HOO, VIL, ROB and NE) and number of tress in the forest and grassland after 1–3 growing seasons (GS1, GS2 and GS3)

We recorded very slight winter frost damage of all poplar clones growing in both sites with the exception of HOO from the forest (Fig. 1). In contrast, there were strong late spring frost damages, mainly of ERI in both plantations and of HOO in the forest (Fig. 1). Leaf rust infection scores differed between GS2 and GS3 (Fig. 1). Leaves of HOO from forest were the most infected by rust, whereas leaves of VIL were infected to a similar extent in both sites. On the other hand, ERI in both plantations and HOO on grassland were only very slightly infected by leaf rust.

The differences in mean tree height between the grassland and forest (Fig. 3) were negligible (forest ≈ grassland, $p \geq 0.05$). However, in ANOVA the effects of clone and site as well as clone × site interaction on tree height were significant ($p = 0.006$) after GS3 (Table 2). DBH and stem volume of trees growing on the grassland was higher than in the forest ($p < 0.0001$ and $p < 0.001$, respectively) and the effect of clone and site on both was evident (Fig. 3; Table 2). The larger DBH of poplar clones growing on the grassland in comparison with the forest was more pronounced after GS3 than after GS2, with the exception of VIL, which reached similar values in both sites after GS3 ($p > 0.05$). The same applied to stem volume (Fig. 3; Table 2).

Table 2 Significance of differences in response variables between poplar clones (ERI, HOO, VIL, ROB, and NE) from forest and grassland

Variables	ERI	HOO	VIL	ROB	NE	ANOVA effects (p values)[a]		
						Clone (C)	Site (S)	C × S
Total chlorophyll	**	**	**	**	**	<0.0001	<0.0001	<0.0001
Projected leaf area	*	**	**	**	**	<0.0001	<0.0001	<0.0001
Leaf dry mass	ns	**	ns	**	**	<0.0001	<0.0001	<0.0001
Specific leaf area	ns	ns	**	ns	**	<0.0001	0.007	<0.0001
Soluble carbohydrates	*	ns	**	**	**	<0.0001	<0.0001	<0.0001
Starch	**	ns	**	**	**	<0.0001	<0.0001	<0.0001
Total non-structural carbohydrates	**	ns	**	**	**	<0.0001	<0.0001	<0.0001
Winter frost damage	ns	**	ns	ns	ns	<0.0001	<0.0001	<0.0001
Spring frost damage	**	**	ns	**	ns	<0.0001	<0.0001	<0.0001
Rust leaf damage (in GS 2)	ns	**	ns	*	**	<0.0001	<0.0001	<0.0001
Rust leaf damage (in GS 3)	ns	**	ns	ns	**	<0.0001	<0.0001	<0.0001
Tree height (after GS 2)	ns	ns	ns	ns	ns	<0.0001	ns	ns
Tree height (after GS 3)	ns	*	ns	*	ns	<0.0001	0.002	0.006
Tree diameter at breast height (after GS 2)	*	**	**	**	**	<0.0001	<0.0001	ns
Tree diameter at breast height (after GS 3)	**	**	ns	**	**	<0.0001	<0.0001	0.012
Stem volume (after GS2)	*	**	**	**	*	<0.0001	<0.0001	ns
Stem volume (after GS3)	**	*	ns	**	**	<0.0001	<0.0001	0.025
Activity of acid phosphomonoesterases	**	**	**	**	**	0.017	<0.0001	<0.0001
Activity of alkaline phosphomonoesterases	**	**	**	**	**	<0.0001	<0.0001	<0.0001

Significance levels: * $p < 0.05$; ** $p < 0.01$; *ns* indicates $p \geq 0.05$

[a] Results of ANOVA for effects of poplar clone (C), site (S) and C × S interaction

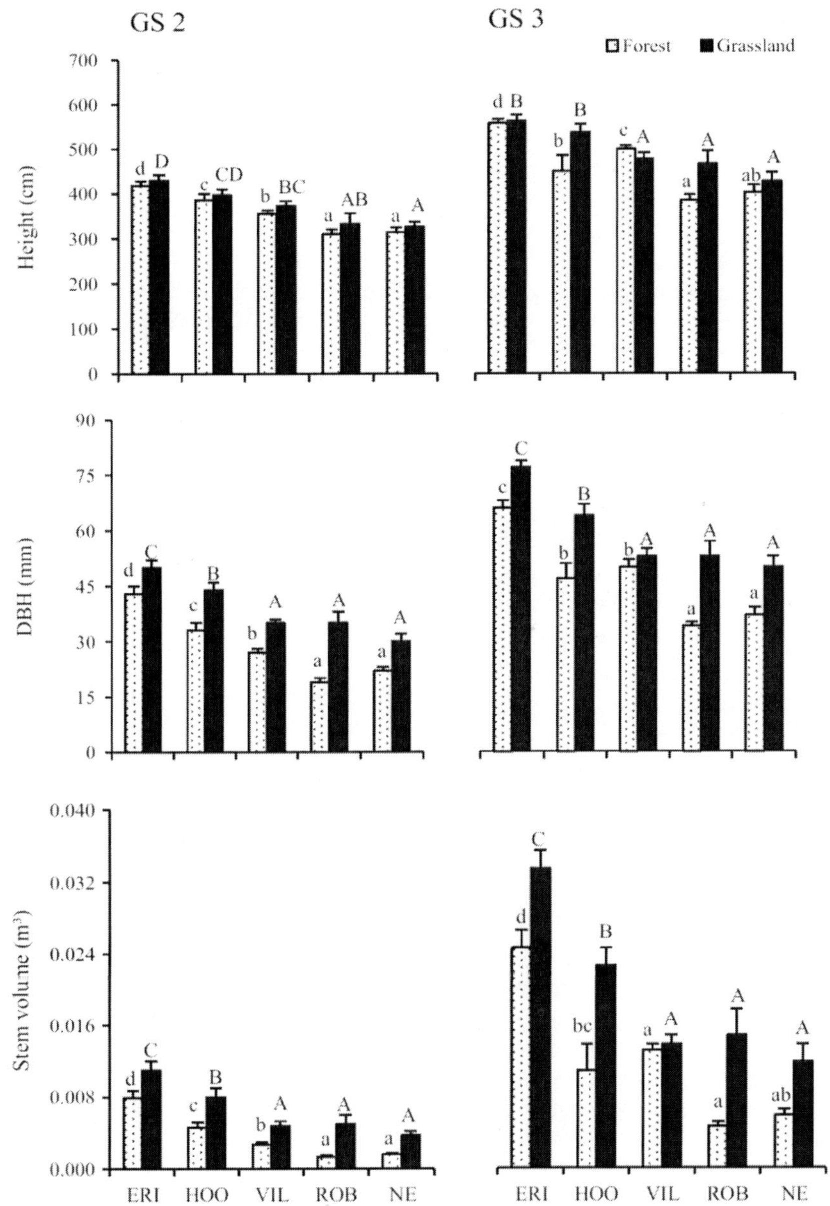

Figure. 3 Tree height, diameter at breast height (130 cm, DBH) and stem volume after second and third growing seasons (GS2 and GS3) for poplars in the forest and grassland. Other explanations as in Fig. 2

Soil element concentrations and enzymatic activity after GS3

In comparison to the rhizosphere in the forest, the rhizosphere soil in the grassland had lower pH and concentrations of C_{Total}, N_{Total}, Ca_{Total}, Ca–EDTA, Mg_{Total}, Mg–EDTA, S_{Total}, S–SO_4, Fe and C:N ratio but higher concentrations of N–NO_3, N–NH_4, P_{Total}, P–PO_4, K_{Total}, K–EDTA and Fe–EDTA in November of GS3 (Table 1). When pH and essential mineral elements were compared before poplar planting (in March 2009) and after GS3 (November 2011) then the direction of changes in pH and C_{Total}, P–PO_4, K–EDTA, Ca_{Total}, S–SO_4 concentrations in the forest and grassland were similar but different for N_{Total}, N–NO_3, N–NH_4, K_{Total}, S_{Total} and Fe–EDTA (Table 1). The pH in the rhizosphere of individual poplar clones on the grassland was lower than in the forest (Table 3). Differences were also noted in concentration of all the analyzed elements, with the exception of the C:N ratio for VIL, the concentration of N–NH_4 for VIL, ROB and NE, and Mg–EDTA and S–SO_4 for NE, which were similar in both sites (Table 3). Activity of Acid–Pase was higher ($p < 0.0001$) and Alkali–Pase was lower ($p < 0.0001$) in the rhizosphere soil of the grassland than in the forest after GS3 (Fig. 3 in supplementary material; Table 2).

DISCUSSION

Morphophysiological characteristics of leaves and fine roots

It is assumed that PLA, SLA and N concentration are leaf characteristics linked to net assimilation rate, tree growth and productivity (Broeckx et al. 2012; Johansson and Karacic 2011). PLA is often independent, while SLA and leaf N levels are dependent on environmental conditions and genotypes (Al Afas et al. 2007; Marron et al. 2007). In our study, PLA, SLA and N concentrations in young leaves varied between sites and clones, and the highest values were achieved by the best growing clone ERI (Figs. 1, 3; Table 2, and Table 1 in supplementary material). SLA was correlated with N in leaves of trees on the grassland ($r = 0.70$; $p < 0.0001$) and in the forest ($r = 0.37$; $p = 0.019$). N concentrations in young and mature leaves ranged from 2.5 to 3.4 % on the grassland and from 1.9 to 2.7 % in the forest (Table 1 in supplementary material), i.e. were within the range reported for *Populus* leaves (Pellis et al. 2004). Although N was a non-limiting factor on both plantations, the significantly lower N level in leaves in the forest than on the grassland ($p < 0.0001$ for young leaves and $p = 0.005$ for mature leaves) may reflect their lower photosynthetic capacity, depending on the N concentration (Al Afas et al. 2007; Weih and Rönnberg-Wästljung 2007).

Table 3 Total concentrations of macro-and micronutrients and their bioavailable forms (N–NO$_3$, N–NH$_4$, P–PO$_4$, S–SO$_4$, K–EDTA, Ca–EDTA, Mg–EDTA, Fe–EDTA) and pH in the rhizosphere soil of poplar clones (ERI, HOO, VIL, ROB and NE) in the forest and grassland

Elements	Forest					Grassland				
	ERI	HOO	VIL	ROB	NE	ERI	HOO	VIL	ROB	NE
pH (H_2O)	7.54 ± 0.01 a	7.58 ± 0.03 ab	7.58 ± 0.01 ab	7.60 ± 0.01 b	7.59 ± 0.03 ab	5.26 ± 0.17 A	5.30 ± 0.05 A	5.42 ± 0.08 B	5.53 ± 0.08 C	5.51 ± 0.08 D
pH (1 n KCL)	7.23 ± 0.01 a	7.24 ± 0.02 a	7.24 ± 0.02 a	7.28 ± 0.02 b	7.29 ± 0.01 b	3.75 ± 0.23 A	3.88 ± 0.1 B	3.96 ± 0.04 C	3.96 ± 0.04 C	4.03 ± 0.07 D
C_{Total} (%)	5.51 ± 0.08 c	3.85 ± 0.02 a	4.13 ± 0.08 b	4.3 ± 0.1 b	4.1 ± 0.1 b	1.41 ± 0.01 C	1.4 ± 0.1 BC	1.19 ± 0.04 A	1.34 ± 0.04 B	1.18 ± 0.02 A
C:N	14.5 ± 0.4 ab	14.1 ± 0.4 ab	13.3 ± 0.7 a	13.2 ± 0.7 a	14.8 ± 0.4 b	9.5 ± 0.2 A	9.2 ± 0.4 A	**11.7 ± 0.4** C	9.8 ± 0.3 A	10.5 ± 0.4 B
N_{Total} (%)	0.37 ± 0.002 c	0.27 ± 0.002 a	0.31 ± 0.01 b	0.33 ± 0.01 b	0.28 ± 0.01 a	0.16 ± 0.001 C	0.15 ± 0.01 BC	0.101 ± 0.001 A	0.14 ± 0.01 B	0.11 ± 0.004 A
N–NH_4 (mg kg^{-1})	1.8 ± 0.1 ab	1.5 ± 0.1 a	1.7 ± 0.1 a	2.1 ± 0.1 b	1.7 ± 0.1 a	2.9 ± 0.1 B	2.6 ± 0.1 AB	**2.1 ± 0.2** A	**2.5 ± 0.2** A	**2.2 ± 0.2** A
N–NO_3 (mg kg^{-1})	17.1 ± 0.3 d	8.2 ± 0.1 a	8.7 ± 0.3 a	15.9 ± 0.3 c	12.1 ± 0.3 b	42.3 ± 0.6 D	33.7 ± 0.8 C	26.3 ± 0.3 A	32.3 ± 0.2 C	30.5 ± 0.8 B
P_{Total} (mg kg^{-1})	600 ± 50 b	390 ± 30 a	590 ± 70 b	520 ± 60 ab	460 ± 40 a	1430 ± 100 C	1300 ± 80 C	1020 ± 80 B	1220 ± 70 BC	730 ± 20 A
P–PO_4 (mg kg^{-1})	2.4 ± 0.2 c	1.74 ± 0.1 b	1.8 ± 0.2 b	1.7 ± 0.1 b	1.3 ± 0.1 a	30 ± 1 C	27 ± 1 B	26 ± 1 B	25 ± 1 AB	23 ± 1 A
K_{Total} (mg kg^{-1})	740 ± 50 c	620 ± 20 a	640 ± 10 ab	710 ± 20 bc	580 ± 20 a	1900 ± 30 C	1820 ± 40 C	1620 ± 40 B	1530 ± 40 A	1510 ± 20 A
K-EDTA (mg kg^{-1})	53 ± 2 c	35 ± 1 a	60 ± 2 d	45 ± 1 b	32 ± 1 a	450 ± 20 D	370 ± 15 C	350 ± 14 C	310 ± 14 B	111 ± 20 A
Mg_{Total} (mg kg^{-1})	1190 ± 30 b	1120 ± 60 ab	1090 ± 80 ab	1050 ± 20 ab	1020 ± 10 a	1000 ± 20 B	890 ± 60 B	700 ± 30 A	725 ± 17 A	735 ± 20 A
Mg-EDTA (mg kg^{-1})	75 ± 2 c	56 ± 2 a	65 ± 2 b	72 ± 2 bc	54 ± 6 a	40 ± 2 B	36 ± 2 AB	39 ± 1 B	34 ± 1 A	**33 ± 2** A
Ca_{Total} (mg kg^{-1})	23,300 ± 900 a	30,000 ± 1000 a	30,000 ± 2000 a	34,000 ± 3000 a	30,600 ± 800 a	650 ± 20 C	610 ± 10 BC	580 ± 20 B	520 ± 10 A	510 ± 20 A
Ca–EDTA (mg kg^{-1})	6300 ± 500 a	7600 ± 50 b	7400 ± 100 b	7500 ± 200 b	6600 ± 100 a	590 ± 20 C	556 ± 9 B	538 ± 10 B	490 ± 10 A	468 ± 20 A
S_{Total} (%)	0.03 ± 0.002 a	0.02 ± 0.001 a	0.02 ± 0.001 a	0.02 ± 0.001 a	0.02 ± 0.001 a	0.01 ± 0.0001 A	0.01 ± 0.0003 A	0.01 ± 0.0001 A	0.01 ± 0.0003 A	0.01 ± 0.001 A
S–SO_4 (mg kg^{-1})	20.3 ± 0.7 c	17.1 ± 0.3 b	22.4 ± 0.3 d	17.4 ± 0.3 b	12.2 ± 0.3 a	13.9 ± 0.2 B	11.5 ± 0.1 B	9.4 ± 0.2 A	9.2 ± 0.2 A	**11.2 ± 0.2** B
Fe_{Total} (mg kg^{-1})	6300 ± 200 c	5400 ± 200 b	5700 ± 200 b	5600 ± 200 b	4800 ± 100 a	4900 ± 200 C	4700 ± 100 BC	3800 ± 400 A	4100 ± 100 AB	3600 ± 100 A
Fe–EDTA (mg kg^{-1})	44 ± 1 b	43 ± 1 b	45 ± 3 b	39 ± 1 a	39 ± 1 a	148 ± 5 C	142 ± 5 C	118 ± 5 B	104 ± 4 A	107 ± 5 AB

Mean values ± SE of 9 independent samples for ERI, VIL, ROB, and NE clones from grassland and forest and 9 and 5 samples for HOO from grassland and forest, respectively, collected in November of the last growing season (GS3). Values designed with the same letters are not significantly different (LSD-test; $p \geq 0.05$). Bold font indicates lack of significant differences between the forest and grassland ($p \geq 0.05$)

Lower chlorophyll concentrations in leaves from forest than from grassland ($p < 0.0001$) (Fig. 1) can also indicate their lower photosynthetic capacity. Highly significant correlations were found between chlorophyll and leaf N and Fe concentrations in the forest ($r = 0.86$; $p < 0.0001$ and $r = 0.78$; $p < 0.0001$, respectively), earlier also reported in other studies (Lombard et al. 2010; Weih and Rönnberg-Wästljung 2007). Chlorophyll concentrations in leaves from the grassland were correlated with leaf Fe and Mg content ($r = 0.88$; $p < 0.0001$ and $r = 90$; $p < 0.0001$, respectively) but not with leaf N content ($r = 0.27$; $p > 0.05$). Mg and Fe play crucial roles in the chlorophyll molecule and photosynthesis, and SPAD readings were closely related to levels of these foliar elements (Klooster et al. 2012; Lombard et al. 2010). The unreliable relationship between leaf chlorophyll and N concentrations on the grassland may be a consequence of the impact of N concentration and distribution within the leaf and also other factors on SPAD readings (Klooster et al. 2012; Lombard et al. 2010). The lowest chlorophyll concentrations were noted for clone ERI (Fig. 1), which showed the best growth parameters (Fig. 3). Carbon concentrations in leaves were similar for all poplar clones at both locations (Table 1 in supplementary material). This confirms earlier suggestions about the lack of relationship between chlorophyll and C content and tree growth (Marron et al. 2007).

On the grassland, about 90 % of fine roots of planted poplar clones were found in the top soil (0–5 cm soil depth; excluding the leaf litter) and the value was particularly high for ERI clone. In the forest site, nearly 80 % of fine roots were at 10–15 cm soil depth, like in other studies with deciduous trees (Terzaghi et al. 2013). The surface location of fine roots on the grassland increases their access to rainwater and nutrients from litter fall. This may result in higher concentrations of N and Mg in fine roots on the grassland than in the forest (Table 2 in supplementary material) despite the lower N and Mg concentrations in the soil on the grassland (Tables 1, 3). On the other hand, the fine roots in the top soil on the grassland could be vulnerable to pathogenic bacteria, fungi and insects living in/on the leaf litter. However, in the course of the study, there was no visible damage increase on fine roots from the grassland in comparison with the forest. Further, high K (Table 2 in supplementary material) and low soluble sugars concentrations in the fine roots from the grassland (Fig. 2 in supplementary material) may also indicate that fine roots were not attractive to pathogens and insects (Amtmann et al. 2008; Römheld and Kirkby 2010). The growth of fine roots in the top soil may lead to frost damage in case of severe winter, especially without snowfall. During our study, no such an event was observed.

Higher concentrations of N and C in fine roots from the grassland than from the forest were linked with lower values of the C:N ratio (Table 2 in supplementary material). The concentrations of C in fine roots are connected with their construction cost and N levels are connected with metabolic activity and root longevity, while the fine root C:N ratio can reflect their lifespan (Terzaghi et al. 2013). Assuming the above, our results seem to indicate a shorter lifespan and higher fine root turnover rate on the grassland than in the forest.

Survival, frost hardiness, rust resistance, growth and stem-wood production

Of all the planted poplars, only clone ERI had the 100 % survival rate in both sites after GS3, which confirms its high ability to survive in various environments and growing conditions (Coyle et al. 2006a; Kaczmarek et al. 2013). In comparison with the other clones, leaves of ERI developed earlier in the spring and very quickly expanded the leaf blade, which became discolored and fell off in late autumn. However, a consequence of fast leaf development can be leaf damage due to late spring frosts. The sharp temperature drop in May after an early spring warm period (Fig. 1 in supplementary material) damaged expanding leaves of ERI, especially in the forest site (Fig. 1). However, damaged leaves were rapidly replaced with new ones, whose PLA, weight and SLA already in mid-June reached higher values than in the other poplar clones (Fig. 1). Clone ERI was also characterized by high leaf rust resistance and frost hardiness, like in earlier research (Coyle et al. 2006b. Clone ERI achieved the highest values of DBH and stem volume in comparison with other poplar clones after GS3 (Fig. 3). Stem volume for ERI (0.034 ± 0.002 m^3) in suboptimal growing conditions on the grassland was 2.7–5.0-fold greater after GS3 than for ERI in studies with fertilization and irrigation (Coyle et al. 2006a; Kaczmarek et al. 2013). Although these differences to some extent could be results of the growth environment, they clearly indicate the great potential for growth and stem-wood production of ERI clone on the grassland.

The HOO clone achieved good growth parameters on the grassland but it seemed inappropriate for growth in the forest (Fig. 3), where HOO was highly susceptible to *Melampsora* rust infection (Fig. 1). The consequences of leaf rust damage can include a decrease in photosynthetic efficiency and frost hardiness (Aylott et al. 2008), stunted growth and even mortality of young trees (Coyle et al. 2006b. Severe frost from 23 to 27 December 2010, when temperature dropped to −23.5 °C (Fig. 1 in supplementary material), killed most of the rust-infected trees. Also serious late spring frost damages were noted in rust-infected HOO trees in the forest (Fig. 1). Earlier reports indicate that the level of foliar rust damage depends on site conditions (Aylott et al. 2008), while others suggested that only certain poplar clones are more susceptible to rust in a given area (Coyle et al. 2006b. Foliar rust damages of HOO as well as ROB and NE were generally higher in the forest than on the grassland (Fig. 1; Table 2). The sites greatly differed in soil conditions (Table 1) and the lower incidence of the disease on the grassland could be due to the high concentration of K in the soil, leaves and fine roots (Table 3and Tables 1, 2 in supplementary material) of poplars growing there, as high K diminishes the incidence of diseases (Amtmann et al. 2008; Römheld and Kirkby 2010). A lower ability to survive as well as poorer growth were observed in VIL after GS3 (Fig. 3). That clone was resistant to frost and moderately susceptible to rust infection but significantly attacked by the insect pest *Saperda populnea* (L.). Curvature of the trunk of infested trees disturbed their growth performance, and the weakened trees were easily broken by the wind. VIL was previously recommended for wood plantations for energy (Kačík et al. 2012; Rosso et al. 2013), but considering our results, especially because of its susceptibility to *S. populnea*, it is not advised for use. Poplar clones ROB and NE are also often used for bioenergy plantations (Baum et al. 2012; Broeckx et al. 2012). However, results of this study show that ROB and NE rather should not be selected for stem-wood productivity under suboptimal soils, because of their damage by spring frost, low disease resistance and low growth parameters. The native European white poplar (*Populus alba*) and poplar clones BOE and BEA had also very poor survival rates (0–20 %, data not shown). On the other hand, these poplars grow well in the *Populetum* collection of ID PAS. The high survival rate and growth potential of *Populus alba*, BOE and BEA were the basis for their frequent applications in short-rotation coppice systems (Dillen et al. 2013; Johansson and Karačić 2011; Rosso et al. 2013; Truax et al. 2012). However, in our plantation, management of these poplars proved to be unsuitable for long-term stem-wood production.

Elemental analysis and enzymatic activity of the soil

Soil texture, pH and concentrations of essential mineral elements differed between forest soil and grassland before poplar planting. Differences remained also after 3 years of poplar growth, with the exception of concentrations of inorganic N (Table 1). Ammonium ($N–NH_4$) and nitrate ($N–NO_3$) concentrations were lower in grassland soil than in the forest before poplar planting and higher after 3 years of poplar growth in the grassland. This may indicate a rapid OM decomposition and nitrogen mineralization and nitrification (Stauffer et al. 2014) on the grassland after GS3, probably partly due to greater litter inputs through better growth of poplars and short lifespan and higher fine root turnover on the grassland than in the forest sites. The hypothesis of rapid OM biodegradation is also supported by a higher concentration of N in leaves (2.8 ± 0.1 % on the grassland and 2.4 ± 0.1 % in the forest; $p = 0.005$) and lower C:N ratios (20.8 ± 0.6 vs. 35.2 ± 0.8; $p < 0.0001$) in fine roots, which might effectively stimulate the growth of soil microorganisms (Mao et al. 2011).

The comparison of the chemical properties of forest soil before poplar planting and after GS3 indicates that poplar planting led to a reduction in alkaline pH and Ca_{Total} and an increase in OM, C_{Total}, N_{Total}, K_{Total}, $P–PO_4$, K–EDTA and $S–SO_4$ concentrations in rhizosphere soil after GS3 (Tables 1, 3).

The greater N_{Total} concentration probably resulted from an increase in organic N, as the concentrations of inorganic N in the soil did not change after GS3. Plants can use organic N through enzymatic hydrolysis of proteins and peptides (Nacry et al. 2013) of soil OM, and its concentrations increased after GS3. A significant ($p = 0.023$) increase in the soil C:N ratio to ~14 after GS3 implies appropriate N availability in soil (Peichl et al. 2012). Moreover, concentrations of N_{Total}, N–NH_4 and N–NO_3 in the soil, as well as N in poplar leaves and roots (Tables 1, 2 in supplementary material) were within the ranges reported in the literature (Pellis et al. 2004). The very low concentrations of bioavailable P (P–PO_4) before poplar planting increased about 1.9-fold in rhizosphere soil after GS3, but without changes in P_{Total}. The latter might be due to the presence of stable, non-bioavailable forms of P in the high Ca_{Total} and alkaline soil pH (Shen et al. 2011). Despite the increase in the concentrations of inorganic P after GS3, the concentration of P–PO_4 in the rhizosphere can still be classified as very low. Nevertheless, the P concentrations in leaves and roots (Tables 1, 2 in supplementary material) were even higher than those given in the literature (Ciadamidaro et al. 2014). The increase in P–PO_4 in the forest after GS3 could be the result of the catalytic transformation of organic P esters, due to a high activity of Alkali–Pase (303 ± 6 μg *p*–NP g^{-1} DM h^{-1}, Fig. 3 in supplementary material), effectively produced by soil bacteria, fungi and animals but not secreted by plant roots, and partly also of Acid–Phase (130 ± 3 μg *p*–NP g^{-1} DM h^{-1}, Fig. 3 in supplementary material), originating mainly from plant roots with rhizosphere microorganisms (Nannipieri et al. 2012).

Afforestation of the grassland led to a decrease in pH and Ca_{Total} and Ca–EDTA and an increase in OM, C_{Total}, P_{Total}, P–PO_4, K–EDTA, S–SO_4 and Fe–EDTA in the rhizosphere soil after GS3 (Table 1). No changes in N_{Total} and N–NH_4 were accompanied by a nearly sevenfold increase ($p < 0.0001$) in concentrations of N–NO_3. Nitrate is easily available as highly mobile anions in the top soil and may be taken up by plant roots. Poplars have a strong preference for N–NO_3-dominated soil solutions (Woolfolk and Friend 2003) at low soil pH (Lafleur et al. 2013). An about 1.5-fold increase ($p = 0.0002$) in the C:N ratio in the rhizosphere soil and its value near 10 after GS3, together with high nitrate concentrations, may indicate accelerated OM biodegradability as well as N mineralization and nitrification process, in which ammonium is converted to nitrate by nitrifying bacteria (Stauffer et al. 2014). Nitrification may acidify the soil (Colin-Belgrand et al. 2003). Therefore, it could be speculated that the decreased pH from 6.2 before poplar planting to 5.4 in the rhizosphere soil after GS3 (Table 1) may be partly attributed to increased nitrification. Soil pH could be also lowered by the release of acid exudates by plant roots and rhizosphere microorganisms (Jones 1998). Rhizosphere acidification might improve the mobilization and availability of soil elements (i.e. Ca, Mg, K, Fe and P) (Hinsinger et al. 2011; Lafleur et al. 2013) and thereby enhance nutrient uptake. The consequence of the latter may be the relatively high concentration of elements in the leaves and roots of planted poplar (Tables 1, 2 in supplementary material), similar to the values reported by others (Ciadamidaro et al. 2014; Kabata-Pendias and Pendias 2001), despite very low concentrations of e.g. Mg, Ca and Fe in the rhizosphere soil after GS3 (Tables 1, 2). The increase in bioavailability of nutrients could better satisfy the high nutrient demand of poplar for biomass growth, but at the same time could lead to depletion of base cations in soil (Lafleur et al. 2013), which was seen for Ca (Table 1), especially in the rhizosphere of ERI on grassland after GS3 (Table 3). In addition, the increased availability of P–PO_4 through dissolving P-minerals (Hinsinger et al. 2011) is probably not sufficient for growing poplar, as indicated by the high activity of Acid–Phase (221 ± 4 µg *p*–NP g^{-1} DM h^{-1}, Fig. 3 in supplementary material), hydrolyzing the soil organic P complexes (Amtmann et al. 2008; Nannipieri et al. 2012).

CONCLUSIONS

This study shows that clone ERI is the best candidate for potential stem-wood production among all the poplar clones planted in the forest and grassland irrigated for many decades with PSW. On the grassland the stem volume for ERI was 2.8-fold higher than in the slowest growing NE after GS3. It has been shown recently that for poplar growth, clonal selections can be made after GS3 (Kaczmarek et al. 2013). Therefore it can be assumed that clone ERI in a long-life plantation (10–15 years) will continue to have the highest stem volumes and thereby wood productivity of all the planted poplar clones in the forest and grassland. Climatic instability in the temperate climate of Poland (Central Europe) and resistance to leaf rust were important factors affecting survival and growth of poplar clones planted in the forest and grassland. The stronger leaf rust resistance of poplar planted in the grassland than in the forest and higher frost hardiness and less severe spring frost damage could have partly resulted from higher concentrations of K in the soil, leaves and fine roots of poplars growing in the grassland, and the above properties undoubtedly were responsible for the better growth of poplars planted in the grassland than in the forest. The beneficial effect of higher K in soil and crops on resistance to diseases and frost damage were already reported earlier (Amtmann et al. 2008; Römheld and Kirkby 2010).

Due to poplar planting on the grassland, soil quality parameters (OM, C_{Total}, N–NO_3, P_{total}, P–PO_4, K–EDTA, S–SO_4, Fe–EDTA concentrations) in the topsoil after GS3 were significantly better than before poplar planting and hence afforestation promoted revitalization of grassland soil. The increase in soil fertility can be a consequence of decomposition of high-quality leaves of the fast-growing poplars, but also of the short lifespan of fine roots and their high turnover rates. Accelerated litter decomposition, together with increased soil C concentrations following poplar planting on the grassland, may indicate increased soil C sequestration after GS3. The nutrient enrichment of the soil under poplar was linked with soil acidification to pH 5.4, but it still remained in the optimum pH range (5.0–7.5) for poplar growth (Zalesny Jr and Bauer 2007). Simultaneously, Ca soil concentrations declined. Therefore there is a need for liming (preferably dolomitic lime) to promote further growth of the poplars and thus higher wood production on the grassland.

Land application of food processing wastewater represents its simple and cost-effective use (Kretschmer et al. 2002). However, as wastewater can be loaded with toxic trace elements or harmful organic substances, long-term application imposes a risk of groundwater contamination and threatens plant growth (Kretschmer et al. 2002). Aryal and Reinhold (2015) conclude that poplar trees meet the criteria of 'perfect' plants to be grown on fields irrigated with food processing wastewater and propose *Populus deltoides* × *P. nigra* DXN 34 poplars for planting on lands treated with fruit and vegetable processing wastewater. For improvement of soil fertility and bioenergy crops on lands irrigated with PSW, we suggest use of the ERI clone.

ACKNOWLEDGMENTS

This work was supported by the National Centre for Research and Development, Poland (Grant No. N R14 0058 06). We thank the staff of the Potato Industry Company "Trzemeszno" for their kind cooperation. Thanks are also due to the Institute of Meteorology and Water Management, National Research Institute, Warsaw, Poland, for climatic data. We gratefully acknowledge everyone who helped us with field work, plant and soil analysis. Comments and suggestions on this manuscript provided by anonymous reviewers are appreciated.

REFERENCES

1. Al Afas N, Marron N, Ceulemans R (2007) Variability in *Populus* leaf anatomy and morphology in relation to canopy position, biomass production, and varietal taxon. Ann For Sci 64:521–532

2. Amtmann A, Troufflard S, Armengaud P (2008) The effect of potassium nutrition on pest and disease resistance in plants. Physiol Plantarum 133:682–691
3. Arienzo M, Christen EW, Quayle W, Kumar A (2009) Review of the fate of potassium in the soil-plant system after land application of wastewaters. J Hazard Mater 164:415–422
4. Aryal N, Reinhold DM (2015) Reduction of metal leaching by poplars during soil treatment of wastewaters: small-scale proof of concept studies. Ecol Eng 78:53–61
5. Aylott MJ, Casella E, Tubby I, Street NR, Smith P, Taylor G (2008) Yield and spatial supply of bioenergy poplar and willow short-rotation coppice in the UK. New Phytol 178:358–370
6. Baum S, Bolte A, Weih M (2012) High value of short rotation coppice plantations for phytodiversity in rural landscapes. GCB Bioenergy 4:728–738
7. Broeckx LS, Verlinden MS, Ceulemans R (2012) Establishment and two-year growth of a bio-energy plantation with fast-growing *Populus* trees in Flanders (Belgium): effects of genotype and former land use. Biomass Bioenerg 42:151–163
8. Buelow MC, Steenwerth K, Parikh SJ (2015) The effect of mineral-ion interactions on soil hydraulic conductivity. Agr Water Manag 152:277–285
9. Chang M, Zhou S-G, Lu N, Ni J-R (2008) Starch processing wastewater as a new medium for production of *Bacillus thuringiensis*. World J Microbiol Biotechnol 24:441–447
10. Charmley E, Nelson D, Zvomuya F (2006) Nutrient cycling in the vegetable processing industry: utilization of potato by-products. Can J Soil Sci 86:621–629
11. Ciadamidaro L, Madejón E, Robinson B, Madejón P (2014) Soil plant interactions of *Populus alba* in contrasting environments. J Environ Manage 132:329–337
12. Colin-Belgrand M, Dambrine E, Bienaimé S, Nys C, Turpault MP (2003) Influence of tree roots on nitrogen mineralization. Scan J For Res 18:260–268
13. Coyle DR, Coleman MD, Durant JA, Newman LA (2006a) Survival and growth of 31 *Populus*clones in South Carolina. Biomass Bioenerg 30:750–758
14. Coyle DR, Coleman MD, Durant JA, Newman LA (2006b) Multiple factors affect pest and pathogen damage on 31 *Populus* clones in South Carolina. Biomass Bioenerg 30:759–768
15. Dillen SY, Djomo SN, Al Afas N, Vanbeveren S, Ceulemans R (2013) Biomass yield and energy balance of a short-rotation poplar coppice with multiple clones on degraded land during 16 years. Biomass Bioenerg 56:157–165
16. Gruenewald H, Brandt BKV, Schneider BU, Bens O, Kendzia G, Hüttl RF (2007) Agroforestry systems for the production of woody biomass for energy transformation purposes. Ecol Eng 29:319–328
17. Hamilton A, Stagnitti F, Xiong X, Kreidl S, Benke K, Maher P (2011) Wastewater irrigation: the state of play. Vadose Zone J 6:823–840
18. Hinsinger P, Brauman A, Devau N, Gerard F, Jourdan C, Laclau J-P, Le Cadre E, Jaillard B, Plassard C (2011) Acquisition of phosphorus and other poorly mobile nutrients by roots. Where do plant nutrition models fail? Plant Soil 348:20–61
19. Johansson T, Karačić A (2011) Increment and biomass in hybrid poplar and some practical implications. Biomass Bioenerg 35:1925–1934
20. Jones DL (1998) Organic acids in the rhizosphere: a critical review. Plant Soil 205:25–44
21. Kabata-Pendias A, Pendias H (2001) Trace elements in soils and plants, 3rd edn. CRC Press, Boca Raton
22. Kačík F, Ďurkovič J, Kačíková D (2012) Chemical profiles of wood components of poplar clones for their energy utilization. Energies 5:5243–5256
23. Kaczmarek DJ, Coyle DR, Coleman MD (2013) Survival and growth of a range of *Populus*clones in central South Carolina USA through age ten: do early assessments reflect longer-term survival and growth trends? Biomass Bioenerg 49:260–272
24. Kalra YP, Maynard DG (1991) Methods manual for forest soil and plant analysis. For Can, Northwest Reg, North For Cent, Edmonton, Alberta. Inf Rep NOR-X-319

25. Klooster WS, Cregg BM, Fernandez RT, Nzokou P (2012) Growth and physiology of deciduous shade trees in response to controlled-release fertilizer. Sci Hortic 135:71–79
26. Kretschmer N, Ribbe L, Gaese H (2002) Wastewater reuse for agriculture. Technol Resour Manag Dev Sci Contrib Sustain Dev 2:37–64
27. Lafleur B, Paré D, Claveau Y, Thiffault E, Bélanger N (2013) Influence of afforestation on soil: the case of mineral weathering. Geoderma 202–203:18–29
28. Laganière J, Angers DA, Paré D (2010) Carbon accumulation in agricultural soils after afforestation: a meta-analysis. Glob Change Biol 16:439–453
29. Lichtenthaler HK, Wellburn RR (1983) Determination of total carotenoids and chlorophylls *a*and *b* of leaf extracts in different solvents. Biochem Soc Trans 11:591–592
30. Lombard K, O'Neill M, Mexal J, Ulery A, Onken B, Bettmann G, Heyduck R (2010) Can soil plant analysis development values predict chlorophyll and total Fe in hybrid poplar? Agrofor Syst 78:1–11
31. Lorenc-Plucińska G, Walentynowicz M, Niewiadomska A (2013) Capabilities of alders (*Alnus incana* and *A. glutinosa*) to grow in metal-contaminated soil. Ecol Eng 58:214–227
32. Lu N, Zhou S-G, Zhuang L, Zhang J-T, Ni J-R (2009) Electricity generation from starch processing wastewater using microbial fuel cell technology. Biochem Eng J 43:246–251
33. Mao R, Zeng DH, Li LJ (2011) Fresh root decomposition pattern of two contrasting tree species from temperate agroforestry systems: effects of root diameter and nitrogen enrichment of soil. Plant Soil 347:115–123
34. Marron N, Dillen SY, Ceulemans R (2007) Evaluation of leaf traits for indirect selection of high yielding poplar hybrids. Environ Exp Bot 61:103–116
35. Mateo JJ, Maicas S (2015) Valorization of winery and oil mill wastes by microbial technologies. Food Res Int 73:13–25
36. Mocek A, Drzymała S (2010) Genesis, analysis and soil classification. Poznan University of Life Sciences, Poznan **(in Polish)**
37. Nacry P, Bouguyon E, Gojon A (2013) Nitrogen acquisition by roots: physiological and developmental mechanisms ensuring plant adaptation to a fluctuating resource. Plant Soil 370:1–29
38. Nannipieri P, Giagnoni L, Renella G, Puglisi E, Ceccanti B, Masciandaro G, Fornasier F, Moscatelli MC, Marinari S (2012) Soil enzymology: classical and molecular approaches. Biol Fert Soils 48:743–762
39. Öborn I, Andrist-Rangel Y, Askekaard M, Grant CA, Watson CA, Edwards AC (2005) Critical aspects of potassium management in agricultural systems. Soil Use Manage 21:102–112
40. Peichl M, Leava NA, Kiely G (2012) Above- and belowground ecosystem biomass, carbon and nitrogen allocation in recently afforested grassland and adjacent intensively managed grassland. Plant Soil 350:281–296
41. Pellis A, Laureysens I, Ceulemans R (2004) Growth and production of a short rotation coppice culture of poplar I. Clonal differences in leaf characteristics in relation to biomass production. Biomass Bioenergy 27:9–19
42. Römheld V, Kirkby EA (2010) Research on potassium in agriculture: needs and prospects. Plant Soil 335:155–180
43. Rosenwinkel K-H, Austermann-Haun U, Meyer H (2005) Industrial wastewater sources and treatment strategies. In: Jördening H-J, Winter J (eds) Environmental biotechnology: concepts and applications. Wiley, Weinheim, pp 49–77
44. Rosso L, Facciotto G, Bergante S, Vietto L, Nervo G (2013) Selection and testing of *Populus alba* and *Salix* spp. as bioenergy feedstock: preliminary results. Appl Energy 102:87–92
45. Shen J, Yuan L, Zhang J, Li H, Bai Z, Chen X, Zang W, Zang F (2011) Phosphorus dynamics: from soil to plant. Plant Physiol 156:997–1005Central

46. Stauffer M, Leyval C, Brun JJ, Leportier P, Berthelin J (2014) Effect of willow short rotation coppice on soil properties after 3 years of growth as compared to forest, grassland and arable land uses. Plant Soil 377:423–438
47. Szuba A, Wojakowska A, Lorenc-Plucińska G (2013) An optimized method to extract poplar leaf proteins for two-dimensional gel electrophoresis guided by analysis of polysaccharides and phenolic compounds. Electrophoresis 22:3234–3243
48. Terzaghi M, Montagnoli A, Di Iorio A, Scippa GS, Chiatante D (2013) Fine-root carbon and nitrogen concentration of European beech (*Fagus sylvatica* L.) in Italy Prealps: possible implications of coppice conversion to high forest. Front Plant Sci 4:1–8
49. The European Starch Industry Association (2014) The European Starch Industry. http://www.aaf-eu.org/. Accessed 31 Sept 2014
50. Truax B, Gagnon D, Fortier J, Lambert F (2012) Yield in 8 year-old hybrid poplar plantations on abandoned farmland along climatic and soil fertility gradients. For Ecol Manag 267:228–239
51. Verlinden MS, Broeckx LS, Zona D, Berhongaray G, De Groote T, Camino Serrano M, Janssens IA, Ceulemans R (2013) Net ecosystem production and carbon balance of an SRC poplar plantation during its first rotation. Biomass Bioenergy 56:412–422
52. Vu KD, Tyagi RD, Valéro JR, Surampalli RY (2010) Batch and fed-batch fermentation of *Bacillus thuringiensis* using starch industry wastewater as fermentation substrate. Bioprocess Biosyst Eng 33:691–700
53. Weih M, Rönnberg-Wästljung A-N (2007) Shoot biomass growth is related to the vertical leaf nitrogen gradient in *Salix* canopies. Tree Physiol 27:1551–1559
54. Woolfolk WRM, Friend AL (2003) Growth response of cottonwood roots to varied NH_4:NO_3ratios in enriched patches. Tree Physiol 23:427–432
55. Zalesny RS Jr, Bauer E (2007) Selecting and utilizing *Populus* and *Salix* for landfill covers: implications for leachate irrigation. Int J Phytoremed 9:497–511
56. Zalesny RS Jr, Hall RB, Zalesny JA, McMahon BG, Berguson WE, Stanosz GR (2009) Biomass and genotype × environment interactions *of Populus* energy crops in the Midwestern United States. Bioenergy Res 2:106–122

CHAPTER 3

Land Use Affects the Soil C Sequestration in Alpine Environment, NE Italy

Diego Pizzeghello[1,*], Ornella Francioso[2], Giuseppe Concheri[1], Adele Muscolo[3] and Serenella Nardi[1]

[1]*Dipartimento di Agronomia, Animali, Alimenti, Risorse Naturali e Ambiente (DAFNAE), Università degli Studi di Padova, 35020 Legnaro (Padova), Italy*

[2]*Dipartimento di Scienze Agrarie, Università degli Studi di Bologna, 40127 Bologna, Italy*

[3]*Dipartimento di Gestione dei Sistemi Agrori e Forestali, Università Mediterranea di Reggio Calabria, 89060 Gallina (Reggio Calabria), Italy*

ABSTRACT

Soil carbon sequestration is strongly affected by soil properties, climate, and anthropogenic activities. Assessing these drivers is key to understanding the effect of land use on soil organic matter stabilization. We evaluated land use and soil depth influencing patterns of soil organic matter stabilization in three types of soil profiles located under the same pedogenetic matrix and alpine conditions but with different vegetation cover. The stock in soil organic carbon in the mean 0–20 cm layer increased from prairie (31.9 t ha^{-1}) to prairie in natural reforestation (42 t ha^{-1}) to forest (120 t ha^{-1}), corresponding to increments of 1.3-fold prairie, for prairie in natural reforestation, and of 3.8-fold prairie for forest. The forest showed the highest humic carbon (21.7 g kg^{-1}), which was 2.8 times greater than the prairie in natural reforestation and 4 times higher than the prairie. ^{13}C-NMR spectroscopic measurements suggested a different C pattern. The prairie in natural reforestation and the prairie were characterized by a higher content in O,N-alkyl C with respect to the forest. Alkyl C and aromatic C in the prairie in natural reforestation and prairie did not show relevant differences while they decreased with respect to the forest. Carboxyl and phenolic C groups were markedly higher in forest and prairie than prairie in natural reforestation. Alkyl C, carboxyl C, and phenolic C prevailed in the Ah horizons whereas aromatic C and O,N-alkyl C were dominant in the B horizons. Overall, the marked distribution of O,N-alkyl C and alkyl C in humic substances (HS) indicates a low degree of humification. Nevertheless, in forest, the relatively high presence of aromatic C designated HS endowed with a relatively high humification degree. Thus, our results might suggest that in the alpine environment of NE Italy differences in soil organic matter (SOM) stocks and characteristics are affected by land use and anthropic activities.

Keywords: alpine; forest; humic; soil organic carbon; land use; ^{13}C-NMR

1. INTRODUCTION

Forests play an important role in the global carbon (C) cycle by capturing CO_2 from the atmosphere, converting it into biomass, and emitting it back into the atmosphere or fixing it into stable soil organic matter pools [1]. The amount of soil C is the result of the balance between the rate of organic C input in leaf and root biomass and its mineralization. In boreal forests, and in particular in alpine environment, SOM is quantitatively important because it favours the thick superficial organic horizons called humus [2] with different ecological value. The extent in C pool in forests depends on soil properties, climate, and anthropogenic activities [3], while the stabilization of soil organic matter (SOM) is mediated by soil structure and their interactions [4], which protect organic matter from decomposition [5]. The influence of tree species on C pool has been widely studied [6,7,8,9,10,11], however, a comprehensive understanding of the magnitude of tree influence across site types has not yet been well elucidated.

C storage in soil ecosystems can also be affected by stand management practices and land use [12,13]. Brown and Lugo [14] showed that, following deforestation, sites lost up to 60%–70% of the initial C contained in mature forests. Carbon loss is due to the rapid decomposition of organic matter, with consequent C released as CO_2 to the atmosphere, or by erosion processes in which C may be deposited somewhere else in the landscape, with only some lost as CO_2. Land management plays an important role as source and/or sink of C [15].

Therefore, soil organic C (SOC) stock can be modified by land use because of different inputs of organic matter, different canopy structures, as well as different physical and chemical properties of soil. There is a considerable amount of evidence showing that forests converted to cultivated cropland decreased their organic layer, soil C content, and cation exchange capacities [16]. In a study by Li et al., [17] over 30 years, the SOC storage did not decrease significantly after a land use change from primary rain forest to fallow land, however, it did increase significantly (49.3%) due to conversions to natural secondary forest and rubber plantations.

SOM characteristics is also influenced by seasonal changes, management practices and/or land uses and litter composition [18,19,20]. Djukic et al. [21] reported that the content of O-alkyl C increased with cooler and moisture climates, while aromatic C increased under warmer conditions. As the chemical composition of SOM can have significant impacts on its turnover affecting in turn the magnitude and the temporal pattern, the ecosystem responses to environmental changes [19,20,21,22] has not been well clarified yet. Moreover, there are a number of questions about the SOM formed in alpine soils. Because the SOM decomposition is considered as one of the largest C fluxes in the global terrestrial, an evaluation of different land use/vegetation cover may be a valid approach to better understand the key mechanisms to preserve it.

In this study three representative soil profiles from the mountain environment of NE Alpes (Italy), developed under the same pedogenetic, and climate conditions and different vegetation cover were described and analysed. The SOC balance was estimated and its more recalcitrant fraction was analysed by using 13C NMR spectroscopy. Our assumption was that the soil characteristics of our sites were comparable in the past, and that the current differences are due to different land use/vegetation cover.

2. MATERIAL AND METHODS

2.1. Study Site

The study was conducted on the Ampezzane Dolomites, which are located near Cortina d'Ampezzo in the province of Belluno (BL), on Northestern Alpes, Italy (Figure 1). The Ampezzane Dolomites include the massifs of Sorapiss, Croda da Lago, Nuvolau, and Cristallo, and show geomorphological features of past glacial conditions. The principal valley, the Cortina d'Ampezzo Valley, spans from 1224 m to 3244 m of altitude, and represents the site of this study (latitude 46°30′ N, longitude 12°08″ E). The climate is continental with a mean annual temperature of 4–5 °C, and the mean annual precipitation is of about 1000 mm, with a maximum in July, and prolonged periods without rain and snow in summer and in winter. The morphology is gently undulating, due to the widespread geological formation, which is made up of limestone yellowish layers, fine sandstones, and marne. Because of the easy erodibility, mass movements and flows are frequent. These materials originate Haplic Luvisols [22] which are generally rich in clay and relatively deep. The arboreal vegetation is comprised of forests characterized by the presence of *Picea excelsa* Lam. Link, *Abies alba* Mill, *Fagus sylvatica* L., and *Pinus cembra* L. Spruce is the most common tree species because of man who has favoured it eliminating beech and Silver fir. The remarkable pioneer attitude of spruce led to the formation of secondary spruce forests or new generation of spruce forests over large areas of abandoned pastures. These forests are replaced by grasslands, where the slopes are milder, and *Pinus mugo* T., where the slope is greater.

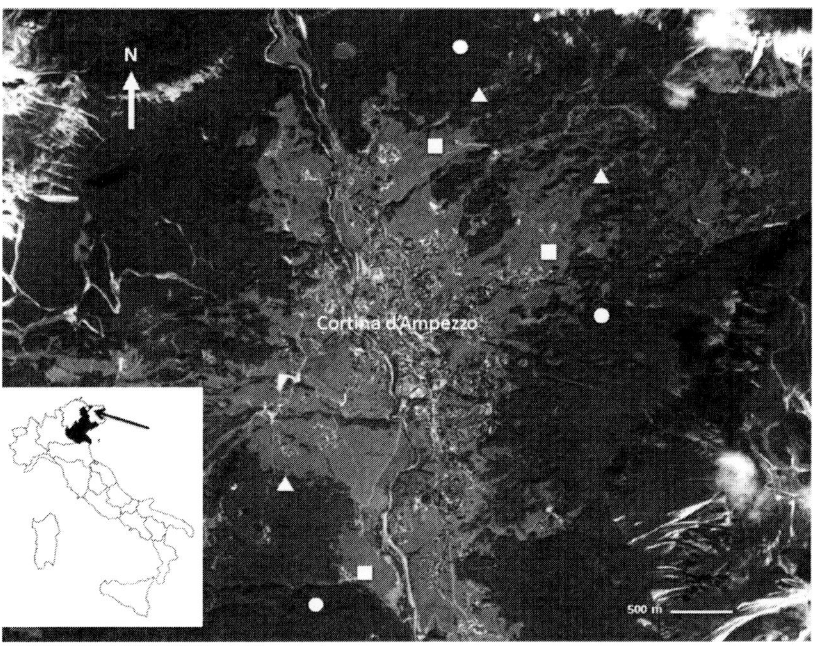

Figure 1. Map of Cortina d'Ampezzo Valley (Northestern Italy) indicating the study locations (circle, forest; triangle, reforestation; square, prairie).

2.2. Soil Sampling

In June, three 20 m × 20 m plots were set up based on three major land use types in this region: (i) prairie (P), (ii) reforestation by larch (R), and (iii) multi-layered forest (F) composed by spruce (*Picea abies* (L.) H. Karst), larch (*Larix decidua* Mill., and Scots pine (*Pinus sylvetris* L.); and three replicates per land use type were made (Figure 1). The plots were considered true replicates given that the distance between each other exceeded the spatial dependence of most soil chemical variables [23]. From the documentation present at the municipal offices of the town, we can state that the prairies have existed for several centuries, whereas the wooded area is at least of 200 years old.

Three sampling points in each plot were selected. We excavated the soil down to a depth where we found consolidated rock material. In order to estimate the stock of SOC with soil depth, intact soil cores were sampled at 20-cm intervals with 100-cm^3 stainless steel cylinders to quantify bulk density. Disturbed soil samples were also used for physical and chemical analysis. Data sheets for the soil profiles were compiled following the guidelines for the description of soils published by [24]. Forest and reforestation area soils were classified [22] as Haplic Luvisol, and prairie soils were classified as Aric Anthrosol.

2.3. Laboratory Analysis

The samples were air-dried. The proportion of soil skeleton was determined by weight after sieving the bulk samples at 2 mm. Soil pH was measured potentiometrically using 1:2.5 soil/water extracts. Total carbonate content (C) was determined by the calcimeter method and gravimetric loss of CO_2. Particle size analysis was performed according to the hydrometer method, using sodium hexametaphosphate as a dispersant [25]. Organic C was assayed by measuring dichromate oxidation [26] and total nitrogen (TN) according to the Kjeldahl method [27]. The bulk density was calculated by drying the soils at 105 °C and subsequently weighing the soil samples. Finally, the SOC stock (SOCs) for each sampling site was calculated using the following formula:

$$SOCs = \sum_{i}^{m} SOCi \times BDi \times Di \times (1 - Ski) \times 10 \quad (1)$$

where *SOCs* was the stock of *SOC*, SOC_i is the *SOC* concentration of the *i*th layer (g kg^{-1}), BD_i is the bulk density of the *i*th layer (g cm^{-3}), D_i is the thickness (m), and Sk_i is the skeletal (%) of this layer.

Humic substances were extracted with 0.1 N NaOH (1:10 *w/v*) by shaking the suspension for 16 h at room temperature in a N_2 atmosphere. The mixture was centrifuged at 5000 rpm for 30 min, and the supernatants were dialyzed in Visking tubes (18,000 mol wt cut-off) against distilled water to pH 6.0. Subsequently, the solution was filtered through Whatman No. 42 filter paper, and then through a column of Amberlite IR 120 H^+ form [28].

^{13}C-NMR spectra were performed by dissolving ca. 300 mg of humic extract with 0.5 mL of 2 M NaOH + 1.5 mL D_2O for one day at room temperature under N_2. The suspension was then filtered through acid-washed glass wool, which was then washed with 1 mL of D_2O, and the filtrates were then combined. The ^{13}C-NMR spectra were recorded at 125.7 MHz on a Bruker AMX-500 spectrometer (Bruker, Karlsruhe, Germany), using inverse-gated decoupling experiments for quantitative intensity distribution. The chemical shift was expressed in ppm on a scale relative to external sodium 3-trimethylsilvylpropiniate at 0 ppm. According to Candler et al. [29], each NMR spectrum was divided into the following five regions: alkyl-C (0–48 ppm), O,N-alkyl C (48–105 ppm), aromatic-C (105–145 ppm), phenolic-C (145–165 ppm), and carboxyl-C (165–200 ppm). The areas of the chemical shift regions were determined from the integral curves and were expressed as a fraction of total area.

2.4. Isotopic Determination

Carbon-13 to carbon-12 ratios ($\delta^{13}C$) were determined by a method similar to a procedure reported by Craig [30]. The system consisted of a Carlo Erba NA 1500 elemental analyzer (Carlo Erba, Milano, Italy) interfaced with a SIRA II-VG mass spectrometer (Micromass, Manchester, UK) for stable isotope ratio determination. The weight of sample for each determination was ca. 1.5 mg. Results are expressed in parts per million ‰ calculated according to equation $\delta‰ = (R_{sample}/R_{standard} - 1) \times 1000$, where R_{sample} and $R_{standard}$ are the isotopic ratios of the sample and reference, respectively. The international standard is CO_2 obtained from Peedee belemnite limestone from South Carolina with a conventional value $\delta = 0$. The Craig correction was applied to the raw parts per thousand values for the ^{17}O contribution to m/z of 45 [30]. The experiments were carried out in triplicate. The precision of the isotopic analysis was better than ±0.01‰. The overall reproducibility of the experiments, including the combustion, was ±0.2‰.

2.5. Statistical Analyses

All examined variables were tested for normality and homoscedasticity by Shapiro–Wilk's and Levene's tests, respectively, and transformed when necessary to satisfy assumptions required by parametric statistics. Data were the means of three independent replicates, and the standard deviations did not exceed 5%. Correlations between variables were determined using Spearman's coefficient. Two-way analysis of variance (ANOVA) was used to evaluate the soil physical-chemical variables under different land use types and depths. The Student–Newman–Keuls test was applied to compare the differences between group means [31]. Analyses were carried out in SPSS, version 22 (IBM SPSS, Chicago, IL, USA).

3. RESULTS

3.1. Morphology

Morphological descriptions of the three representative profiles belonging to the three types of land uses are reported (Table 1). Observations revealed that the formation of horizons was mostly layer dependent. What appeared related to the nature of the parent layer is the thixotropy, a property that was developed in quasi all horizons showing a sub-angular blocky structure. Nonetheless, in forest soil, properties appeared to not be completely inherited from the parent material, and in the Ah horizon, a friable structure prevailed. However, from chemical and physical analysis, pH had a narrow range of variability from forest (8.1) to reforestation and prairie (ca. 7.7) (Supplementary Materials Table S1), BD was high in reforestation and forest (ca. 1.2 g cm^{-3}) (Table S1) whereas, sand, silt, clay, and carbonates did not change with land use (Table S1). pH, BD, clay, and carbonates increased with depth ($p \leq 0.05$) (Table S2).

3.2. Soil Organic C and N Concentration and Stock under Different Land Uses and Depths

ANOVA results showed that SOC and TN were significantly affected by the land use (p = 0.001) and depth (p = 0.001). Forest had the highest concentrations in SOC (Figure 2), showing a mean value 2.2-fold higher than those in prairie (56.5 versus 25.7 g kg^{-1}) (p = 0.05) (Table S1). No differences were found between prairie and reforestation. SOC decreased throughout the profiles (Figure 2) from the upper Ah to the lower Ck layers (74.6–7.1 g kg^{-1}) ($p = 0.05$) (Table S1). Thus, forest Ah horizon had a SOC value 2.4-fold higher than reforestation and 3.3-fold higher than prairie (130, 54, 39.5 g kg^{-1}, respectively) (land use × depth interaction significant at $p = 0.001$) (Figure 3). Similar to SOC, the concentration in TN in the Ah horizon of forest was 2-fold higher than prairie and reforestation (5.5, 2.1, 2.4 g kg^{-1}, respectively) (land use × depth interaction significant at $p = 0.001$).

Table 1. Morphological description of the three representative profiles belonging to the three land use types (prairie, P; reforestation, R; forest, F), Cortina d'Ampezzo Valley, Italy.

Land Use	Soil Horizon	Depth cm	Colour [a]	Structure [b]	Roots [c]	Skeleton [d] %	Limit [e]	Observations
Soil under prairie (Aric Anthrosol, IUSS Working Group WRB, 2015)								
P	Ap	0–20	10YR 2/2 (very dark brown)	2,3 f, sbk	3 vf, m	5	c, i	no effervescent
	Bh	20–80	10YR 3/3 (dark brown)	m, sbk	1 vf, f	5	c, u	no effervescent
	CB	80–103	10YR 4/2 (dark greyish brown)	sbk	1 vf	25	c, u	weak effervescent
	Ck	103–>150	10YR 4/1 (dark grey)	ms	-	40	uk	Effervescent
Soil under reforestation (Haplic Luvisol, IUSS Working Group WRB, 2015)								
R	Ah	0–32	10YR 2/2 (very dark brown)	2,3 sbk	3 vf, f	5	c, i	no effervescent
	Bt	32–71	10YR 3/3 (dark brown)	m, sbk	1 f	3	c, u	no effervescent
	CB	71–92	10YR 4/2 (dark greyish brown)	sbk	1 vf	15	c, u	weak effervescent
	Ck	92–>150	10YR 4/1 (dark grey)	ms	-	40	uk	effervescent
Soil under forest (Haplic Luvisol, IUSS Working Group WRB, 2015)								
F	Ah	0–16	10YR 2/2 (very dark brown)	2 fr	3 vf, m	2	c, i	no effervescent
	ABh	16–51	10YR 3/3 (dark brown)	2 m, sbk	2 vf, f	2	c, l	no effervescent
	Bt	57–88	10YR 3/6 (dark greyish brown)	m, sbk	1 f	5	c, u	weak effervescent
	Ck	88–>150	10YR 4/1 (dark grey)	ms	-	50	uk	effervescent

[a] Moist and crushed, according to the Munsell soil colour charts (1954 edition). [b] 1 = weak, 2 = moderate, 3 = strong; f = fine, m = medium, c = coarse; fr = friable; g = granular, sbk = sub-angular blocky, s = massive. [c] 1 = few, 2 = plentiful, 3 = abundant; mi = micro, vf = very fine, f = fine, m = medium. [d] By mass. [e] c = clear; i = irregular; u = undulating; uk = unknown.

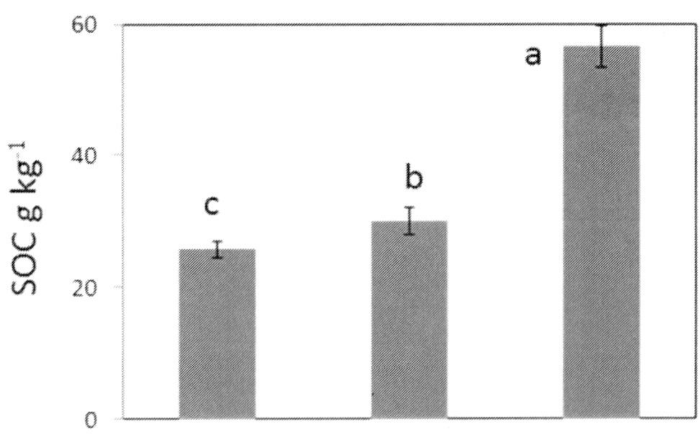

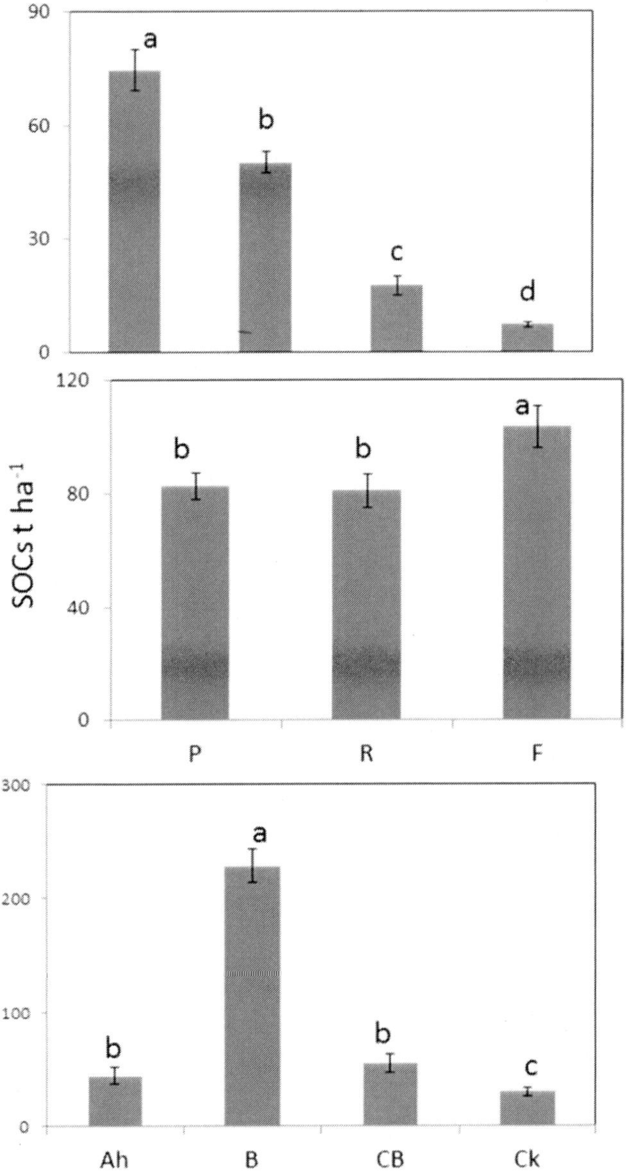

Figure 2. Soil organic carbon concentration and stock per land use (forest, F; reforestation, R; prairie, P) and soil depth. Data followed by the same letter are not statistically different from each other (Student's test, $p = 0.05$).

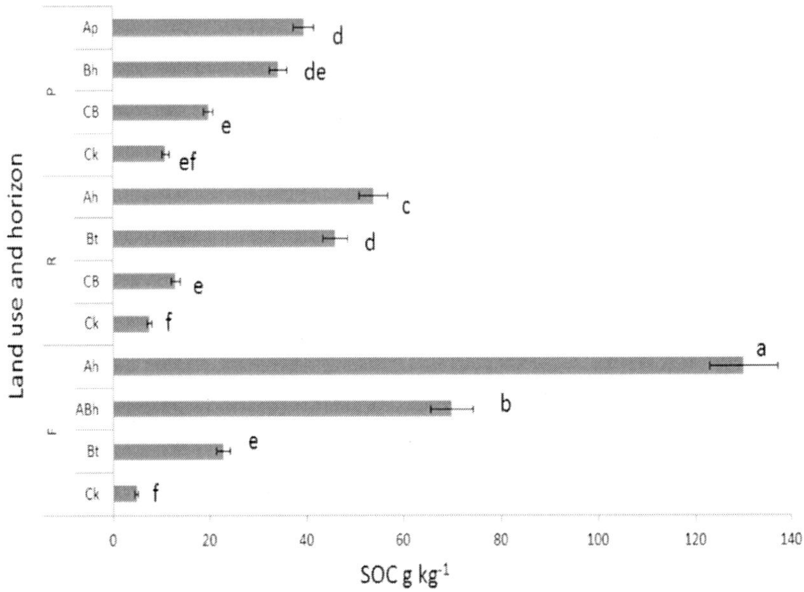

Figure 3. Soil organic carbon concentration according to the land use (forest, F; reforestation, R; prairie, P) per soil depth interaction. Data followed by the same letter are not statistically different from each other (Student's test, $p = 0.05$).

Concerning the stock in organic carbon (SOCs) (t ha^{-1}) (Figure 2), forest soil had a higher mean value (103 t ha^{-1}) than prairie and reforestation (*ca*. 80 t ha^{-1}) ($p = 0.05$) (Table S1), whilst the B horizon was the more endowed (228 t ha^{-1}) among layers ($p = 0.05$) (Table S2). Thus, the ABh horizon of forest had the maximum amount in SOCs, reaching a value of 264 t ha^{-1} (Figure 4). Calculating the mean upper 20 cm layer, SOCs increased from the prairie and reforestation to the forest, showing values of 31.9, 42 and 120 t ha^{-1}, respectively. This corresponded to increments of 1.3-fold prairie for reforestation, and 3.8-fold prairie for forest.

3.3. Humic Substance Features under Different Land Uses and Depths

Forest had a higher C:N ratio (20.0) than prairie and reforestation (ca. 17.3) ($p = 0.05$) (Table S1), and a HC content of 21.7 g kg^{-1}, which was greater than reforestation (2.8-fold) ($p = 0.05$), and prairie (4-fold) ($p = 0.05$) (Table S1). Both C:N ratio and humic C (HC) content decreased with depth, showing values from 21.8 and 31.1 g kg^{-1} in the surface layer to 11.5 and 0.8 g kg^{-1} at depth ($p = 0.05$) (Table S2).

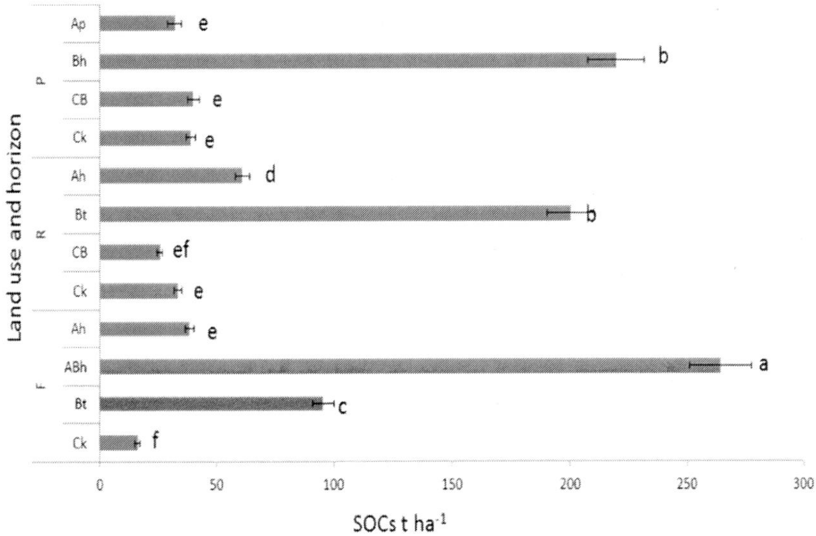

Figure 4. Soil organic carbon stock according to the land use (forest, F; reforestation, R; prairie, P) per soil depth interaction. Data followed by the same letter are not statistically different from each other (Student's test, $p = 0.05$).

^{13}C-NMR spectra of humic substances revealed a different pattern in C with O,N-alkyl > alkyl > aromatic > carboxyl > phenolic. With respect to land use, reforestation and prairie had a higher content in O,N-alkyl C (49.43% and 46.80%, respectively) with respect to forest (39.38%) ($p = 0.05$) (Table 2). Comparing prairie with reforestation no differences were found in alkyl (22.85% and 23.76%) and aromatic C (16.91% and 16.23%), whilst both increased in forest (25.97% and 20.20%) ($p = 0.05$) (Table 2). Carboxyl and phenolic C were higher in forest (11.76% and 4.03%, respectively) and prairie (11.3% and 4.72%) than in reforestation (8.40% and 2.93%) ($p = 0.05$) (Table 2). Respect to depth, alkyl, carboxyl, and phenolic C prevailed in the Ah horizons, whereas aromatic and O,N-alkyl C were dominant in the B horizons (Table 2). The $\delta^{13}C$ of HS was falling in a narrow range with lower values in prairie (−25.9‰) than in reforestation and forest (−24.9 and −25.1‰), whilst B horizons had higher $\delta^{13}C$ values (−24.48‰) than Ah horizons (−26.12‰) ($p = 0.05$).

Table 2. Distribution of C determined ^{13}C-NMR and $\delta^{13}C$ in the three land use types, Cortina d'Ampezzo Valley, Italy.

Land Use		Chemical Shift of C (ppm) %					$\delta^{13}C$ ‰
	Hor	Carboxyl	Phenolic	Aromatic	O,N-Alkyl	Alkyl	
P	Ah	11.63 b*	5.43 a	13.73 d	45.80 b	25.80 c	−26.80 d
	B	11.13 b	4.01 b	20.10 b	47.80 b	19.90 e	−24.96 b
		11.3 A	4.72 A	16.91 B	46.80 A	22.85 B	−25.88 B
R	Ah	10.91 b	4.00 b	13.93 d	43.73 bc	27.50 b	−25.90 c
	B	5.90 d	1.86 c	18.53 bc	55.13 a	20.03 e	−23.93 a
		8.40 B	2.93 C	16.23 B	49.43 A	23.76 B	−24.91 A
F	Ah	16.16 a	4.23 b	15.53 c	36.86 d	28.93 a	−25.66 c
	B	7.36 c	3.83 b	24.86 a	41.90 c	23.00 d	−24.56 b
		11.76 A	4.03 B	20.20 A	39.38 B	25.97 A	−25.11 A
	Ah	12.90 *a*	4.55 *a*	14.40 *b*	42.13 *b*	27.41 *a*	−26.12 *b*
	B	8.13 *b*	3.23 *b*	21.16 *a*	48.27 *a*	20.98 *b*	−24.48 *a*

* In the same column differences among land use (capital letters), depth (italicized letters), and land use × depth interaction (lowercase) were at $p = 0.05$ by Student-Newman-Keuls test.

The correlations between soil physical and chemical parameters, SOC and TN were weakly correlated with pH ($r = -0.39$, $p = 0.019$ and $r = -0.40$, $p = 0.015$) and clay ($r = -0.50$, $p = 0.002$ and $r = -0.52$, $p = 0.001$), whilst SOC was highly and negatively correlated with TN ($r = -0.98$, $p < 0.001$), HC ($r = -0.95$, $p < 0.001$), and with BD ($r = -0.75$, $p < 0.001$).

Moreover, HC showed high correlation value with alkyl C ($r = 0.82$, $p = 0.001$), O,N-alkyl C ($r = -0.83$, $p = 0.001$), and carboxyl C ($r = 0.76$, $p = 0.001$). $\delta^{13}C$ was strongly correlated with clay ($r = 0.90$, $p = 0.001$), SOC stock ($r = 0.79$, $p = 0.001$), HC:OC ratio ($r = 0.75$, $p = 0.001$), phenolic C ($r = 0.88$, $p = 0.001$), aromatic C ($r = 0.71$, $p = 0.001$), alkyl C ($r = 0.71$, $p = 0.001$), and carboxyl C ($r = 0.67$, $p = 0.002$).

4. DISCUSSION

Our results showed land use is reflected on SOC concentration and stock, thus forest had high SOC concentrations and stocks with respect to reforestation and prairie. This confirms that in boreal ecosystems forests have higher litter input than prairies [32,33]. Nevertheless, the amount of organic C in soil is known to be affected also by soil properties such as bulk density, and high SOC promotes larger soil pores, which in turn decreases bulk density [34]. In fact, BD was inversely related with SOC, and it was significantly affected by land use. Indeed, forest had high SOC and relative low BD, whereas prairie had low SOC and high BD. A similar trend was reported comparing middle-aged forestland with shrubland [32]. However, while concentration in SOC was highest in the top soil layers, the stock in SOC had its maximum in the Bh, Bt, and ABh horizons. The increased SOCs in the B horizon may be due to the contribution of root litter and downward SOC transport by bioturbation and/or soil water percolating. SOC accumulation was also strictly related with soil pH as already found [35,36].

Behind the stock in SOC, in this study we found that land use affected the SOM evolution and production of more stable C forms, such as humus. Indeed, the relatively low C:N and production in humic carbon found in prairies indicated that SOM progressed towards humification, while the high C:N and HC of forests indicate that the SOM moved towards the humification-immobilization process. Intermediate position between prairie and forest was instead found in reforestation.

By studying the organic matter composition via ^{13}C-NMR technique, peptide, carbohydrate, and alkyl C were found to be the main components of the HS in forest soils [36]. Indeed, the alkyl and aromatic C increased from the prairies to forests. Probably, the inputs of waxes from resinous litter could have increased the proportion of alkyl groups in forest soils. Indeed, the SOC under grass vegetation has been reported to be poorer in alkyl-C compared to the SOC under forest [19,37,38]. Overall, the marked distribution of O,N-alkyl C and alkyl C in HS suggest a low degree of humification that could be partly due to anthropic activity. Nonetheless, in forest, the relatively high presence of aromatic C indicated a higher humification degree than prairie and reforestation. ^{13}C-NMR results also showed marked changes due to soil depth. In particular, the labile components such as alkyl and carboxylic C decreased with increasing depth, while the more persistent aromatic C components dominated. The observed pattern may be associated with progressing litter decomposition and increasing humification. These results agree with previous findings [21] that found along an alpine elevation gradient that labile organic compounds dominated the upper soil layers, whereas persistent materials accumulated in deep layers. Similar results were previously demonstrated for humic acids [39]. Moreover, high pH implies favourable conditions for organic matter decomposition leading to humification and accumulation of aromatic structures [21-40]. Indeed, in our study, the aromatic components of HS positively correlated with the soil pH, while the carboxyl C showed an opposite trend. δ^{13} C values can give important information about origin and decomposition patterns of the organic matter. Stable carbon isotopes are important markers of OM and are widely used in studies of ecology [41]. Our data proved that organic matter extracted from the three different land uses was produced by C_3 vegetation, as expected for an alpine environment such as that of Cortina, which experienced a temperate or colder climate. Moreover, the slightly lower (more negative) and higher (less negative) δ^{13} C values in Ah and B horizons, respectively, depended on the different chemical composition of organic compounds. In fact, during decomposition processes some isotopic fractionation does occur, due to the preference accorded by microorganisms to ^{12}C with respect to the heavier ^{13}C [42]. A similar progressive increase of $\delta^{13}C$ with depth has been reported for the bulk SOM by other authors [43,44], and thus it may be inferred that the stronger decomposition of the extractable SOM in the deep horizons reflects that of the bulk SOM. Land use might influence the stabilization of CO_2 levels increasing C sink in soils [15].

It must be remembered that in forestry contexts as in this work, the concept of land use is closely linked to a different plant cover which consequently implies a different management. It is therefore difficult to determine whether the differences found in the studied soils are due to the different use of land or different coverage. Plant cover, stand age and site physical and topographic variables are known to affect soil carbon content and characteristics [6,9], but this study showed that the alpine forest carries a large amount of organic carbon and humic carbon, which is respected both at reforestation and prairie. Therefore, by virtue of a management point of view aimed at decreasing or counteracting the increase in the amount of CO_2 in the atmosphere, result important to preserve and where possible increase the surface dedicated to such uses of soil. Forest, obviously, appears to be the much more efficient ecosystem than prairie and reforestation in term of carbon storage. In terms of management it should also be encouraged the formation of thick B horizons which are useful as site of humic carbon, the more stable and more fertile carbon in soil. However, it is still important to address the assumption we made that the prairie and forest types were comparable in the past, and that the current soil attribute differences were due to the presence of forest. However, because these locations are so long established (both forest and prairie have been in that condition for 200 years or longer), we really do not know the origin of those vegetation types. It is certainly possible that there were initial differences in soils or other conditions that favoured the establishment of forest on the plots that are now forest, partially explaining the difference in the detected soil characteristics.

5. CONCLUSIONS

In this study, we found in alpine environment of NE Italy that SOC concentration and stock changed consistently with land use type and soil depth. Each horizon of the forest soil had greater amounts of SOC with respect to prairie and natural reforestation. In particular, in the 20 cm surface layer of a soil forest, carbon stock was 4 times higher in comparison to prairie, demonstrating that change in land use or disturbance may be the cause of soil carbon loss. Spectroscopic analysis provided evidence of a marked distribution of O,N-alkyl C and alkyl C in HS, suggesting a low degree of humification that could be partly due to anthropic activity. Nevertheless, in forest, the relatively high presence of aromatic C indicated that HS endowed a high humification degree. In short, the results of this study provide evidence that land use change from prairie to secondary forest plantation improves the humification process, by promoting soil carbon storage in the form of refractory C, such as humus. Therefore, it is possible to suggest that the conversion of prairie to forestland may represent a good practice for C sequestration.

ACKNOWLEDGMENTS

The authors would like to remember and thank Dino Dibona, pedologist, naturalist, and connoisseur of the mountain.

REFERENCES

1. Fang, X.; Zhang, C.; Wang, Q.; Chen, X.; Ding, J.; Karamage, F. Isolating and Quantifying the Effects of Climate and CO_2 Changes (1980–2014) on the Net Primary Productivity in Arid and Semiarid China. *Forests* 2017, *8*, 60.
2. Zanella, A.; Jabiol, B.; Ponge, J.F.; Sartore, G.; De Waal, R.; Van Delft, B.; Cools, U.; Katzensteiner, K.; Hager, H.; Englisch, M. A European morpho-functional classification of humus forms. *Geoderma* 2011, *164*, 138–145
3. Grüneberg, E.; Ziche, D.; Wellbrock, N. Organic carbon stocks and sequestration rates of forest soils in Germany. *Glob. Chang. Biol.* 2014, *20*, 2644–2662
4. Six, J.; Conant, R.T.; Paul, E.A.; Paustian, K. Stabilization mechanisms of soil organic matter: Implications for C-saturation of soils. *Plant Soil* 2002, *241*, 155–176
5. Kaiser, K.; Guggenberger, G. Mineral surfaces and soil organic matter. *Eur. J. Soil Sci.* 2003, *54*, 219–236

6. Pizzeghello, D.; Nicolini, G.; Nardi, S. Hormone-like activities of humic substances in different forest ecosystems. *New Phytol.* 2002, *155*, 393–402
7. Augusto, L.; Ranger, J.; Binkley, D.; Rothe, A. Impact of several common tree species of European temperate forests on soil fertility. *Ann. For. Sci.* 2002, *59*, 233–253
8. Vesterdal, L.; Schmidt, I.K.; Callesen, I.; Nilsson, L.O.; Gundersen, P. Carbon and nitrogen in forest floor and mineral soil under six common European tree species. *For. Ecol. Manag.* 2008, *255*, 35–48
9. Carletti, P.; Vendramin, E.; Pizzeghello, D.; Concheri, G.; Zanella, A.; Nardi, S.; Squartini, A. Soil humic compounds and microbial communities in six spruce forests as function of parent material, slope aspect and stand age. *Plant Soil*2009, *315*, 47–65
10. Vesterdal, L.; Clarke, N.; Sigurdsson, B.D.; Gundersen, P. Do tree species influence soil carbon stocks in temperate and boreal forests? *For. Ecol. Manag.* 2013, *309*, 4–18
11. Augusto, L.; De Schrijver, A.; Vesterdal, L.; Smolander, A.; Prescott, C.; Ranger, J. Influences of evergreen gymnosperm and deciduous angiosperm tree species on the functioning of temperate and boreal forests. *Biol. Rev.*2015, *90*, 444–466
12. Muscolo, A.; Bagnato, S.; Sidari, M.; Mercurio, R. A review of the roles of forest canopy gaps. *J. For. Res.* 2014, *25*, 725–736
13. James, J.; Harrison, R. The Effect of Harvest on Forest Soil Carbon: A Meta-Analysis. *Forests* 2016, *7*, 308
14. Brown, S.; Lugo, A.E. Tropical secondary forests. *J. Trop. Ecol.* 1990, *6*, 1–32
15. Arnhold, S.; Otieno, D.; Onyango, J.; Koellner, T.; Huwe, B.; Tenhunen, J. Soil properties along a gradient from hillslopes to the savanna plains in the Lambwe Valley, Kenya. *Soil Tillage Res.* 2015, *154*, 75–83
16. Seyoum, B. Assessment of soil fertility status of Vertisols under selected three land uses in Girar Jarso District of North Shoa Zone, Oromia National Regional State, Ethiopia. *Environ. Syst. Res.* 2016, *5*, 18
17. Li, Y.; Xia, Y.; Lei, Y.; Deng, Y.; Chen, H.; Sha, L.; Cao, M.; Deng, X. Estimating changes in soil organic carbon storage due to land use changes using a modified calculation method. *iForest* 2015, *8*, 45–52
18. Bravo, F.; Lucà, M.; Mercurio, R.; Sidari, M.; Muscolo, A. Soil and forest productivity: A case study from Stone pine (*Pinus pinea* L.) stands in Calabria (southern Italy). *iForest* 2011, *4*, 25–30
19. Montecchio, D.; Francioso, O.; Carletti, P.; Pizzeghello, D.; Chersich, S.; Previtali, F.; Nardi, S. Thermal analysis (TG-DTA) and DRIFT spectroscopy applied to investigate the evolution of humic acids in forest soil at different vegetation stages. *J. Ther. Anal. Calorim.* 2006, *83*, 393–399
20. Muscolo, A.; Panuccio, M.R.; Mallamaci, C.; Sidari, M. Biological indicators to assess short-term soil quality changes in forest ecosystems. *Ecol. Ind.* 2014, *45*, 416–423
21. Djukic, I.; Zehetner, F.; Tatzber, M.; Gerzabek, M. Soil organic-matter stocks and characteristics along an Alpine elevation gradient. *J. Plant Nutr. Soil Sci.* 2010, *173*, 30–38
22. IUSS Working Group WRB. *World Reference Base for Soil Resources 2014, Update 2015. International Soil Classification System for Naming Soils and Creating Legends for Soil Maps*; World Soil Resources Reports No. 106; FAO: Rome, Italy, 2015
23. Marriott, C.A.; Hudson, G.; Hamilton, D.; Neilson, R.; Boag, B.; Handley, L.L.; Robinson, D. Spatial variability of soil total C and N and their stable isotopes in an upland Scottish grassland. *Plant Soil* 1997, *196*, 151–162
24. FAO. *Guidelines for Soil Description*, 4th ed.; FAO: Rome, Italy, 2006; p. 110
25. Gee, G.W.; Bauder, J.W. Particle-size analysis. In *Methods of Soil Analysis, Part 1. Physical and Mineralogical Methods*, 2nd ed.; Agronomy Monograph; American

Society of Agronomy/Soil Science Society of America: Madison, WI, USA, 1986; Volume 9, pp. 383–411
26. Walkley, A.; Black, I.A. An examination of Degtjareff method for determining soil organic matter and a proposed modification of the chromic acid titration method. *Soil Sci.* 1934, *37*, 29–38
27. Kjeldahl, J. A new method for the determination of nitrogen in organic matter. *Z. Anal. Chem.* 1883, *22*, 366–382
28. Stevenson, F.J. *Humus Chemistry: Genesis, Composition, Reactions*; John Wiley and Sons: New York, NY, USA, 1994
29. Candler, R.; Luping, C.; Michaelson, G.J.; Malcolm, R.L. Characterization of fractionated soil organic matter from two Alaskan arctic soils by 13C NMR and titration analysis. In *The Role of Humic Substances in the Ecosystems and in Environmental Protection*; IHSS Polish Society of Humic Substances: Wroclaw, Poland, 1997; pp. 301–307
30. Craig, H. Isotope standards for carbon and oxygen and correction factors for mass spectrometric analysis of carbon dioxide. *Geochim. Cosmochim. Acta* 1957, *12*, 133–149
31. Sokal, R.R.; Rohlf, F.J. *Biometry*, 1st ed.; Freeman: San Francisco, CA, USA, 1969
32. Wang, T.; Kang, F.; Cheng, X.; Han, H.; Ji, W. Soil organic carbon and total nitrogen stocks under different land uses in a hilly ecological restoration area of North China. *Soil Tillage Res.* 2016, *163*, 176–184
33. Gelaw, A.M.; Singh, B.R.; Lal, R. Soil organic carbon and total nitrogen stocks under different land uses in a semi-arid watershed in Tigray, North Ethiopia. *Agric. Ecosyst. Environ.* 2014, *188*, 256–263
34. Lister, T.W.; Burger, J.A.; Patterson, S.C. Role of vegetation in mitigating soil quality impacted by forest harvesting. *Soil Sci. Soc. Am. J.* 2004, *68*, 263–271
35. Motavalli, P.P.; Palm, C.A.; Parton, W.J.; Elliott, E.T.; Frey, S.D. Soil pH and organic carbon dynamics in tropical forest soils: Evidence from laboratory and simulation studies. *Soil Biol. Biochem.* 1995, *27*, 1589–1599
36. Pizzeghello, D.; Berti, A.; Nardi, S.; Morari, F. Phosphorus-related properties in the profiles of three Italian soils after long-term mineral and manure applications. *Agric. Ecosyst. Environ.* 2014, *189*, 216–228
37. Beyer, L. The chemical composition of soil organic matter in classical humic compound fractions and in bulk samples—A review. *J. Plant Nutr. Soil Sci.* 1996, *159*, 527–539
38. Golchin, A.; Clarke, P.; Baldock, J.A.; Higashi, T.; Skjemstad, J.O.; Oades, J.M. The effects of vegetation and burning on the chemical composition of soil organic matter in a volcanic ash soil as shown by 13C NMR spectroscopy. I. Whole soil and humic acid fraction. *Geoderma* 1997, *76*, 155–174
39. Tatzber, M.; Stemmer, M.; Spiegel, H.; Katzlberger, C.; Haberhauer, G.; Mentler, A.; Gerzabek, M.H. FTIR-spectroscopic characterization of humic acids and humin fractions obtained by advanced NaOH, $Na_4P_2O_7$, and Na_2CO_3 extraction procedures. *J. Plant Nutr. Soil Sci.* 2007, *170*, 522–529
40. Quideau, S.A.; Anderson, M.A.; Graham, R.C.; Chadwick, O.A.; Trumbore, S.E. Soil organic matter processes: Characterization by 13CNMR and 14C measurements. *For. Ecol. Manag.* 2000, *138*, 19–27
41. Guo, Q.; Zhu, G.; Chen, T.; Yang, J.; Yang, J.; Peters, M.; Wei, R.; Tian, L.; Han, X.; Hu, J. Spatial variation and environmental assessment of soil organic carbon isotopes for tracing sources in a typical contaminated site. *J. Geochem. Explor.* 2017, *175*, 11–17
42. Ågren, G.O.; Bosatta, E.; Balesdent, J. Isotope discrimination during decomposition of organic matter: A theoretical analysis. *Soil Sci. Soc. Am. J.* 1996, *60*, 1121–1126
43. Bol, R., Harkness, D.D.; Huang, Y.; Howard, D.M. The influence of soil processes on carbon isotope distribution and turnover in the British uplands. *Eur. J. Soil Sci.* 1999, *50*, 41–51

44. Certini, G.; Agnelli, A.; Corti, G.; Capperucci, A. Composition and mean residence time of molecular weight fractions of organic matter extracted from two soils under different forest species. *Biogeochemistry* 2004, *71*, 299–316

CHAPTER 4

Agroforestry and Management of Trees in Bunya County, Mayuge District, Uganda

Monica Kyarikunda,[1] Antonia Nyamukuru,[2] Daniel Mulindwa,[3] and John R. S.Tabuti[2,4]

[1]*Bishop Stuart University, P.O. Box 09, Mbarara, Uganda*

[2]*Sustainable Use of Plant Diversity (SUPD), P.O. Box 16794, Wandegeya, Uganda*

[3]*Department of Environmental Management, College of Agriculture and Environmental Sciences, Makerere University, P.O. Box 7062, Kampala, Uganda*

[4]*College of Agricultural and Environmental Sciences (CAES), Makerere University, P.O. Box 7062, Kampala, Uganda*

ABSTRACT

Woody plant resources continue to disappear in anthropogenic landscapes in Uganda. To slow down further loss of these resources requires the collaboration of farmers in tree planting in agroforestry systems. Tree planting interventions with the collaboration of farmers require a good understanding of tree management practices as well as trees that best satisfy farmers' needs. We carried out this research to determine (1) the most preferred tree species and reasons why they are preferred, (2) the species conservation statuses, and (3) existing tree management practices and challenges to tree planting. Fourteen priority species valued because they yield edible fruits and timber have been prioritised in this study. Farmers are interested in managing trees but are constrained by many factors, key among which is scarcity of land and financial capital to manage tree planting. Trees are managed in crop fields and around the homestead. From farmers' reports, the highly valued species are increasing in the landscape. In conclusion, the potential to manage trees in agroforestry systems exists but is hampered by many challenges. Secondly, the liking of trees that supply edible fruits seems to support the welfare maximisation theory which ideally states that rural people manage trees with the aim of having regular access to products that satisfy their household needs and not for income generation.

1. INTRODUCTION

Woody species (i.e., tree and shrubs) are threatened with local extinction and are disappearing fast, most especially in tropical Africa [1, 2]. In Uganda tree loss is rapidly growing and is estimated at 1.8% per year [3]. This rate of tree loss is fastest in anthropogenic landscapes outside protected areas where it is estimated to occur at a rate of 2.27% per annum. For this reason the forest estate outside protected areas reduced from 3.46 million to 2.3 million ha between 1990 and 2005, a difference of about 1.16 million ha. Mayuge District, where this study was conducted, is one of the districts that has had the greatest loss of woody species and has lost almost all its forest cover [3].

The reasons that lead to loss of trees are chiefly land use change. In most parts of the world including Uganda land cover has been converted from woodlands or forests to create land for crop and livestock agriculture. The underlying cause for most of the land use change and consequently woody cover loss, in Uganda, is human population growth. The rapid human population growth estimated at more than 3% per annum has increased demand for land for cultivation and settlement, as well as woody products. In Uganda people exploit woody resources as fire wood, for timber and poles for construction purposes [4]. As an example of the level of woody species demand, over 95% of Uganda's households depend on firewood and charcoal for their energy needs and 91% of all roundwood consumed is consumed as fuelwood [5]. It is this extensive consumption of tree products that leads to overexploitation and land cover change.

The rapid loss of trees and shrubs threatens livelihoods that are dependent on those resources. Human populations, globally, depend very much on plant resources for their sustenance and economy [1], for example, the use of fuelwood for cooking. In addition plants (woody or otherwise) also generate ecosystem services that are invaluable to humans in ways such as rainfall generation or provision of nectar for pollinators. Because of these reasons, the rate of woody species loss must be slowed in Uganda.

To address the challenge of fast tree cover loss in Uganda will require increases in rates of tree planting. At present Uganda has made important strides in tree planting on lands outside protected areas through interventions for establishment of tree plantations. However, most of these interventions such as the Sawlog Production Growers Scheme (SPGS) have targeted large plantation farmers and have ignored small holder farmers. Successful increase in tree/shrub cover requires the participation of small holder farmers because they own most of the land in Uganda: approximately 70% of all land in Uganda is privately owned by small holder farmers [6]. This means therefore that small holder farmers are invaluable stakeholders in interventions for increasing tree/shrub cover in Uganda.

To involve farmers, we need to understand their attitudes and perceptions concerning tree management, how they currently manage species, and what challenges hinder tree growing or which opportunities exist for growing trees. Farmers make decisions whether to participate or not in interventions such as tree growing based on considerations of how such interventions maximise their welfare and also fit into their livelihood strategy and available resources [7–9]. Overall these decisions are also influenced by historical events such as circumstances that have directed resource availability. According to Miller et al., [8], analysing people's decisions can reveal key variables that affect their behaviours and which can therefore be used to design successful and sustainable interventions.

Additionally, farming systems include a wide variety of agroforestry practices. These tend to vary from place to place in both extent and intensity of management [10] in line with the varying agroecological and socioeconomic factors prevailing in those areas. This validates the need to make agroforestry studies location-specific and avoid too much generalization especially by development organizations [11]. This study was undertaken in Mayuge District to address an otherwise ignored topic and to increase understanding about the perception and behaviour of farmers towards the potential of adoption of agroforestry systems to ameliorate the environmental challenges of Uganda. We specifically determined (1) the most preferred tree species and reasons why they are preferred, (2) the species conservation statuses, and (3) existing tree management practices and challenges to tree planting.

2. STUDY AREA AND METHODS

Bunya County is found in Mayuge District, Eastern Uganda, at 00°12′N, 33°32′E, and approximately 120 km from Kampala the Capital City of Uganda. The district neighbours other major towns, namely, Iganga and Jinja. This means that there is high demand for tree products.

The district has a tropical climate with a bimodal rainfall pattern with peaks in March–May and October-November; the rainfall ranges between 900 and 1200 mm per annum. The district is endowed with fertile soils and hence has high potential for agriculture. The topography is relatively flat with high ridges and isolated hills, wavy low lands, and open water reserves [12]. The soils are skeletal, shallow, or poorly drained and are characterized mostly as Nitisols and Luvisols.

The vegetation of Mayuge has adversely been affected by rapid population growth with its attendant problems of deforestation, environmental degradation, and pollution [12]. Between 1990 and 2005, the district lost 97% of its forest cover, starting with 15,000 ha in 1990; only 450 ha were remaining by 2005. In the same period agriculture increased by more than 200% from 660 ha in 1990 to 2006 ha in 2005 [3].

The population of the district is estimated at approximately 473,000 people and the population density at 440 people/km^2 [13]. The main source of livelihood is subsistence farming. Sixty-two percent of the population is 19 years of age or less. Most of the agriculture in the district is on the subsistence level. However, there is commercial sugar cane growing that is increasing and many residents are cutting down trees to establish sugar cane plantations (Makerere takes climate change project to Mayuge to encourage farmers to plant trees (http://caes.mak.ac.ug/news/news-archieve/1235-makerere-takes-climate-change-project-to-mayuge-to-encourage-farmers-to-plant-trees.html)). The economic activities include crop cultivation (e.g., maize, cassava, groundnuts, cocoa, cotton, coffee, beans, sweet potatoes, and millet) and fishing. All these demographic and socioeconomic characteristics point to a high potential to overexploit tree resources and to convert land cover to agriculture.

2.1. Methods Field work for this study was carried out between July and September 2012. The study adopted an ethnobotanical approach. We conducted interviews using structured open and close ended questionnaire in face to face interviews with household heads.

Respondents were selected using multistage sampling procedure based on administrative units. In Uganda 5 or 6 villages make a parish, a similar number of parishes make a subcounty, and a similar number of subcounties are collected to make a county. Using random sampling procedure, three subcounties were selected from each county and, from each of these, one parish was randomly selected. Lastly, from each selected parish one village was selected. Thirty households were then selected from each of the three villages to make a total of 90 respondents (44 of these were female).

The interviews were designed to collect information on the most preferred species. Free listing techniques were used, where informants were requested to list ten species that came first to mind. This technique was important because the aim was to focus on the most important species and people tend to mention important species first [14]. Interviews also investigated respondent perceptions of trees conservation status; tree management practices, ownership of tree resources (tree tenure systems), tree products marketing, and threats and constraints to the management of woody plant species; and also the personal profiles of the respondents. Respondents were asked to give the current state of woody tree species by listing down species that are locally increasing or reducing and factors contributing to these changes. Every interview lasted about 30 minutes. At the end of the interview every household was given a token of appreciation to compensate for time spent in interviews. Although rewarding respondents is a controversial issue among researchers, some have recommended it on grounds of ethics and efficiency [15], and we share a similar opinion.

Voucher specimens were collected and taken to Makerere University Herbarium (MHU) where identification and archiving were done. The survey data was summarised into frequencies/percentages using Microsoft Excel 2007. To test for relationships within the data set, frequencies were computed and chi-square tests were carried out on cross tabulations. Preferences were ranked based on frequencies considering the species with higher frequencies to be the most preferred.

A use value was computed for each species using the expression below:

$$V_s = \sum \frac{f_i}{n}, \tag{1}$$

where V_s is the use value for a species s, f_i is the frequency of mentioning a given use of species s by respondents, and n is the total number of respondents in the study (see Nyamukuru et al. [16] for a description).

3. RESULTS

3.1. Respondents' Socioeconomic Characteristics

The respondents interviewed in this study were mostly peasant farmers (82%) who earn their livelihood through subsistence farming of food crops that include maize, cassava, sorghum, and sweet potatoes and also cash crops like sugar cane. The respondents are mostly Christians of mixed denominations: Protestants, Catholics, and Pentecostals (Table 1). They are affiliated to 13 tribes, dominated by the Basoga (59%) and the Japadholas (14%). Other tribes included Bagisu, Banyole, Atesots, Badama, Baluya, Bakedi, and Bagwere. They have little or no formal education, not exceeding secondary level or 13 years of education.

Table 1: Respondents' sociodemographic details .

Attribute	Percentage
Age	
<30	26
30–49	40
50–69	30
>70	4
Education	
None	31
Primary	42
Secondary	23
Tertiary	3
Sex of household head	
Female headed	23
Male headed	77
Tribe	
Musoga	59
Japadhola	14
Muganda	6
Other tribes	21
Religion	
Moslem	43
Protestant	32
Catholic	20
Pentecostal	4

3.2. Preferred Woody Species and Their Values

Fourteen tree species are valued by the people of Bunya County. The species have been prioritised based on computed use values and reports by respondents that they are grown by the people of Bunya and that they will continue to grow them in the future and also because the species were mentioned by 30% or more of the respondents (Table 2). An additional three species Eucalyptus spp., Senna siamea, and Senna spectabilis, though not shown in Table 2, because they had very low total use values (=0.20), are considered: Eucalyptus was liked by farmers and 15% had plans to plant it, while the Senna species were being planted by 24% of the farmers.

Table 2: Priority woody species of Bunya and their uses. Shown are use values and component category use values. Also shown are proportions of respondents mentioning the species and those who have planted/are planning to plant the species. The total use values are larger than component category use values because category use values less than 0.1 have been excluded.

Species	Total use value	Use value by category: Edible fruit	Timber	Construction poles and wood	Firewood	Income	Respondents mentioning species (%)	Respondents who have planted species (%)	Respondents planning to plant species (%)
Artocarpus heterophyllus Lam.	1.30	0.93	0.13	0.02	0.07	0.06	99	72	31
Mangifera indica L.	1.23	0.93	0.03		0.11	0.07	99	87	31
Milicia excelsa (Welw.) C. Berg	1.03	0.01	0.57	0.32	0.04	0.02	73	42	15
Persea americana Mill.	1.01	0.80	0.01		0.03	0.06	86	67	24
Maesopsis eminii Engl.	1.00		0.46	0.23	0.19	0.06	67	46	55
Ficus natalensis Hochst.	0.83	0.01	0.13	0.09	0.11		33	51	10
Markhamia lutea (Benth.) K. Schum.	0.68		0.13	0.41	0.06	0.02	50	43	6
Canarium schweinfurthii Engl.	0.57	0.37	0.08	0.02	0.06	0.01	47	22	3
Citrus sinensis (L.) Osb.	0.56	0.41			0.04	0.04	30	17	13
Pinus spp.	0.50	0.01	0.22	0.09	0.01	0.08	39	23	43
Coffea canephora Froehner	0.50	0.04			0.01	0.44	30	53	16
Carica papaya L.	0.46	0.40				0.03	44	44	8
Albizia coriaria	0.37		0.17	0.07	0.03		26	23	1
Theobroma cacao L.	0.18	0.07				0.11	51	11	20

These priority species are multipurpose that yield products that include edible fruits, timber, construction, and firewood. The most frequently mentioned species were Mangifera indica L. (93%), Artocarpus heterophyllusLam. (99%), and Persea Americana Mill. (86%), and these were mostly valued for their edible fruits (see Table 2). Milicia excelsa (Welw.) C. Berg and Pinus spp. were preferred for their timber. Some of these especiallyCoffea canephora Froehner are sold to earn incomes. For example, all species except Carica papaya L. can be used as firewood, but the topmost three for firewood are Maesopsis eminii Engl., Senna spp., and Eucalyptusspp.

Some species are known to deliver the following ecosystem services: providing shade for crops (e.g., Ficus sycomorus L.); creating wind breaks;

contributing to rain fall formation; enriching soils (fixing nitrogen in the soil), keeping soils moist, and controlling soil erosion; and adding scenic beauty. These values were mentioned by few respondents leading to low computed use values. For this reason, these environmental services as well as the species known to provide them are not included in Table 2. All the priority species are introduced and naturalised apart from M. excelsa, M. eminii, and Canarium schweinfurthii Engl. that are indigenous and Pinusspp. that are exotic.

Some tree products are sold to earn incomes; however, only 7% of the respondents sell tree products to get income, and they sell mostly edible fruits. Other products commonly sold and the species that they are gathered from are shown in Table 3. The products are sold at farmers homes (87%), but, some like timber are sold in designated markets (12%). It was not easy to determine the frequency for selling of products, apart from edible fruits that are sold twice a year on average. Everybody including children can sell tree products, although adults in the home are more involved.

Table 3: Species and products sold in the study area. The numbers are frequencies of respondents who say that products from species are sold for money.

	Edible fruits	Timber	Fuelwood	Living tree	Seedlings	Charcoal	Seeds	Poles	Round wood
Coffea canephora Froehner	27		1		1				
Persea americana Mill.	18		1		1				
Artocarpus heterophyllus Lam.	17	1							
Maesopsis eminii Engl.		10	1	3	1			1	1
Mangifera indica L.	12		2		1	1			
Milicia excelsa (Welw.) C. Berg		9		1				1	
Carica papaya L.	5		1				1		

Females mentioned slightly more species than males; however, there was no statistical association between species and gender (χ^2,$P > 0.05$). Additionally, women tend to preferentially mention species that are used for firewood, medicine, and food compared to males who mentioned mostly species used for timber, charcoal, and those used in construction.

3.3. Farmer Perceptions on Conservation Status of Tree Species.

Overall, woody species according to respondents are declining in abundance mostly because of destructive exploitation to get products like timber or charcoal (80%), from clearing land for agriculture (9%) or pests (9%). The conservation status, specific to prioritised woody species, was determined in Bunya County by asking respondents to list species that are becoming scarce (hard to find) and those that are increasing or becoming abundant. A comparison was made on the perception of whether a species was increasing or declining based on a comparison of the numbers of respondents stating that a species was declining or increasing. And from this the greater of the two frequencies was taken to be the conservation status of the species (Table 4). For this reason the species Markhamia lutea (Benth.) K. Schum., M. indica, and A. heterophyllus are considered to be abundant, while M. excelsa and Albizia coriaria are considered to be scarce (Table 4).

Table 4: Conservation status of tree species.

Species	Scarce	Abundant	Status
Markhamia lutea (Benth.) K. Schum.	14	19	Abundant
Mangifera indica L.	12	55	Abundant
Artocarpus heterophyllus Lam.	10	51	Abundant
Pinus spp.	10	30	Abundant
Maesopsis eminii Engl.	8	52	Abundant
Persea americana Mill.	2	26	Abundant
Coffea canephora Froehner	4	19	Abundant
Ficus natalensis Hochst.	9	18	Abundant
Senna spp.	1	12	Abundant
Acacia spp.	1	9	Abundant
Carica papaya L.	1	9	Abundant
Milicia excelsa (Welw.) C. Berg	67	3	Scarce
Albizia coriaria Oliv.	34	1	Scarce
Citrus sinensis (L.) Osb.	10	4	Scarce
Eucalyptus spp.	9	2	Scarce
Canarium schweinfurthii Engl.	5		Scarce

Mayuge in the recent past has experienced increased sugar cane growing and this has forced people to convert large tracts of land to this crop. Other factors were pests and diseases, lack of knowledge on how to propagate woody species, and lack of planting materials or the fact that they are not planted or protected by farmers in the area. Abundant species are on the other hand increasing because they are planted or are protected by farmers when found growing naturally. The abundance of woody species was attributed to active husbandry by planting and protection as well on the intrinsic nature of the species in that they propagate easily. This is shown in Table 5 for the most popular species that include M. indica, M. eminii, A. heterophyllus, Pinus spp., and C. canephora.

3.4. Local Tree Husbandry Practices.

Farmers in Bunya County are interested in tree planting and are actively involved. All respondents who participated in this study had planted a tree in the last 5 years and were willing to plant more in future. Trees are propagated from seed, seedlings, wildings, and cuttings. Seedlings are prepared by farmers or purchased from nursery gardens managed by farmers (Table 6). All family members, including children, are involved in tree planting. Trees are maintained in crop fields, as well as around the home, in the home-garden (or backyard) or the compound/courtyard (Table 6). Planting as live fence in farm boundaries was not observed.

Table 5: Reasons to explain increases in species availability in Bunya County. The species in this table are not necessarily the most valued.

Species	Planted	Protected	Propagates easily	Drought resistant	Grows naturally	Matures fast	Pest resistant	Dispersed by birds	Seedlings available
Mangifera indica L.	16	14	5	1					
Maesopsis eminii Engl.	29		3			1	1	1	
Artocarpus heterophyllus Lam.	17	10	4	1					
Pinus spp.	23								1
Coffea canephora Froehner	13	1	1						
Persea americana Mill.	10	4	3						
Ficus natalensis Hochst.	6	3	4						
Markhamia lutea (Benth.) K. Schum.	1	3	6		1	1			
Canarium schweinfurthii Engl.	3	4	1						
Carica papaya L.	2	3	1						
Senna spp.	1	3	1						
Moringa oleifera Lam.	1	1							
Burtdavya nyasica	1	1							
Total	131	47	29	2	1	2	1	1	1

Table 6: Common tree management practices, planting materials, where trees are maintained, and source of planting materials as reported by respondents.

Tree husbandry practices	Number	Planting materials	Number
Planting	28	Seeds	38
Protection against damage	26	Seedlings	31
Pruning	18	Wildings	27
Root suckering	3	Cuttings	4
Location where trees are grown		*Source of planting material*	
Crop field	42	My land	97
Backyard	29	Others (local administration = 2; bought, neighbour)	4
Compound or courtyard	27		
Hedge	2		
Along the road	1		

The people besides having an intrinsic interest in tree planting also receive encouragement from many different sources, chief among which is the radio, and from other farmers and village or other subnational meetings (Table 7).

3.5. Challenges That Affect Tree Planting.

Farmers in Bunya County who did not plant trees were asked to give reasons that hinder them from planting trees and, according to them, the major reasons were pests (termites), drought, lack of land, diseases (that attack mostly coffee and oranges), and lack of seeds and seedlings (Table 8). Other factors that hinder tree planting mentioned by respondents were limited capital to buy planting materials, garden tools and chemicals for spraying, human and animal damage, limited labour, and poor propagation methods as well as government policies on tree harvesting.

Table 7: Sources of information for tree planting.

Sources of information	Frequency
Radio	37
Seminars[1]	16
Other farmers	16
Relatives	5
School	3
Church and mosque	3
Others[2]	4
Environmental officers/representatives	1
Busoga Forest Company	1
Television	1

[1] Seminars include community meetings at subcounty (10 respondents), village (3), and others (3).
[2] Others include books, environmental officers/representatives, Busoga Forest Company, and television.

4. DISCUSSION

4.1. Priority Woody Species and Their Conservation Statuses.

The 14 species prioritised in this study are valued mostly for their edible fruits, firewood, timber, and construction wood/poles. These benefits justify increased investment in the development of agroforestry systems that will contribute to food security [17]. Their role in income generation is low however; the computed use values were low for income generation and few people were involved in marketing of tree products. This seems to support the welfare maximisation theory described by Scherr [7]. According to this theory, rural people manage trees with the aim of having regular access to products that satisfy their household needs and not for income generation. Delayed return on investment and underdeveloped markets are cited as major barriers to agroforestry development [17]. The abundant tree species such as Artocarpus heterophyllus, Mangifera indica, and Pinus spp. have edible products and high quality timber and are sometimes sold to generate income for communities. They thus play an important role in household economies [17]. Over and above this, useful species are likely to create favourable attitudes that contribute to their active management for their conservation among farmers within agroforestry systems. Outside of agroforestry systems, the attribute of being useful may instead lead to overexploitation and sometimes destructive harvesting.

Table 8: Constraints to tree growing.

Constraints	Frequency
Pests, mostly termites and rarely moles (*Nfuko*)	45
Drought	42
Lack of land	19
Tree diseases (coffee, oranges, and avocado are mostly affected)	9
Lack of seeds and seedlings	9
Lack of funds to purchase inputs, for example, fertilizers	8
Limited labour	6
Domestic animals	4
Destruction or theft of seedlings	4
Lack of market for tree products	2
Government ban to stop people from cutting their own trees	1
Some trees, for example, Mvule, cannot be intercropped with other crops	1

Different attributes have made many species easy to maintain in the community. The species that are perceived to be abundant all have fast rates of growth, are easy to propagate, and have high survival rates [18], and their planting materials are readily available. The other factor is that the species are actually grown or spared when found growing in the area.

Similar factors for tree availability were reported in Balawoli County [19]. Furthermore, these species are mostly crop trees for which the technology to propagate, grow, and manage them is well developed. Indeed, the prioritised species are valued in many other areas in Uganda and outside Uganda. For instance, M. indica is highly valued all over Africa. It is highly likely that exchange and transfer of technologies between sites will be easy when the species are valued in many different locations compared to if the species were valued in only a small area like Bunya County and when the technology to grow them is not developed. This is good and increases the potential to manage them in agroforestry systems.

Some tree species are perceived to be declining in abundance: those that are slow growing species and those that are lowly valued by farmers. They are not actively managed by planting. The exceptions are Citrus sinensisand Eucalyptus spp. which, though fast growing, are declining in extent and numbers. Features contrary to those mentioned for abundant species, such as slow growth, are some of the reasons that dissuade farmers from growing tree species. Another factor for the decline is that the species are harvested destructively to get firewood, charcoal, or poles for building.

5. TREE MANAGEMENT CHALLENGES

Two of the key reasons for not planting trees are linked to lack of knowledge on pest and disease management and can be addressed through training on pest and disease management. Other challenges such as lack of funds to purchase inputs, for example, fertilizers, have been recognized by the Ugandan government as a key bottleneck to tree planting and maintenance (NPA, 2010). More to that, sugar cane growing in Mayuge has taken most of the land that would be available for tree growing. These factors reduce the appeal for farmers to maintain tree species in agroforestry systems and will need addressing by the responsible institutions such as NAADS (National Agricultural Advisory Services) or NFA (National Forestry Authority) and others.

The challenge of land scarcity can be addressed by encouraging and incentivising farmers with large pieces of land to plant more trees on their farms. Elsewhere, in Kenya, farmers faced with land scarcity decided to use land borders or hedgerows, where trees least interfere with food crop growing [20]. This is an option worth investigating in Bunya County, where, presently, hedgerow is the least used land unit for tree growing.

Concerning seedlings, community level tree nurseries need to be established for easing access to seedlings of priority tree species. Care must be taken not to give farmers free seedlings as it has been observed that some farmers who receive free seedlings abandon them [21]. The government of Uganda should operationalise the National Tree Planting Act and relevant policies to provide finances to farmers to acquire capital for inputs. This will address the challenge of lack of capital and finances to purchase farm inputs and implements.

Over and above these factors, and notwithstanding the fact that the trees are presently valued for their subsistence value, increases in tree planting will not happen unless decent markets for tree products are created. The most important stimulus for active tree growing is presence of markets for tree products (ICRAF website). Residents of Bunya County are willing to collaborate with the government and other relevant stakeholders in maintaining priority woody species. This can be possible by providing them with inputs including seeds and seedlings for planting and chemicals for spraying, conservation education, and capital to buy equipment and land. Lack of land was a major hindrance to tree planting and conservation in the areas visited.

6. CONCLUSIONS AND RECOMMENDATIONS

Opportunities for cultivating trees within Bunya exist because tree species are valued for their products. In this study we have generated a short list of 14 priority species that implementers can choose from when promoting tree planting. The priority species are valued mostly for their edible fruits, timber, firewood, and poles especially for construction purposes. Interventions to plant trees should focus on the priority species identified here, because these are the ones that farmers are interested in managing. According to Franzel et al. [22], tree planting interventions with farmers are likely to succeed when farmers' needs and preferences are properly assessed and respected by project implementers.

Trees are maintained in crop fields and home-gardens as well as in the compound and are propagated from seedlings, wildings, and cuttings. There is little evidence of commercial nursery tree operators. This shows that farmers cannot afford the cost of seedlings.

Many challenges that impact on tree growing have been identified in this study. The most significant in this study is that the market for nonwood tree products is very weak. If markets for tree products remain poor, farmers will not dedicate their best effort to tree planting and maintenance.

Farmers need training and sensitization on better methods of tree planting and management and the role of trees in climate change abatement. There is need to study markets of tree products in this area and other parts of Uganda to understand this value chain with a view to strengthening tree products markets and to provide a decent return on tree investment. There is also need to conduct an ecological study to determine the floristic composition and stocking rate of woody species in Bunya County. This will inform the need to plant trees and also the land use units on the farm where trees can be planted.

ACKNOWLEDGMENTS

This study was supported by a Rockefeller Foundation funded project (no. 2009CL1323). In a special way, the authors would like to extend their sincere thanks to all the respondents for providing this information and to the local district administration for all the help they provided to the authors. Mr. Samuel Kakungu, the Local Council Chairperson of Buwaaya village, moved with the authors from home to home showing them households they needed to interview.

REFERENCES

1. FAO, State of the Worlds Forests, Food and Agriculture Organization of the United Nations, Rome, Italy, 2009.
2. M. C. Hansen, P. V. Potapov, R. Moore et al., "High-resolution global maps of 21st-century forest cover change," Science, vol. 342, no. 6160, pp. 850–853, 2013.
3. NFA, National Biomass Study: Technical Report 2005, National Forestry Authority, 2009.
4. NEMA (National Environment Management Authority), State of The Environment Report for Uganda 2010, NEMA, 2010.
5. R. Drigo, WISDOM (Woodfuel Integrated Supply/Demand Overview Mapping) – East Africa, FAO, Rome, Italy, 2006.
6. J. Obua, J. G. Agea, and J. J. Ogwal, "Status of forests in Uganda," African Journal of Ecology, vol. 48, no. 4, pp. 853–859, 2010.
7. S. J. Scherr, "Meeting household needs: farmer tree-growing strategies in western Kenya," in Tree Management in Farmers Strategies: Responses to Agricultural Intensification, J. E. M. Arnold and A. P. Dewees, Eds., pp. 141–173, Oxford University Press, Oxford, UK, 1995.
8. B. W. Miller, P. W. Leslie, and J. T. McCabe, "Coping with natural hazards in a conservation context: resource-use decisions of Maasai households during recent and historical droughts," Human Ecology, vol. 42, no. 5, pp. 753–768, 2014.
9. S. S. Meijer, D. Catacutan, O. C. Ajayi, G. W. Sileshi, and M. Nieuwenhuis, "The role of knowledge, attitudes and perceptions in the uptake of agricultural and agroforestry innovations among smallholder farmers in sub-Saharan Africa," International Journal of Agricultural Sustainability, vol. 13, no. 1, pp. 40–54, 2015.
10. S. J. Scherr, "The role of extension in agroforestry development: evidence from western Kenya," Agroforestry Systems, vol. 18, no. 1, pp. 47–68, 1992.
11. N. T. Hansen, A. Raebild, and H. H. Hansen, "Management of trees in northern Ghana - when the approach of development organizations contradicts local practices," Forests Trees and Livelihoods, vol. 21, no. 3-4, pp. 241–252, 2012.

12. Mayuge District, Mayuge District Local Government Five Year Development Plan (2010/2011–2014/2015), Mayuge District, 2011.
13. UBOS, National Population and Housing Census 2014: Subcounty Report—Eastern Region, Uganda Bureau of Statistics, Kampala, Uganda, 2016.
14. M. Lykke, "Local perceptions of vegetation change and priorities for conservation of woody-savanna vegetation in Senegal," Journal of Environmental Management, vol. 59, no. 2, pp. 107–120, 2000.
15. J. A. Van Den Brakel, R. Vis-Visschers, and J. J. G. Schmeets, "An experiment with data collection modes and incentives in the dutch family and fertility survey for young moroccans and turks," Field Methods, vol. 18, no. 3, pp. 321–334, 2006.
16. Nyamukuru, A. Nabatanzi, S. Mpiira, and J. R. S. Tabuti, "Locally preferred woody species and their management in Kiruhura and Arua districts, Uganda," Ethnobotany Research and Applications, vol. 14, pp. 49–61, 2015.
17. FAO, "Advancing agro forestry on the policy agenda: a guide for decision makers," in Agro Forestry Working Paper No. 1, G. Buttoud, O. Ajayi, G. Detlefsen, F. Place, and E. Torquebiau, Eds., p. 37pp, Food and Agriculture Organization of the United Nations, Rome, Italy, 2013.
18. B. Katende, A. Birnie, and B. Tengnäs, Useful Trees and Shrubs for Uganda, Identification: Propagation and Management for Agricultural and Pastoral Communities, 2010. National Development Plan (2010/11–2014/15), Regional Soil Conservation Unit. National Planning Authority, Kampala, The Republic of Uganda, 1995.
19. J. R. S. Tabuti, "Important woody plant species, their management and conservation status in Balawoli sub-county, Uganda," Ethnobotany Research & Applications, vol. 10, pp. 269–286, 2012.
20. M. M. Backes, "The role of indigenous trees for the conservation of biocultural diversity in traditional agroforestry land use systems: The Bungoma case study: In-situ conservation of indigenous tree species," Agroforestry Systems, vol. 52, no. 2, pp. 119–132, 2001.
21. M. Nieuwenhuis and N. O'Connor, "Challenges and opportunities for small-scale tree nurseries in the East African highlands," Unasylva, vol. 51, no. 203, pp. 56–60, 2000.
22. S. Franzel, H. Jaenicke, and W. Janssen, "Choosing the right trees: setting priorities for multipurpose tree improvement," ISNAR Research Report 8, International Service for National Agricultural Research, The Hague, The Netherlands, 1996.

CHAPTER 5

Understanding patterns of tree adoption on farms in semi-arid and sub-humid Ethiopia

Miyuki Iiyama . Abayneh Derero . Kaleb Kelemu . Catherine Muthuri . Ruth Kinuthia . Ermias Ayenkulu . Evelyn Kiptot . Kiros Hadgu . Jeremias Mowo . Fergus L. Sinclair

World Agroforestry Centre (ICRAF), Nairobi, Kenya

Japan International Research Center For Agricultural Sciences (JIRCAS), Tsukuba, Japan

Ethiopian Environment and Forest Research Institute, Addis Ababa, Ethiopia

Ethiopian Institute of Agricultural Research, Addis Ababa, Ethiopia

Bangor University, Wales, UK

ABSTRACT

Trees on farms are a widespread feature of landscapes across a large part of Ethiopia with an important role in enhancing the resilience of smallholder livelihoods through the provision of ecosystem services. Despite their importance, little is known about what trees are planted or retained from natural regeneration by different types of farmers that results in the pattern of tree cover found in the region. We address this knowledge gap through analysis of household survey data from semi-arid and sub humid areas of Oromia regional state. A set of composite variables that represent distinctive patterns of tree cover on farms were derived from principal component analysis and Pearson correlation analysis. This revealed two major tree adoption strategies: farmer managed natural regeneration (FMNR) of trees to meet subsistence needs as well as contributing to other ecosystem services; and, high value agroforestry (HVAF) involving planted trees used largely to produce fruits, timber and fodder. Regression analysis further identified fine-scale variation in ecological and socio-economic factors that affect which of these two broad strategies are adopted by farmers. Favorable climatic conditions coupled with institutional arrangements to control free grazing were pre-conditions for HVAF, whereas poor biophysical potential and sloping land provided a positive incentive for farmers to adopt FMNR. Farmers with preferences for tree species with multiple utilities and locational flexibility favored FMNR while adoption of HVAF was more asset-driven. Our findings reveal that farmers integrate many native and exotic tree species on their farms to meet their variable farm conditions, needs and asset profiles in stark contrast to most tree promotion efforts that focus on a few, usually exotic, tree species. We recommend that future agroforestry promotion should embrace a diversity of tree species appropriate to matching the fine scale variation in ecological conditions and farmer circumstances encountered in the field.

Keywords: Characterization Farmer managed natural regeneration (FMNR) Ecosystem services Livelihoods

INTRODUCTION

Ethiopia's economy is heavily dependent on agriculture which supports 83 % of the population mainly through production of rain-fed grain, predominantly teff, maize and wheat as well as livestock, principally cattle, sheep and goats (Deressa et al. 2009). During the last century, the expansion of agriculture to support the growing population was in general achieved by massive deforestation that has been followed by soil degradation (Bewket 2002; Gelaw et al. 2014). Over much of the country this has transformed forests and woodlands into agricultural land with scattered trees (Tesfaye et al. 2014).

Trees on farms are characteristic of a large part of the Ethiopian agricultural landscape today, while tree species distribution and management intensity varying with agro-ecological conditions principally defined by rainfall, altitude, and soil type (Poschen 1986; Teklay et al. 2007; Gelaw et al. 2014). It is common for farmers to manage natural regeneration of trees (FMNR) within crop fields by protecting seedlings and young trees, mostly native species that have germinated from soil seedbanks (Poschen 1986). Farmers usually retain between 1–20 trees of selected species per hectare and minimize impact on the companion crops through occasional lopping and pollarding of trees (Poschen 1986).

Examples of this practice include *Cordia africana* intercropping with maize in sub-humid zones (Yadessa et al. 2009), *Faidherbia*-based agroforestry in teff-wheat zones (Poschen 1986) and a diverse range of *Acacia* species such as *A. tortilis* and *A. senegal* in fields in low lying savanna regions (Degefu et al. 2011). It is also common for farmers to deliberately plant and manage trees on their farms, such as fast growing timber woodlots predominantly using *Eucalyptus* spp. or fruit orchards (Deininger and Jin 2006).

Through FMNR and active tree planting on farms, the widespread adoption of trees on agricultural land can play an important role in enhancing tree diversity and cover at landscape scale. This can mitigate and reverse deforestation and land degradation with large potential for soil organic carbon and nitrogen sequestration (Bewket 2002; Mekuria and Aynekulu 2013; Gelaw et al. 2014). The positive relationship between higher rural population and tree cover observed by Tiffen et al. (1994) in semi-arid Kenya has been also reported in the Blue Nile basin as tree cover has been restored over the last four decades as a result of local initiatives to plant trees at the household level along with community afforestation and forest protection (Bewket 2002).

Since the early 1990s there has been a surge of research on the adoption of agroforestry innovations in the tropics, motivated by a perceived gap between advances in agroforestry science and the rate of adoption of trees on farms (Mercer 2004). They can be largely categorized into either ex-ante or ex-post adoption studies. Ex-ante studies rely primarily on social and financial analyses of on-farm trials of agroforestry innovations to assess their adoption potential (Franzel and Scherr 2002). In contrast, ex-post studies aim at identifying factors that have affected adoption through analysis of data on the performance of agroforestry options on farm and the types of farmers who have and who have not adopted (Coe et al. 2016). A meta-analysis of 120 ex-post studies concluded that technology adoption was explained by preferences, resource endowments, market incentives, biophysical factors, and risk and uncertainty (Pattanayak et al. 2003). Meijer et al. (2015) argue that intrinsic factors such as knowledge and attitudes of farmers are also critical. Both ex-ante and ex-post studies recognize the multicomponent nature of agroforestry (Mercer 2004). Most of these studies, however, focus on a single "modern" or "new" agroforestry technology, and methodologically treat the adoption as a binary choice of planting a specific tree species or not, driven by financial and economic factors (Mercer 2004). For example in Ethiopia, several studies examined factors affecting the uptakes of small-scale *Eucalyptus* woodlots by smallholders (Deininger and Jin 2006; Jagger and Pender 2003). Costs and returns of investment emerged critical in determining decisions to plant trees along with tenure security (Jagger et al. 2005; Duguma 2013).

In contrast, research on adoption has rarely studied indigenous agroforestry systems in the tropics that have evolved uniquely within the locally specific landscape contexts they are embedded in, often over long time periods (Sinclair and Walker 1999). Biggelaar and Gold (1996), based on a case study from Rwanda, argue that the adoption of indigenous agroforestry systems are driven by farmers' preferences for specific tree species with multiple utilities and locational flexibility rather than solely driven by financial and economic factors. Indeed, many farmers in Ethiopia, including those who have not practiced any form of intensive tree planting, implement FMNR to manage a set of indigenous species scattered on farm, not only for goods such as fuel, fodder, and fruits, but also for ecosystem services such as shade and soil amelioration (Poschen 1986). A household normally keep indigenous trees across farmlands for multiple utilities that they deem manageable and valuable in order to optimize capture and use of scarce environmental resources (Negash 2007).

In reality, different agroforestry practices such as various forms of tree planning and indigenous practices such as FMNR co-exist (Nyaga et al. 2015). Unfortunately, there have been few empirical studies in Ethiopia or further afield in sub-Saharan Africa (SSA) to understand patterns of tree cover in agricultural landscapes as a whole rather than focusing on a single technology. It is imperative to address this gap both in terms of knowledge and methodology if present landscapes are to be understood and their future resilience ensured.

The objectives of this paper are two-fold. Firstly, to present a novel systematic method for characterizing complex patterns of tree cover on farms, including both indigenous practices and tree planting as a commercial investment in terms of their structure (species composition), function (utilities) and socioeconomic aspects (management intensity and commercial goals). Secondly, to identify a fine scale variation in factors that affects their adoption.

METHODS

We used household survey data from both semi-arid and sub-humid agroecosystems in Ethiopia to characterize tree cover on farms by deriving proxy variables reflecting adoption intensities as well as multi-dimensionality of utilities using a multi-variate analytical method. We then examined associations between distinctive patterns of tree adoption on farms and both ecological and socio-economic factors that determine their adoption and can be used to match agroforestry interventions to the contexts in which they are appropriate.

Study area, data collection

The Oromia National Regional State accounts for 34 % of the total area of Ethiopia (The National Regional Government of Oromia 2016) and with a population of over 27 million people (Federal Democratic Republic of Ethiopia Population Census Commission 2008), is the largest state in terms of land area and population in the country. It is a region of great physiographic diversity whose landscapes include high and rugged mountain ranges, undulating plateaus, panoramic gorges, deep incised river valleys and rolling plains (Ethiopian Government Portal, accessed January 7, 2016). The lowlands of the eastern sub-region have an arid climate. The intermediate highlands of central and western Oromia have a hot tropical rainy climate, while the highlands have a warm temperate, rainy climate. The distribution of mean annual rainfall varies from place to place and from year to year, decreasing in all directions from the western highlands (1600–2400 mm) towards the eastern and south eastern arid lowlands (less than 400 mm) (The National Regional Government of Oromia 2016). The present research was conducted in East Shewa Zone that falls within the semi-arid agroecology, and East Wollega and West Shewa Zones that are in the sub-humid agroecology (Fig. 1).

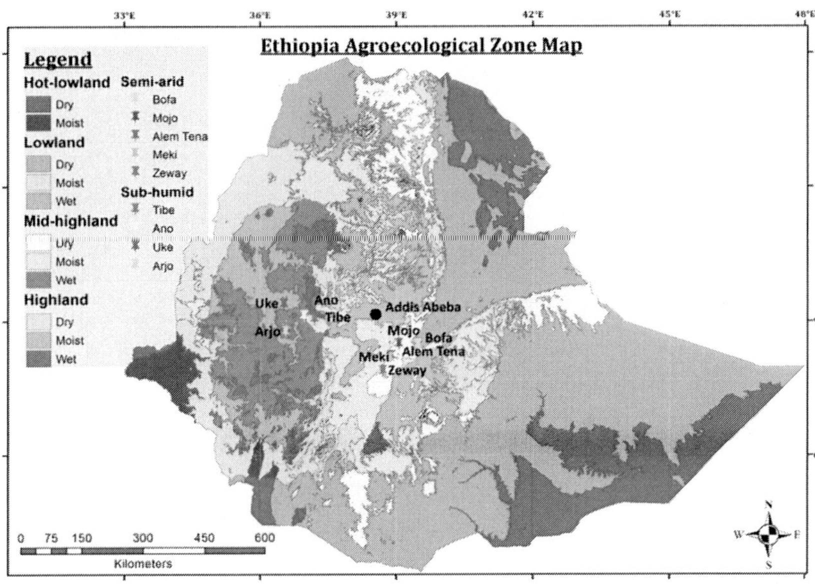

Figure. 1 Agroeclogical map of Ethiopia with locations of selected sites

The semi-arid sites mostly fall in the lowlands of the Central Rift Valley with an altitude less than 1500 m, then rise up to 2300 m at the mountain fringes of the Rift. Grain crop and livestock farming are dominant in the study area. Some diversity is observed in terms of the combination and management of tree-crop systems (Endale 2014); from teff-wheat plus *Faidherbia albida* (syn. *Acacia albida* Delile A.Chev, locally called gerbi) to maize-beans-sorghum plus *Acacia ssp.* across the north–south transect (Fig. 2a), and from teff-wheat plus *Faidherbia albida* to teff-maize-sorghum plus *Acacia tortilis* Hayne (tadecha, or ajo loc) and *Zizyphus mucronata* Willd. (ourqura) across the west-east transects, while the livestock system is dominantly communal grazing of cattle and goats on farmland (Fig. 2b).

Figure. 2 Typical agroforestry practices observed in semi-arid and sub-humid zones of Oromia State, Central Ethiopia. **a** Faidherbia albida amongst teff in semi-arid Oromia. **b**Ziziphus mauritania and Acacia abysinica in maize fields in semi-arid Oromia. **c** Cordia africana in a maize field in sub-humid Oromia. **d** Various trees around a homestead in sub-humid Oromia

The sub-humid sites are characterized by rugged landscapes, with hills and valleys. Dominant crops are maize, sorghum, teff, nug (*Guizotia abyssinica*—an oil crop), with their relative importance varying with altitude and micro-climate. In contrast, diversity and types of trees species observed are heterogeneous across the north–south and east-south transects as well as by altitude (Teshome 2014). *Croton macrostachyus* Hochst. (bakanisa) is dominant in home compounds and farm boundaries in southwestern mid-highlands, *Ficus spp.* such as *Ficus sycomorus* (oda) and *Ficus vasta* Forssk (kiltu) in eastern mid-highlands, and *Cordia africana*Lam. (wadessa) in eastern parts, while small woodlots of *Eucalyptus camaldulensis* Dehnh. (bargamo dima) and scattered *Acacia abysinica Hochst.* (lafto) on farm are commonly observed (Fig. 2c). Across the east–west transect, fruits trees such as pawpaw, mangoes, and coffee are observed in home compound, along rivers or localities with irrigation structures available (Fig. 2d).

The administrative hierarchy of Ethiopia is the state-zone-woreda. Woreda is equivalent to a district, within which there are a number of kebeles or villages. Five and four woredas were chosen from the semi-arid and sub-humid zones respectively, to reflect the transition of observed diversity in tree-crop systems. All five sites selected in the semi-arid zones were located in the dry-midlands. In the sub-humid zones, two sites were in the wet mid-highlands, while the other two were in the moist midland and wet lowland respectively (Fig. 1). The selection of a kebeles was done in consultation with woreda administrative officers to be representative of each of the five woredas. A kebele is the smallest administrative unit in Ethiopia, in which households are registered and recorded. According to the 2007 Census, the population of a rural kebeles in Oromia varied widely from less than 100 to over a thousand households (Central Statistical Agency 2010), while the sizes reported by local officials for the surveyed kebeles ranged from about 300 to over a thousand households (Table 1). The minimum sample size of 568 out of 6135 households in all the selected kebeles was calculated using the formula of Israel (1992) for ±4 % precision level, while the minimum 10 % of the households in each of the surveyed kebeles were targeted for interviews.

The socio-economic survey focused on collecting qualitative information about farmers' perceptions relating to the status of trees on farms as well as the income/asset status of households, while quantitative information about trees on farms was captured through a tree inventory for about a third of the households surveyed. The inventory allows correction of any bias in householders' recollection regarding quantitative variables such as the number of trees on farm. The socio-economic data were collected between November and December 2012. A total of 687 households were interviewed across the two agroecologies. In most cases, the respondents were household heads or spouses who, combined, accounted for over 80 % of surveyed households (Table 1). The data is made publicly available on DATAVERSE (http://hdl.handle.net/1902.1/21219 UNF:5:EwSwq3/7ysbpqpawnencvg == World Agroforestry Centre [Distributor] V21 [Version]).

Table 1 Sampling information and characteristics of the respondents of the surveyed households

Agro-ecology (zones)	Semi-arid (East Shewa Zone)					Sub-humid (East Wollega and West Shewa Zones)				Total
Site name Site code	Bofa Semi-arid 1	Mojo Semi-arid 2	Alemtena Semi-arid 3	Meki Semi-arid 4	Zeway Semi-arid 5	Arjo Sub-humid 1	Ano Sub-humid 2	Bako Sub-humid 3	Uke Sub-humid 4	
Sampling information										
Woreda (district)	Boset	Lume	Bora	Dugda	Adami Tulu Jido Kombolcha	Jima Arjo	Gobu Sayo	Bako Tibe	Guto Gida	
The number of kebeles	33	35	18	36	43	20	8	28	21	
The name of the kebele surveyed	Sara Areda	Ejersa Jero	Berta Sami	Mukiye Lamaan	G/W/Booramoo	Wayu Kumba	Ongobo Bakanisa	Oda Haro	Uukkee Badiya	
The number of households in the Surveyed kebele[a]	1002	314	463	314	672	739	1005	719	907	6135
The number of households surveyed	120	37	55	48	80	76	104	75	92	687
% of the households surveyed in the kebele	12.0	11.8	11.9	15.3	11.9	10.3	10.3	10.4	10.1	11.2
Respondents' characteristics										
Household head (%)	70	65	67	60	65	64	66	68	64	
Spouse (%)	20	27	25	21	25	16	26	20	27	
Other family member (%)	10	8	7	19	10	20	8	12	9	

[a] The estimates of the number of households reported by local officers when the authors visited kebele offices in November 2012

Research questions and hypotheses

Patterns of tree cover on farm in Ethiopia, as elsewhere in SSA, are complex and heterogeneous in terms of (1) mixture of species, (2) utilities derived from these species, (3) management intensity and niches occupied, under (4) specific biophysical and socio-economic circumstances (Nyaga et al. 2015). These dimensions correspond to criteria used to classify agroforestry practices: (1) structural, the nature and arrangement of components; (2) functional, the role and output derived from them; (3) socioeconomic, the type of management; and (4) agroecological, the zone where a system exists or is adoptable (Sinclair 1999).

Our major research question was how to characterize patterns of tree cover found on farms. More specifically, we propose a systematic method to identify patterns of tree cover on farms in terms of (1) species composition, (2) multiple utilities, (3) management intensity and niches occupied, and then identify fine-scale variation both ecological and socio-economic contextual factors that determine their adoption.

Our first major hypothesis was that farm households adopt portfolios of tree species that maximize utilities from goods and services derived from them, which, in turn, determine the appropriate management intensity and niches within the farm for the trees to occupy. We assume that the planting of tree species that produce high value fruits, fodder and timber is driven commercially, as they are deliberately planted and more intensively managed as an investment (Franzel and Shurr 2002). In contrast, we assume that native tree species naturally regenerated in agroforestry practices are managed less intensively but still serve multiple utilities, not only for direct consumption such as fuel and construction materials but also for ecosystem services such as shade and as windbreaks, soil fertility enhancement and soil erosion control (Poschen 1986; Biggelaar and Gold 1996).

Our second major hypothesis was that patterns of trees on farm are bounded by biophysical and socio-economic conditions, which are classified into five major categories; biophysical factors, preferences, resource endowments, risk and uncertainty, and market incentives, following the definitions by Pattanayak et al. (2003). We discuss each of these, in turn.

Biophysical factors relate to influences on the physical production process associated with farming (Pattanayak et al. 2003). In our study area, East Shewa in the semi-arid zone is less agro-ecologically favored in terms of rainfall and soil than West Shewa/East Wollega in the sub-humid zone. A wide variety of tree species, including exotics, are likely to thrive in sub-humid conditions where intensive planting and management are feasible. In contrast, in semi-arid conditions, moisture stress constrains the survival of planted seedlings and well adapted native species are most likely to survive. While the five chosen semi-arid sites are all located in dry mid-highlands and relatively homogeneous in terms of topography, two of the sub-humid sites are located in higher, sloping conditions than the others, which may affect tree-crop management incentives.

Preferences are placeholders for the broad category of farmer specific influences such as risk tolerance, attitudes to conservation and intra-household homogeneity (Pattanayak et al. 2003). Because farmer preferences are difficult to measure explicitly, socio-demographic proxies such as age, gender, education, and social status are used instead. In an African context, gender of head of household may influence what trees are on a farm, because female headed households may take different decisions about managing trees than male-headed households. Traditional land/tree tenure systems often do not allow women to plant trees according to their own preference unless men approve, even though it is the women who bear the burden of taking care of raising trees and collecting firewood (Deininger et al. 2009). Family size and composition can also matter in terms of sufficiency and quality of labour to manage trees on farm while it is impossible to determine a priori the direction of the influence on adoption of this broad category (Pattanayak et al. 2003).

Risk and uncertainty reflect the unknowns in the market and institutional environment under which decisions are made. Given the long gestation period of investments in farming and forestry, lower risk and uncertainty will in general foster technological adoption (Pattanayak et al. 2003), while the extensive review of experiences from SSA indicate the ambiguity of such impacts which are highly context specific (Place 2009). For this study we include land, parcel characteristics and land-related policy and institutional experiences as proxy variables for risk and uncertainty. In Ethiopia, the land remains state owned while the constitution affirms the right of every adult access to land. The recent effort to improve security of land tenure in Ethiopia includes a land certification through decentralized mechanisms, where the regional government would issue land certificates to individual farmers (ARD Inc. 2004; Deininger et al. 2008, 2009). Experiences of land resettlements and "grabs" where the government designate certain areas for development, such as for irrigation schemes, outside investors or internal redistribution (Deininger et al. 2009; ARD Inc. 2004) can however influence tree planting positively or negatively depending on context. An interesting case was reported in Northern Ethiopia where tree planting was undertaken after land resettlement, as a way to visibly manifest land rights, whereas terracing was done in situations where there was a minimum level of tenure security (Deininger and Jin 2006). For Oromia state, only 39 % of households were reported to receive the certificates several years after the program started in 2003–2004 (Deininger et al. 2008; Holden et al. 2011). Given this situation, the status of land tenure is expected to vary among the surveyed households with uncertain impacts on investment decisions. Communal grazing which is widely practiced in the country can also affect patterns of tree cover on farm. Communal grazing causes soil degradation but also affects the survival of tree seeds and seedlings on farms, which can affect incentives to intensify or extensify tree management on farms (Gebremedhin et al. 2004; Kassahun et al. 2009; Mekuria and Aynekulu 2013; Tesfaye et al. 2014). Under given tenure conditions, access to more land may provide locational flexibility for trees to be managed without competing with crops and livestock, if the availability of labour and other inputs is not constraining. Parcel fragmentation may have mixed effects. It may deter tree management efforts because of increased transaction costs associated with long distances between parcels and result in uneconomic operational holdings (Bizimana et al. 2004). In other contexts fragmentation may offer incentives for farmers to manage trees in some parcels or niches where soil types and slope conditions better match with trees rather than with crop and/or livestock farming, as a risk diversification and management strategy (Blarel et al. 1992).

Resource endowments measure the resources available to the technology adopter for implementing the new technology. Examples of resource endowments include asset holdings such as livestock and savings while we used proxy variables of diverse asset categories and farm/off-farm income categories. Generally, resource endowments are likely to be positively correlated with the probability of adoption. However, it is likely that different endowments will encourage different agroforestry practices (Pattanayak et al. 2003; Iiyama et al. 2008).

Market incentives include factors related to lower costs and/or higher benefits from technology adoption. In general, a factor that is expected to increase the net benefits associated with the technology is likely to be a positive influence on adoption (Pattanayak et al. 2003). In Ethiopia, markets designed specifically for agroforestry are rudimentary compared to those for crop/livestock (personal communication to Ethiopian partners). Therefore, how market access affects patterns of trees on farms are not known.

Data processing and analyses

In order to capture the complex roles of trees on farms, it is useful to develop a method to simultaneously capture multiple dimensions of patterns of tree use on farms, namely, species composition and their associated utilities, niches and management intensities. The socio-economic survey collected data about whether or not a farmer (a) raised seeds/seedlings; (b) planted seeds/seedlings; (c) protected naturally regenerated trees; or (d) had ever had trees on their farm, over the last three years. For the adoption categories with positive answers, farmers were further asked to provide names of tree species. Counts of different species were aggregated for each household with a mean of three different species per household. While 105 and 126 different tree species were counted in the semi-arid zone and sub-humid zone respectively, some species names reported in local languages were difficult to identify or have verified by Ethiopian botanists. At the same time, farmers often mentioned genus names only such as *Acacia* or *Eucalyptus* without specifying species. This potentially introduces problems in counting species diversity from farmers' recall, which needs to be verified by tree inventory.

Detailed questions were then asked about the farmers' reasons for adopting, and the utilities they derived from each tree species. Utilities were categorized into those with predominantly commercial value (timber, fruit, fodder, and medicine), subsistence (construction and tools for domestic purposes), fuel (firewood, charcoal), environmental services (shade, windbreaks, soil fertility, erosion control), fencing (either as live fences or poles), and other utilities (not categorized above). Often farmers provided more than two utilities for one species with the most important ranked as the primary utility and others as secondary. Farmers tend to rank higher or to give priorities to utilities from tangible goods, such as firewood, fruits, and timber, which contribute directly to consumption and income. Nevertheless, farmers tended to derive multiple ecosystem services from trees, including shade (microclimate), erosion control, and soil fertility enhancement.

While ranks represent an ordering of a list of items according to their importance for the particular issue under consideration, the lack of a standard scale makes the task of combining ranks over several farmers difficult (Abeyasekera 2001). Replacing ranks by scores enables variables to be treated like numerical data (The University of Reading Statistical Services Centre 2001). In order to quantitatively capture multiple utilities of trees on farm for different farmers, it is useful to derive scores to reflect such multiple utilities of specific tree species. Scoring exercises are done on the basis of one identified criterion (Abeyasekera 2001), while the choice of scores is not critical, as long as the interpretations of results are insensitive to changes in the actual values attributed to ranks (The University of Reading Statistical Services Centre 2001). In our data, the farmers provided minimum zero to maximum six secondary utilities per tree species for the adoption category of "trees on farm over 3 years". Consequently, we arbitrarily set a score of 0.7 for the primary utility and a score of 0.3 to be divided among the secondary utilities as 0.3 is divisible by any number from one to six. We then calculated weighted utility scores for each tree species mentioned by a household, using the formulae and rules in Box 1.

Box 1 Formulae and rules for calculating and assigning weighted utility scores from ranks

UALL = 1.0xUprimary (fuel, charcoal, fodder…), if counts of Uothers = 0

UALL = 0.7xUprimary (fuel, charcoal, fodder …) and 0.30 for Uothers1…6 (fuel, …), if counts of Uothers = 1

UALL = 0.7xUprimary (fuel, charcoal, fodder …) and 0.15 for Uothers1…6 (fuel, …), if counts of Uothers = 2

UALL = 0.7xUprimary (fuel, charcoal, fodder …) and 0.10 for Uothers1…6 (fuel, …), if counts of Uothers = 3

UALL = 0.7xUprimary (fuel, charcoal, fodder …) and 0.075 for Uothers1…6 (fuel, …), if counts of Uothers = 4

UALL = 0.7xUprimary (fuel, charcoal, fodder …) and 0.06 for Uothers1…6 (fuel, …), if counts of Uothers = 5

UALL = 0.7xUprimary (fuel, charcoal, fodder …) and 0.05 for Uothers1…6 (fuel, …), if counts of Uothers = 6

where UALL, which denotes the utility portfolio of one particular tree species, consists of Uprimary and Uothers which mean the primary utility and secondary utilities respectively.

For example, if a farmer said that fuel was the only one utility derived from his/her *Acacia tortilis* on farm, this species was given a score of 1.0 for fuel (firewood). If another farmer said that *A. tortilis* was primarily for fuel, but also for erosion control as another utility, then the species got 0.7 as the fuel score and 0.3 as the environmental services score. If two other utilities were mentioned, say soil control and fencing, aside from fuel as the primary utility, and then the species got 0.7 for fuel, 0.15 for erosion control and 0.15 for fencing. In this way, the score for one particular species could not exceed 1.0, but with higher numbers of utilities, scores would be subdivided amongst multiple utilities. Our formulae and rules are conceptually similar to the Utility Index (UI) proposed by Biggelaar and Gold (1996) while their index scores could add up to more than 1.0 if farmers reported many utilities for a single species.

Utility scores were aggregated by utility types for each household. Principal Component Analysis (PCA) was then used to derive independent component factors (where each component has zero correlation) from which patterns of tree diversity on farm were derived from species utilities and management intensities as we hypothesized that types of utility determine management intensity.

Another important dimension of patterns of trees on farm are the niches that specific tree species occupy—such as, scattered on farm, on boundaries, as live fences or hedges within fields or as soil conservation bunds, woodlots, in home compounds or fallows. The association between the niches for specific tree species and principal component scores were used to develop utility/management intensity patterns through Pearson correlation analysis.

Once proxy variables for distinctive patterns of trees on farm were derived, regression analyses were employed to determine which agroecological and socio-economic factors affected their adoption. Ordinary Least Squares (OLS) regression was used to estimate the association between the dependent variables—normalized scores calculated for each PCA component representing different patterns of trees on farms—and the explanatory variables. Descriptive statistics of the explanatory variables considered for the regression models are summarized in Table 2.

Table 2 Descriptive statistics of explanatory variables by agroecological zones

Variables	**Unit**	**Semi-arid zone (N = 340)**		**Sub-humid zone (N = 347)**		**F**
		Mean	**SD**	**Mean**	**SD**	
Ecological/topographical condition						
Mid-highland dummy	Wet mid-highland = 1, others = 0	–	–	0.5	0.5	365.40***
Household composition						

Head gender dummy	Male = 1, female = 0	0.9	0.3	0.9	0.3	0.21
Head education level	No formal education = 0, ...	0.7	0.7	0.7	0.7	0.16
Total family size	number	6.7	3.1	6.1	2.5	5.90**
Male ratio	% In total family size	0.5	0.2	0.5	0.2	1.58
Working age member ratio	% In total family size	0.4	0.2	0.5	0.2	0.66
Land/parcel access, characteristics, policy experiences						
Total land size	Hectare	3.7	11.1	1.9	5.8	6.89***
Parcel fragmentation (Simpson index)	Simpson Index	0.5	0.2	0.4	0.3	21.13***
Mean distance of parcels from homesteads	m	1482	1411	1093	1389	13.30***
Proportion of parcels owned with certificate	% Size in total land size	0.8	0.4	0.7	0.4	1.81
Proportion of parcels owned without certificate	% Size in total land size	0.1	0.3	0.1	0.3	1.55
Proportion of parcels rented from others	% Size in total land size	0.1	0.2	0.1	0.3	0.26
Proportion of parcels on sloped/steep	% Size in total land	0.1	0.3	0.5	0.4	188.82***

land	size					
Experience of free communal livestock grazing	Affected = 1, not affected = 0	0.3	0.5	0.6	0.5	49.17***
Experience of land resettlement	Affected = 1, not affected = 0	0.0	0.2	0.0	0.2	0.54
Experience of tenure upgrading	Affected = 1, not affected = 0	0.4	0.5	0.6	0.5	17.09***
Asset/income diversification						
Livestock—local animals	FAO Livestock Unit	6.0	10.0	3.8	3.8	14.64***
Livestock—exotic animals	FAO Livestock Unit	0.1	0.9	0.0	0.5	1.74
Asset—domestic asset value	Estimated value in USD	78	280	31	96	8.84***
Asset—communication asset value	Estimated value in USD	65	97	22	31	60.67***
Asset—transport asset value	Estimated value in USD	60	117	11	70	43.80***
Asset—farm asset value	Estimated value in USD	273	200	238	152	6.77***
Farm income-cereal & pulse	Aggregated scores[a]	3.4	0.9	2.5	1.0	178.63***
Farm income—cash crops	Aggregated scores[a]	0.4	0.8	1.7	1.6	190.82***
Farm	Aggregated	0.0	0.1	0.3	0.7	76.13***

income—roots & tubers	scores[a]					
Farm income—animal	Aggregated scores[a]	2.0	1.6	1.5	1.3	17.48***
Farm income—tree-based	Aggregated scores[a]	0.9	0.9	1.5	1.4	53.24***
Off-farm income—regular business	Aggregated scores[b]	0.4	0.6	0.5	0.6	0.32
Off-farm income—casual	Aggregated scores[b]	0.3	0.5	0.3	0.5	2.29
Off-farm income—remmitance & gift	Aggregated scores[b]	0.1	0.3	0.1	0.3	0.13
Off-farm income—loan	Aggregated scores[b]	0.4	0.6	0.6	0.7	11.83***
Off-farm income—rent	Aggregated scores[b]	0.2	0.4	0.1	0.4	4.34**
Access to extension services, markets, infrastructure						
Participation in field school	Yes = 1, no = 0	0.0	0.2	0.0	0.1	1.12
Participation in training	Yes = 1, no = 0	0.2	0.4	0.2	0.4	3.53*
Participation in field day	Yes = 1, no = 0	0.4	0.5	0.3	0.4	5.95**
Participation in field visit	Yes = 1, no = 0	0.6	0.5	0.6	0.5	0.13
Participation in demonstration farm	Yes = 1, no = 0	0.4	0.5	0.3	0.5	7.83***

Participation in interaction	Yes = 1, no = 0	0.5	0.5	0.6	0.5	3.88**
Participation in community meetings	Yes = 1, no = 0	0.8	0.4	0.9	0.3	4.70**
Participation in training centres	Yes = 1, no = 0	0.9	0.3	0.9	0.3	1.68
Distance to output market	m	4454	344 2	4449	3505	.000
Distance to mortable road	m	569	111 9	1377	2166	37.19***
Distance to tarmac road	m	4830	627 9	16,68 3	21,33 0	92.03***
Distances to markets and roads	Factor scores[c]	(0.3)	0.6	0.2	1.2	32.72***

[a]Aggregated scores for relevant farm produce categories during the last 12 months; 0 = not produced, 1 = produced&consumed, 2 = if produced&sold for cash

[b]Aggregated scores for relevant off-farm income activities involvement during the last 12 months; 0 = not engaged, 1 = engaged and earned income

[c]Factor scores were derived from variables of distances to markets, roads and infrastructure by principal component analysis

RESULTS AND DISCUSSION

Tree adoption strategies

Table 3 presents, regardless of the adoption categories, in the semi-arid zone, *Acacia* species, especially *A. tortilis*, *A. senegal* (kertefa) *A. etbaica* (dodoti) that were commonly mentioned along with *Zizyphus mucronata*, *Faidherbia albida*, and *Balanties aegyptiaca* (bedeno). Table 4 shows that in the sub-humid zone, *Cordia africana*, *Croton macrostachyus*, *Vernonia amygdalina* (ebicha), *Mangifera indica* and *Eucalyptus spp.*, especially *E. camaldulensis* were commonly mentioned. At the same time there was high variability of the proportion of households adopting these tree species across sites within each agroecological zone. For example, the adoption rate of *Acacia tortilis* ranged between 52 and 78 % across sites of the semi-arid zone, while that of *Cordia africana* varied between 38 and 67 % across the sub-humid zone.

Table 3 Proportion of households with the ten most common tree species on farm in the semi-arid sites

Rank.	Tree species name	Vernacular name in the local language(Oromiffa)	Site name	Bofa	Mojo	Alemtena	Meki	Zeway	Total
			Site code	Semi-arid1	Semi-arid2	Semi-arid3	Semi-arid4	Semi-arid5	
			Sample size	120	37	55	48	80	340
1	*Acacia tortilis* Hayne	Tadecha, Ajo[a]		78	70	60	52	74	70
2	*Zizyphus mucronata* Willd.	Qurqura		68	24	31	13	5	35
3	*Faidherbia albida* (syn. *Acacia albida* Delile) A.Chev	Gerbi		13	73	36	40	8	26
4	*Acacia senegal* Willd.	Kertefa		37	5	24	17	23	25
5	*Balanites aegyptiaca* Delile	Bedeno		18	32	16	21	28	22
6	*Acacia etbaica* Schweinf.	Dodoti[b]		18	54	29	4	4	18
7	*Croton macrostachyus* Hochst.	Bakanisa		–	14	38	42	1	14
8	*Melia azedarach* L.	Nimi		19	3	20	13	6	14
9	*Eucalyptus camaldulensis* Dehnh.	Bargamo Dima		13	16	13	4	3	10
10	*Dichrostachys cinerea* (L.) Wight & Arn.	Hatte		4	24	11	4	3	7

The most commonly identified/observed species in the study sites grow over 5 m, thus we define them as trees not shrubs, according to the FAO (2012) definition

[a] *Acacia tortilis* is locally called Tadecha in many parts of semi-arid Oromia, while the same species is called Ajo loc in Zeway

[b] Dodoti can either refer to *Acacia gerrardi* or *Acacia etbaica*, while the tree inventory study by Endale (2014) in the semi-arid sites reported only *Acacia etbaica*. Hence, the references to Dodoti in the semi-arid zones were counted for *Acacia etbaica*

Table 5 presents the descriptive statistics of variables indicating strategies of tree adoption on farms. The semi-arid zone had a lower proportion of households who adopted raising seedlings (4 %, on average 0.11 species per farm) and planting trees (36 %, 0.88 species per farm) than the sub-humid zone (29 %, 0.46 species per farm for raising seedlings, and 71 %, 2.03 species per farm for planting trees), while a higher proportion of households protected (84 %, 2.49 species) naturally regenerated trees on farm than that of the sub-humid zone (73 %, 1.96 species). Overall, 92 % of the surveyed households in the semi-arid zone reported having trees on their farm with a mean of 3.37 tree species, compared to 86 % of households with a mean of 3.50 tree species per farm in the sub-humid zone. In terms of tree niches on farm, higher numbers of species were found scattered in crop fields both for the semi-arid (2.14 species) and sub-humid zones (1.67 species), followed by home compounds and external boundaries or live fences. In both zones, fuel was the most frequent utility (1.22 weighted utility scores in the semi-arid zone and 1.01 in the sub-humid zone). For other utilities, the sub-humid zone households gave higher weighted utility scores for high-value commercial species (0.90 scores) than those in the semi-arid zones (0.23 scores), who in turn gave higher scores for environmental services (0.96 scores) than their sub-humid counterparts (0.70 scores).

Table 4 Proportion of households with the 10 most common tree species on farm in the sub-humid sites

Rank	Tree species name	Vernacular name in the local language (Oromiffa)	Site name Site code Sample size	Arjo Sub-humid1 76	Ano Sub-humid2 104	Bako Sub-humid3 75	Uke Sub-humid4 92	Total 347
1	*Cordia africana* Lam.	Wadessa		38	63	67	48	54
2	*Eucalyptus camaldulensis* Dehnh.	Bargamo Dima		36	46	55	28	41
3	*Croton macrostachyus* Hochst.	Bakanisa		46	51	36	14	37
4	*Vernonia amygdalina* Delile	Ebicha		22	22	35	33	28
5	*Mangifera indica* Wall.	Mango		3	20	17	20	16
6	*Albizia gummifera* C.A.Sm.	Mukarba		33	15	5	4	14
7	*Syzygium guineense* DC.	Badessa		12	12	7	25	14
8	*Ficus sycomorus*	Oda		11	17	3	11	11
9	*Acacia abyssinica* Hochst.	Lafto		13	13	5	–	8
10	*Calpurnia aurea (Lam.)* Benth.	Checka		–	16	9	–	7
10	*Vernonia auriculifera* Hiern	Reji		16	6	5	2	7
10	*Ficus vasta* Forssk.	Kiltu		1	13	8	3	7

The most commonly identified/observed species in the study sites grow over 5 m, thus we define them as trees not shrubs, according to the FAO (2012) definition

Table 5 Extent, form and utility of trees adopted by agroecological zone

	Unit	**Semi-arid zone (cases = 340)**		**Sub-humid zone(cases = 347)**		**F**
		Mean	**Std. Deviation**	**Mean**	**Std. Deviation**	
Trees on Farm						
Proportion of parcels with trees planted	Mean proportion in total parcels	.19	.28	.47	.35	130.15***
Establishment method						

Raised seeds/seedlings during the last 3 years	Proportion of households	.04	.20	.29	.45	84.83***
	Mean number of species types	.11	.58	.46	1.13	27.20***
Planted trees during the last 3 years	Proportion of households	.36	.48	.71	.45	96.24***
	Mean number of species types	.88	1.57	2.03	2.10	65.95***
Protected naturally regenerated trees during the last 3 years	Proportion of households	.84	.37	.73	.45	12.82***
	Mean number of species types	2.49	1.62	1.96	2.04	14.04***
Having trees over 3 years old on farm	Proportion of households	.92	.27	.86	.35	6.89***
	Mean number of species types	3.37	2.18	3.50	2.59	0.47
Tree niches on farm						
Scattered in crop farm	Mean number of species types	2.14	2.01	1.67	2.06	9.17***
External boundary/live fence	Mean number of species types	0.16	0.52	0.36	0.81	15.03***
Hedges within farm/soil conservation bonds	Mean number of species types	0.09	0.48	0.07	0.52	0.18
Woodlot	Mean number of species	0.11	0.64	0.08	0.37	0.61

	types					
Home compound	Mean number of species types	0.68	1.40	1.19	1.67	18.42***
Fallow land	Mean number of species types	0.03	0.44	0.10	0.57	3.65*
Others/NA	Mean number of species types	0.16	0.74	0.02	0.13	11.85
Utilities						
High value commercial species types	Mean scores of weighted utilities	.23	.56	.90	1.28	78.70***
Subsistence species types	Mean scores of weighted utilities	.30	.60	.43	.65	7.59***
Woodfuel species types	Mean scores of weighted utilities	1.22	1.28	1.01	1.17	5.00**
Environmental service species types	Mean scores of weighted utilities	.99	1.12	.70	.92	13.02***
Fence species types	Mean scores of weighted utilities	.47	.74	.36	.65	4.39**
Other species types	Mean scores of weighted utilities	.17	.60	.09	.38	4.18**

*** Significant at 1 %; ** significant at 5 %; * significant at 1 %.

Table 6 shows that some of the variables describing tree adoption on farm in Table 4 were highly correlated. For example, households who produced seedlings were more likely to plant trees, which was then positively associated with utilities from high-value species but negatively with utilities from fuel and environmental services. On the other hand, the number of existing species on farm was positively correlated with all utility types, but especially with woodfuel and environmental services, as well as an establishment by naturally regenerating and protecting trees.

Table 6 Associations amongst variables describing tree adoption patterns

	No. of species whose seeds/ seedlings were raised	No. of species which were planted	No. of species which were naturally regenerated and protected	No. of species over 3 years old on farm	Scores for species types by utility for high commer-cial value	Scores for species types by utility for subsistence use	Scores for species types by utility for woodfuel	Scores for species types by utility for environ-mental services	Scores for species types by utility for fence	Scores for species types by utility other than above
No. of species whose seeds/ seedlings were raised	1									
No. of species which were planted	.383**	1								
No. of species which were naturally regenerated and protected	−.009	−.065	1							
No. of species over 3 years old on farm	−.033	−.012	.409**	1						
Scores for species types by utility for high commercial value	.097*	.105**	.145**	.482**						
Scores for species types by utility for subsistence use	.044	.049	.178**	.478**	.129**	1				
Scores for species types by utility for woodfuel	−.086*	−.103**	.176**	.565**	−.048	.074	1			
Scores for species types by utility for environmental services	−.066	−.099**	.275**	.527**	−.016	.172**	.134**	1		
Scores for species types by utility for fence	−.036	.076*	.248**	.400**	.147**	.153**	.048	−.018	1	
Scores for species types by utility other than above	−.018	.008	.083*	.139**	−.004	.006	−.084*	−.025	−.059	1

** Correlation is significant at the 0.01 level (2-tailed)

* Correlation is significant at the 0.05 level (2-tailed)

Table 7 summarizes the result of the PCA analysis. Four of the extracted components explained about 59 % of the total variance of the original variables included. These four components were interpreted with variables of high factor loadings, as follows. Component 1 had high loadings for the number of natural regenerated and protected species, utilities for subsistence, woodfuel, environmental services and fencing.

Therefore Component 1 was taken to represent farmer managed natural regeneration of trees on farm largely for subsistence, woodfuel, environmental services and fencing (FMNR). Component 2 with higher factor loadings of raising seedlings, planting trees, and utilities for high commercial values was taken to represent active planting high value agroforestry species (HVAF). Component 4 indicates agroforestry practices with focus on environmental services but without fencing, while Component 3 seemingly captured cases in which farmers did not specify utilities derived from specific tree species. Interpretations of Component 1 (FMNR) and Component 2 (HVAF) confirm our assumptions that tree species adopted in indigenous practices are managed less intensively than economically important species but serve for multiple utilities such as fuel and ecosystem services simultaneously (Biggelaar and Gold 1996), while deliberate tree planting is associated with species of high economic utility (Franzel and Sherr 2002).

Table 7 Derived components representing tree adoption strategies

Normalized scores	**Extraction**	**Principal Component**			
		1	**2**	**3**	**4**
No. of species whose seeds/seedlings were raised during the last 3 years	.700	–.192	.689	–.167	.402
No. of species which were planted during the last 3 years	.642	–.203	.737	–.129	.203
No. of species which were naturally regenerated and protected during the last 3 years	.588	.734	.102	.189	.060
Scores for species types by utility for high commercial value	.445	.257	.485	.158	–.344
Scores for species types by utility for subsistence use	.367	.518	.281	.014	.142
Scores for species types by utility for woodfuel	.453	.439	–.246	–.422	.150
Scores for species types by	.626	.558	–.169	–.037	.533

utility for environmental services					
Scores for species types by utility for fence	.654	.456	.308	–.048	–.591
Scores for species types by utility other than above	.823	–.020	.017	.881	.215
Total Variance Explained (% of variance)		18.48	16.95	11.82	11.64
Cumulative (%)		18.48	35.43	47.25	58.89

Figure 3 presents the distribution of the derived principal component scores for households by study sites. Despite variances, the sub-humid sites in general have higher scores for Component 2 (HVAF) than those in semi-arid sites, while the intra-site variability seems larger than the inter-site difference for Component 1 (FMNR) score. The inter-site trends or differences are even less clear for Components 3 (OTHERS) and 4 (ENV-FENCE) with many outliers observed.

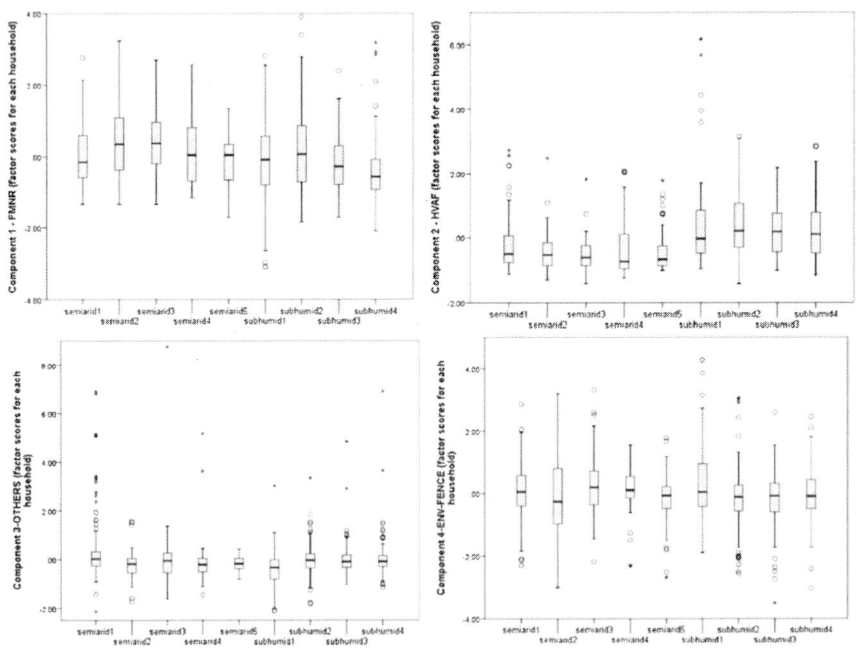

Figure. 3 Box plots of derived principal component factor scores for households by study sites

Table 8 shows the correlations between the derived component scores and niches for trees on farms. Component 1 or FMNR for example had a high association with scattered on farm niches but also significant associations with other niches, including home compounds, hedges in fields and field boundaries. In contrast, Component 2 or HVAF had significant and positive correlations only with home compounds and field boundaries. The relations amongst contextual factors and FMNR confirm our assumption that traditional agroforestry practices are driven by locational flexibility along with multiple utilities, as Biggelaar and Gold (1996) argued from their Rwandan case study. In contrast, the relations amongst contextual variables and HVAF could be explained by the fact that it would be easy for farmers to manage and supervise the growing of commercially valuable tree species in home compounds and fences, whereas scattered trees on farm would be susceptible to low survival rates because of livestock grazing (Gebremedhin et al. 2004; Kassahun et al. 2009; Mekuria and Aynekulu 2013; Tesfaye et al. 2014). This is consistent with the logic behind the homegardens widely observed in southern as well as northwestern regions of Ethiopia, which are known for their rich diversity of economically useful tree species that shade enset (*Ensete ventricosum*, also known as false banana) and/or coffee (Negash 2007; Hylander and Nemomissa 2008; Linger 2014). On the other hand, the association between Component 3 (other, non-classified utilities) and other niches, along with the negative relation of Component 4 (environmental services without fence) and external boundary, would not provide much insight.

Table 8 Association between tree adoption strategies and niches on farm

	Component 1-FMNR	Component 2-HVAF	Component 3-OTHERS	Component 4-ENV-FENCE	Scatter in crop farm	External boundary	Hedges	Woodlot	Home compound	Fallow land	Other niches
Component 1-FMNR	1										
Component 2-HVAF	0.00	1									
Component 3-OTHERS	0.00	0.00	1								
Component 4-ENV-FENCE	0.00	0.00	0.00	1							
Scatter in crop farm	.596**	0.02	−.081*	.079*	1						
External boundary	.131**	.128**	0.01	−.136**	−.083*	1					
Hedges	.169**	0.00	−0.06	0.03	−.077*	0.03	1				
Woodlot	.089*	0.04	0.07	0.02	−.079*	0.02	.080*	1			
Home compound	.313**	.170**	0.01	0.03	−.240**	0.00	−0.01	-0.04	1		
Fallow land	.124**	0.04	−0.04	0.01	−0.05	0.04	−0.02	-0.01	0.01	1	
Other niches	0.05	0.00	.344**	0.01	−.093*	−0.03	−0.02	0.02	0.00	−0.02	1

** Pearson Correlation is significant at the 0.01 level (2-tailed)

* Pearson Correlation is significant at the 0.05 level (2-tailed)

Factors affecting the adoption of different strategies

Table 9 presents the results of the regression analyses. Factors significantly affecting the adoption of FMNR included: being in the semi-arid zone, on mid-land, with larger family size, higher ratios of males and working-age members, larger total land size, larger proportion of parcels on sloped land, higher incomes from cereals, pulses and rent; experiences of communal grazing and tenure upgrading, access to training/training centres, and longer distances to markets. FMNR was negatively correlated with field days/community meetings. In contrast, variables found significantly positively associated with HVAF included: being in the sub-humid zone, having a higher proportion of owned land without certificate, higher farm asset values, higher tree-based farm income, regular business off-farm income, field days and proximity to markets. Variables such as higher transport asset values, farm incomes from roots and tubers, and animals, off-farm income from casual activities and loans, experiences of communal grazing, and community meetings had negative associations with HVAF.

Table 9 Regression analysis of contextual factors on tree adoption strategies

	Component 1-FMNR		**Component 2-HVAF**	
	B	**SE**	**B**	**SE**
(Constant)	−.877	.277***	.247	.261
Biophysical factors				
Zone dummy (semi-arid = 1, sub-humid = 0)	.432	.112***	−.538	.106***
Mid-highland dummy (mid-highland = 1, otherwise = 0)	.375	.132***	.165	.124
Preferences (demographic composition)				
Head gender dummy (male = 1, female = 0)	−.187	.170	−.076	.160
Head education level	−.045	.072	.044	.068
Total family size	.038	.020*	−.009	.019
Male ratio	.624	.284**	.190	.267
Working age member ratio	.421	.252*	−.041	.237
Risk and uncertainties				
Total land size	.121	.052**	−.053	.049

Parcel fragmentation (Simpson index)	.036	.050	.017	.047
Mean distance of parcels from homesteads	.069	.053	.074	.050
Proportion of parcels owned with certificate	.236	.274	.419	.259
Proportion of parcels owned yet no certificate issued	.272	.206	.368	.194*
Proportion of parcels rented from others	.084	.185	.185	.175
Proportion of parcels on sloped/steep land	.099	.049**	.018	.046
Experience of free communal livestock grazing	.242	.048***	−.111	.046**
Experience of land resettlement	−.062	.040	−.056	.038
Experience of tenure upgrading	.118	.050**	−.011	.047
Resource endowments				
Livestock—local animals	.056	.056	−.017	.052
Livestock—exotic animals	−.075	.050	−.048	.047
Asset—domestic asset value	−.019	.044	.012	.041
Asset—communication asset value	−.079	.053	−.058	.050
Asset—transport asset value	.051	.048	−.139	.045***
Asset—farm asset value	.000	.060	.361	.057***
Farm income- cereal & pulse	.112	.050**	.006	.047
Farm income—cash crops	−.012	.051	−.021	.048
Farm income—roots & tubers	.023	.046	−.156	.044***
Farm income—animal	.072	.050	−.082	.047*
Farm income—tree-based	.006	.050	.165	.047***
Off-farm income—regular business	.049	.047	.102	.044**
Off-farm income—casual	−.055	.050	−.103	.047**
Off-farm income—remittance & gift	.014	.044	−.042	.041

Off-farm income—loan	.044	.045	–.108	.042**
Off-farm income—rent	.143	.047***	.052	.044
Market incentives				
Participation in field school	.062	.062	.043	.058
Participation in training	.126	.051**	–.009	.048
Participation in field day	–.094	.053*	.245	.050***
Participation in field visit	–.031	.054	.027	.051
Participation in demonstration farm	.026	.052	.011	.049
Participation in interaction	–.068	.051	–.008	.048
Participation in community meetings	–.116	.049**	–.142	.046***
Participation in training centres	.121	.059**	.050	.056
Distances to markets, roads & services	.135	.046***	.077	.043*
F-value	4.135	.000[b]	6.837	.000[b]
Adjusted R^2	.219		.343	

*** Significant at 1 %; ** significant at 5 %; * significant at 1 %

As we had assumed, biophysical factors were found to influence the adoption of different strategies. There is a higher likelihood of adopting HVAF in the sub-humid zone through planting tree species with timber and fruit utility. In contrast, FMNR was highly associated with the semi-arid zone, probably because harsh climatic conditions may discourage farmers from investing in active tree planting because of low survival rates of seedlings. The significant positive sign of the mid-highland dummy, which applies to two out of four sites in the sub-humid zone, for FMNR, indicates that FMNR is also practiced in the sub-humid zone, especially in the mid-highland parts. Figure 3 also indicates that tree adoption strategies have high variability amongst households within the same agroecological zone.

Controlling biophysical factors, preference, risk and uncertainty, and resource endowment factors have contrasting effects on the adoption of FMNR and HVAF.

Labor availability, especially the proportion of male labor, and land availability, are significant factors for FMNR but not constraints for HVAF, while the gender and education level of the household heads are found not significant for both strategies. As FMNR requires extensive management of trees on farm though occasionally pollarding and lopping, a large land size with a large number of family members supplying labor may provide an advantage for households to allocate more labor that is required for FMNR activities over extensive fields. In turn, HVAF, which is associated with homesteads and boundaries, can be managed without being constrained by land and family labor sizes.

Land access conditions and management/policy experiences which are proxy variables for risk and uncertainty also affected the likelihoods of adopting FMNR and HVAF differently. Land ownership provided a positive incentive to adopt HVAF, even without formal certificates in the Oromia context where the certification program had started but the issuing of certificates had not been fully completed. But, land ownership did not significantly affect adoption of FMNR, which still has a positive association with farmer's experience of upgrading their tenure. It is interesting to note that HVAF had negative association with communal grazing, while with FMNR it was positive. In the Ethiopian context, under communal grazing, neighboring farmers free their cattle and goats to browse on crop fields after communal harvesting. It can be interpreted that promoting intensive agroforestry of actively planted tree species for timber, fruit and income crops requires not only favorable agroecological conditions, but also institutional/policy arrangements to set up physical fencing or institutional arrangements of social fencing to protect young trees.

Asset/income profiles also have contrasting patterns of association with FMNR and HVAF. HVAF is associated positively with farm asset value, as well as tree-based income and regular business off-farm income while negatively associated with animal farm income, roots and tuber farm income and casual off-farm income. This implies that intensive agroforestry requires livelihood assets that can be invested in tree planting that possibly diverts livelihood strategies away from traditional livestock or low-paying casual/farming activities. In contrast FMNR has no significant associations with assets, but positive association with cereal and pulse farm income and off-farm income from rent, thus is consistent with traditional staple crop-based livelihoods.

While access to training and training centres were found positive for FMNR and HVAF, there seems room for improving extension modes for promoting agroforestry adoption, as attendance at community meetings had negative associations with both, and field days had a negative association with FMNR. It is contrary to the expectation to find that longer distances to markets, roads and services are found positive for the adoption not only for FMNR of subsistence nature but also HVAF. The significance of longer distances to market was also reported for adoption of improved wheat in Oromia (Solomon et al. 2014). They explained the unexpected outcome on the basis that farmers nearer to markets would focus on more market oriented crops than wheat whose local market was comparatively underdeveloped. In regard to our finding, though, its significant positive association with tree-based income, opportunities of HVAF may not have been fully exploited in the sub-humid sites which were in general rather isolated from market and infrastructure access during the time of the survey. A plausible explanation for the negative association with proximity to markets might be that farmers nearer to markets in semi-arid sites tend to focus on cash crops such as nug and maize, whose markets are relatively well developed, rather than less well known tree crops.

Overall, the regression results indicated the significant impacts of biophysical factors as well as risk and uncertainties on FMNR and HVAF in often contrasting manners, as Pattanayak et al. (2003) also concluded from their meta-analysis that tree planting behavior is most likely to be significantly influenced by these factors. While favorable climatic conditions are a pre-requisite for HVAF, poorer biophysical potential and sloping land appear to provide a positive incentive to adopt FMNR, which has ecosystem service benefits. The possibility of tenure upgrading provides a positive incentive for FMNR by reducing the risks of land appropriation, while communal grazing is also consistent with the adoption of FMNR. In contrast, the adoption of HVAF is deterred by communal grazing which may discourage tree planting because of increased survival risks of seedlings, confirming Pattanayak et al.'s (2003) finding that the adoption of intensive tree planting is contingent on lower risk.

Pattanayak et al. (2003) found that household preference proxies were significant in only 41 % of tree adoption studies related to tree planting, so they were not as critical as biophysical factors and risk and uncertainty, and that the significance and sign of preference variables were often ambiguous. Our research found that some preference variables were important in influencing adoption of FMNR but not HVAF. A probable explanation is that FMNR, which is driven more by multiple utility and locational flexibility, may be facilitated by the availability of labor as well as land. In contrast, more resource endowment variables were significant for HVAF than for FMNR. Resource endowments are likely to be positively correlated with the probability of adoption of tree planting as an investment driven by economic utility.

CONCLUSION

Analysis of household data revealed two distinct strategies for tree adoption in the semi-arid and humid zones of Ethiopia: farmer managed regeneration (FMNR) and planting high value tree species. We revealed that FMNR is a dominant agroforestry practice not only in the semi-arid zone but also on sloping land in the sub-humid zone, and it is consistent with supporting subsistence staple-crop production through provision of multiple utilities that can alleviate negative biophysical constraints. In contrast, high value agroforestry (HVAF) is practiced more in the sub-humid zone and associated with tree-based farm income, assets and off-farm enterprises. Biophysical conditions and resource endowments are not the only determinants of tree planting strategies. Reducing risk and uncertainty through policy and institutional arrangements is critical to ensure tenure security for people to adopt FMNR on one hand, and to handle externalities of communal grazing and adopt HVAF on the other hand. Preferences, which were represented by household demographic variables, were found more important for the adoption of FMNR as larger labour combined with larger land may facilitate locational flexibility. The impacts of market incentives turned out rather contradictory and ambiguous, which may indicate markets and extension systems to promote agroforestry to enhance tree-based incomes and enterprises were not yet fully functional in the study sites.

Our findings imply the critical importance of understanding farmers' preferences for specific species with multiple utilities and locational flexibility (Biggelaar and Gold 1996) which define their management intensities and niches to make fine-scale recommendations of optimal mixes of species and management options. For example, in the sub-humid zone, managing multi-purpose trees, such as *Cordia africana*, are widely adopted by farmers, as a dominant feature of agricultural landscapes (Yadessa et al. 2009). Interestingly, some farmers consider *Cordia africana* as primarily a timber species to earn income and secondarily for fencing and/or/shade and so deliberately plant the tree in homesteads or along external boundaries. Others that primarily see the species as fuelwood and also a soil amendment protect naturally regenerating trees on their farms. Either way, the tree contributes to enhancing livelihoods and food security, even though preferred management modes and intensities vary depending on farmers' perceptions and preference.

It is clear that FMNR as well as other indigenous practices deserve more attention when designing tree promotion initiatives, as they serve a critical role in alleviating negative production conditions through the provision of ecosystem services. Our findings reveal that farmers integrate many native and exotic tree species on their farms to meet their variable farm conditions, needs and asset profiles in stark contrast to most tree promotion efforts that focus on a few, usually exotic, tree species. We recommend that future agroforestry promotion should embrace a diversity of tree species appropriate to matching the fine scale variation in ecological conditions and farmer circumstances encountered in the field.

ACKNOWLEDGMENTS

This study as a part of the baseline survey of the "Trees for Food Security Project" funded by the Australian Centre for International Agricultural Research (ACIAR) and the CGIAR Research Programme on Forests, Trees and Agroforestry. We acknowledge the Ethiopian Institute of Agricultural Research (EIAR) for supporting the implementation of the data collection through mobilizing enumerators and arranging logistics during the survey. The World Agroforestry Centre's (ICRAF) Research Methodology Group (RMG) contributed greatly to the design of the survey tools, training the local enumerators and data entry.

REFERENCES

1. Abeyasekera S (2001) Analysis Approaches in Participatory Work involving Ranks or Scores. A theme paper written for the DFID-funded NRSP project R7033, Statistical Services Centre, The University of Reading. http://www.reading.ac.uk /ssc/resources/AnalysisApproachesInParticipatoryWorkInvolvingRanksOrScores.p df
2. ARD Inc. (2004) Ethiopia Land Policy and Administration Assessment Final Report with Appendices. Submitted to USAID/Ethiopia. May 2004. USAID Contract No. LAG-00-98-00031-00, Task Order No. 4
3. Bewket W (2002) Land cover dynamics since the 1950s in Chemoga Watershed, Blue Nile Basin, Ethiopia. Mt Res Dev 22(3):263–269
4. Biggelaar CD, Gold MA (1996) Development of utility and location indices for classifying agroforestry species: the case of Rwanda. Agrofor Syst 34:229–246
5. Bizimana C, Nieuwoudt WL, Ferrer SRD (2004) Farm size, land fragmentation and economic efficiency in southern Rwanda. Agrekon 43(2):244–262
6. Blarel B, Hazell P, Place F, John Quiggin J. (1992, May) The economics of farm fragmentation: evidence from Ghana and Rwanda. The World Bank Economic Review, vol 6, no. 2 (May, 1992), pp 233–254. http://www.jstor.org/ stable /3989856
7. Central Statistical Agency (2010) Population Size of Kebeles, Central Statistical Agency, 2010-07. http://www.csa.gov.et/newcsaweb/images /documents/surveys /Population%20and%20Housing%20census/ETH-pop-2007/survey0/data/Doc /Reports/STATISTICAL_OROMIYA/Statistical_Oromiya_PartV.pdf
8. Coe R, Njoloma J, Sinclair FL (2016) Loading the dice in favour of the farmer: reducing the risk of adopting agronomic innovations. Exp Agric (in press)
9. Degefu T, Wolde-meskel E, Frostegård A (2011) Multilocus sequence analyses reveal several unnamed Mesorhizobium genospecies nodulating Acacia species and Sesbania sesban trees in Southern regions of Ethiopia. Syst Appl Microbiol 34:216–226. doi:10.1016/j.syapm.2010.09.006

10. Deininger K, Jin S (2006) Tenure security and land-related investment: evidence from Ethiopia. Eur Econ Rev 50(2006):1245–1277. doi:10.1016 /j.euroecorev .2005.02.001
11. Deininger K, Ali DA, Holden S, Zebenbagen J (2008) Rural land certification in Ethiopia: process, Initial impact, and implications for other African countries. World Dev 36(10):1786–1812. doi:10.1016/j.worlddev.2007.09.012
12. Deininger K, Ayalew D Ali, Alemu T (2009) Impacts of land certification on tenure security, investment, and land markets evidence from Ethiopia. Environment for Development. Discussion Paper Series April 2009. EfD DP 09-11
13. Deressa TT, Hassan RM, Ringler C, Alemu T, Yesuf M (2009) Determinants of farmers' choice of adaptation methods to climate change in the Nile Basin of Ethiopia. Glob Environ Change 19(2009):248–255. doi:10.1016/ j.gloenvcha. 2009.01.002
14. Duguma LA (2013) Financial analysis of agroforestry land uses and its implications for smallholder farmers livelihood improvement in Ethiopia. Agrofor Syst 87:217–231. doi:10.1007/s10457-012-9537-.1
15. Endale Y (2014) Assessment of tree species, diversity distribution patterns and socioeconomic uses on farmland in Oromia Regional State: The case of East Shewa Zone. A thesis submitted to the center for environmental science, presented in partial fulfilment of the requirements for the degree of master of environmental science, Addis Ababa university, Addis Ababa, Ethiopia
16. Ethiopian Government Portal. The Oromia National Regional State http:// www.ethiopia.gov.et/stateoromia. Accessed 17 Jan 2016
17. FAO (2012) FRA 2015 Terms and definition, forest resources assessment working Paper 180. Food and Agriculture Organization of the United Nations, Rome
18. Federal Democratic Republic of Ethiopia Population Census Commission. (2008) Summary and Statistical Report of the 2007 Population and Housing Census. United Nations Population Fund(UNFPA), Addis Ababa. http:// ecastats. uneca.org/aicmd/Portals/0/Cen2007_firstdraft.pdf
19. Franzel S, Scherr SJ (2002) Trees on the farm: assessing the adoption potential of agroforestry practices in Africa. CAB International, Published in association with ICRAF, Nairobi
20. Gebremedhin B, Pender J, Tesfay G (2004) Collective action for grazing land management in crop–livestock mixed systems in the highlands of northern Ethiopia. Agric Syst 82(2004):273–290. doi:10.1016/j.agsy.2004.07.004
21. Gelaw AM, Singh BR, Lal R (2014) Soil organic carbon and total nitrogen stocks under different land uses in a semi-arid watershed in Tigray, Northern Ethiopia. Agric Ecosyst Environ 188(2014):256–263. doi:10.1016/j.agee.2014.02.035
22. Holden ST, Deininger K, Ghebru H (2011) Tenure insecurity, gender, low-cost land certification and land rental market participation in Ethiopia. J Dev Stud 47(1):31–47. doi:10.1080/00220381003706460
23. Hylander K, Nemomissa S (2008) Complementary roles of home gardens and exotic tree plantations as alternative habitats for plants of the Ethiopian Montane Rainforest. Conserv Biol 23(2):400–409. doi:10.1111/j.1523-1739.2008.01097.x
24. Iiyama M, Kariuki P, Kristjanson P, Kaitibie S, Maitima J (2008) Livelihood diversification strategies, incomes and soil management strategies: a case study from Kerio Valley, Kenya. J Int Dev 20:380–397. doi:10.1002/jid.1419
25. Israel GD (1992) Sampling the Evidence of Extension Program Impact. Program Evaluation and Organizational Development, IFAS, University of Florida. PEOD-5.October
26. Jagger P, Pender J (2003) The role of trees for sustainable management of less-favored lands: the case of eucalyptus in Ethiopia. For Policy Econ 5:83–95
27. Jagger P, Pender J, Gebremedhin B (2005) Trading off environmental sustainability for empowerment and income: Woodlot devolution in Northern Ethiopia. World Dev 33(9):1491–1510. doi:10.1016/j.worlddev.2004.10.010

28. Kassahun A, Snyman HA, Smit GN (2009) Soil seed bank evaluation along a degradation gradient in arid rangelands of the Somali region, eastern Ethiopia. Agric Ecosyst Environ 129:428–436. doi:10.1016/j.agee.2008.10.016
29. Linger E (2014) Agro-ecosystem and socio-economic role of home garden agroforestry in Jabithenan District, North-Western Ethiopia: implication for climate change adaptation. Linger Springer Plus 2014, 3:154. http://www.springerplus.com/content/3/1/154
30. Meijer SS, Catacutan D, Ajayi OC, Gudeta W, Sileshi GW, Nieuwenhuis M (2015) The role of knowledge, attitudes and perceptions in the uptake of agricultural and agroforestry innovations among smallholder farmers in sub-Saharan Africa. Int J Agric Sustain 13(1):40–54. doi:10.1080/14735903.2014.912493
31. Mekuria W, Aynekulu E (2013) Exclosure land management restores soil properties of degraded communal grazing lands in northern Ethiopia. Land Degrad Dev 24:528–538
32. Mercer DE (2004) Adoption of agroforestry innovations in the tropics: a review. Agrofor Syst 61(1):311–328
33. Negash M (2007) Trees Management and Livelihoods in Gedeo's Agroforests, Ethiopia. For Trees Livelihoods 17:157–168
34. Nyaga J, Barrios E, Muthuri CW, Öborn I, Matiru V, Sinclair FL (2015) Evaluating factors influencing heterogeneity in agroforestry adoption and practices within smallholder farms in Rift Valley, Kenya. Agric Ecosyst Environ 212:106–118
35. Pattanayak SK, Mercer DE, Sills E, Yang JC (2003) Taking stock of agroforestry adoption studies. Agrofor Syst 57:173–186
36. Place F (2009) Land tenure and agricultural productivity in Africa: a comparative analysis of the economics literature and recent policy strategies and reforms. World Dev 37(8):1326–1336. doi:10.1016/j.worlddev.2008.08.020
37. Poschen P (1986) An evaluation of the Acacia albida-based agroforestry practices in the Hararghe highlands of Eastern Ethiopia. Agrofor Syst 4:129–143
38. Sinclair FL (1999) A general classification of agroforestry practice. Agrofor Syst 46:161–180
39. Sinclair FL, Walker DH (1999) A utilitarian approach to the incorporation of local knowledge in agroforestry research and extension. In: Buck LE, Lassoie JP, Fernandes ECM (eds) Agroforestry in sustainable agricultural systems. Lewis Publishers, New York, pp 245–275
40. Solomon T, Tessema A, Bekele A (2014) Adoption of improved wheat varieties in Robe and DigeluTijo Districts of Arsi Zone in Oromia Region, Ethiopia: a double-hurdle approach. Afr J Agric Res 9(51):3692–3703. doi:10.5897/AJAR2014.9047
41. Teklay T, Nordgren A, Nyberg G, Malmer A (2007) Carbon mineralization of leaves from four Ethiopian agroforestry species under laboratory and field conditions. Appl Soil Ecol 35(2007):193–202. doi:10.1016/j.apsoil.2006.04.002
42. Tesfaye A, Negatu W, Brouwer R, Van der Zaag P (2014) Understanding soil conservation decisions of farmers in the Gedeb Watershed, Ethiopia. Land Degrad Dev 25:71–79. doi:10.1002/ldr.2187
43. Teshome DS. (2014) Assessment of tree species, diversity distribution patterns and socioeconomic uses on agricultural landscapes: A case of Western Oromia, Ethiopia. M.Sc. Thesis submitted to the Graduate Studies Jimma University College of Agriculture and Veterinary Medicine, in partial fulfillment of the requirements for the degree of Master of Science in Natural Resource Management (Forest and Nature Conservation). Jimma University, Jimma, Ethiopia
44. The National Regional Government of Oromia, Regional Statistical Abstract, Physical & Socio-Economic Profile and Economic Account Statistics, Physical and Socio-Economic Profile Of Oromia, http://www.oromiabofed.org/images/stories/rstatistics/Physical_and_Socio_Economic_Profile_Of_Oromia.zip. Accessed 7 Jan 2016
45. The University of Reading Statistical Services Centre. (2001) Approaches to the Analysis of Survey Data, Biometrics Advisory and Support Service to DFID,

Statistical Services Centre, The University of Reading, UK. http://www.reading.ac.uk/ssc/resources/ApproachesToTheAnalysisOfSurveyData.pdf
46. Tiffen M, Mortimore M, Gichuki F (1994) More people, less erosion. Environmental recovery in Kenya. ACTS Press, Nairobi
47. Yadessa A, Itanna F, Olsson M (2009) Scattered trees as modifiers of agricultural landscapes: the role of waddeessa (*Cordia africana* Lam.) trees in Bako area, Oromia. Ethiopia. Afr J Ecol 47(Suppl. 1):78–83

CHAPTER 6

Influence of Soil Types on Establishment and Early Growth of Populus trichocarpa

Henrik Böhlenius[1*], Rolf Övergaard[1], Sandra Jämtgård[2]

[1]*Department of Southern Swedish Forest Research Centre, Swedish University of Agricultural Sciences, Alnarp, Sweden*

[2]*Department of Forest Ecology and Management, Swedish University of Agricultural Sciences (SLU), Umeå, Sweden*

ABSTRACT

When planning poplar plantations there are many factors to consider that can influence young plant development. In addition to soil preparation, vegetation management,protection against browsers, and plant-type, it is important to have information on how soil properties influence young plant development. Compa- risons of establishment and growth in different soils are complicated as experimental sites often are geographically distant where temperature, precipitation and vegetation can vary, thus complicate the analysis. In this case study, a new approach of studying growth of different soil at one experimental site are undertaken by translocating agricultural soils, a silty clay loam and a sandy soil, and common forest soil types, forest moraine soil and forest peat to a single experimental site, thus guaranteeing identical climatic conditions. In all soils, transplanted *P. trichocarpa* cuttings initially developed in to plants. After 4 weeks though, plants grown in forest peat stopped their growth while plants in the other soils gradually continued their growth with no evidence that soil types influenced above ground plant development. Unlike above- ground growth, root growth and morphology were influenced by soil texture with more root growth occurring if soils have sandy texture. These findings give advice to some of the limitations when planning for establishment of poplar plantations either at agricultural or at forest land.

Keywords: Poplar Establishment, Soil Texture, Cutting Growth, Nutrient, Agricultural Soil, Forest Soil

1. INTRODUCTION

Soil texture and drainage are two of the most important factors for successful poplar plantation establishment and growth (Baker & Broadfoot, 1979) . Poplars have the ability to initiate adventitious roots from the stem and therefore hardwood cuttings can be planted in order to establish plantations (DeBell & Harrington, 1997; Hartmann, 1975; Hofmann-Schielle et al., 1999) . Poplars are generally considered to prefer well drained alluvial soils with sufficient moisture and nutrients and an intermediate soil texture (sand/loam) (Baker & Broadfoot, 1979) . In northern climates, sandy soils can favor the growth of *Populus x wettsteinii* (hybrid aspen) as these soils warm earlier in spring, but this advantage may be offset by the risk of drought conditions later in the growing season (Bergante et al., 2010; Scotti et al., 2010; Tullus et al., 2011) . Heavy soils with clay, clay loam, and silty clay loam textures are considered less favorable for poplar growth(Stanturf et al., 2001) . However, once established, high growth of poplars can occur on heavier soils (Johansson & Karačić, 2011) . Pinno and Belanger (2009) found that poplar growth is best in soils with more silt and clay. However, Pinno et al. (2009) found that maximum poplar tree growth occurred at a sand content between 55 and 70% with pH near 6. In other tree species, for example *Citrus volkameriana*, soil texture only slightly affects growth (Bouma & Bryla, 2000) . For *Salix nigra*, biomass production was found to be clearly affected by soil characteristics, with higher production in coarse-grained (sand) than in finer-grained sediments (silt/clay): fine textured soil negatively affected leaf size, leaf mass and leaf area (Schaff et al., 2003) . Different soil types can influence plant development through differences in available nitrogen (N) resources leading to a shift in carbon allocation from root to shoot (Albaugh et al., 1998; Beets & Whitehead, 1996;Coleman et al., 2004) , resulting in differences in fine root production (Burton et al., 2000; Gower et al., 1992; Kern et al., 2004; Keyes & Grier, 1981) and changes in leaf size and total leaf area(Knops, 2000) . To date, plantations of poplars are mainly located on marginal abandoned agricultural land (Christersson, 2008; Christersson, 2010; Tullus et al., 2011) . On forest land however, our knowledge of establishment and growth of poplars are limited but there is a large potential in available area if establishment would be possible.

How young poplar growth is influenced by a specific soil type is hard to predict and comparisons are also difficult because sites with different soils are generally geographically distant, so that factors such as precipitation and temperature vary, thus complicating the analysis. In order to undertake accurate comparisons of establishment and growth in different soil types, distances between experimental sites needs to be minimized. In this paper, we created tree cases with different soils and investigated how poplar cutting development is influenced by texture (sand and clay) and origin (agricultural and forest). We used soils with different characteristics and from different sites to grow poplar cuttings under similar conditions to study the effects of soil characteristics.

2. MATERIALS AND METHODS

2.1. The Experimental Site and Collection of Soils

The experiment was performed in the experimental garden at the Swedish University of Agricultural Sciences in Alnarp, Sweden (Lat N 55°39'33" Lon E 13°4'46"). In 2006, holes measuring 2 × 2 m and 1 m deep were dug at the experimental site. The soils (Table 1) were collected in 10 cm layers to a depth of 1 m and placed in the holes in their original order. To separate the experimental soils from the surrounding soil, a polyethylene plastic woven carpet (Mypex,) was placed along the walls of the pits. The only treatments applied during the period from soil collection to the beginning of the experiment were non-chemical vegetation control and irrigation, no fertilizers were added.

2.2. Soil Characteristics

Three different soils were chosen for the experiment: two agricultural soils with different textures and one moraine soil, typical of Swedish forest land and a forest peat also commonly found in Swedish forest land (Table 1). The soil texture was determined by sieve analysis, the humus content by Loss on Ignition (LOI) (Table 1), soil Nitrogen (N), Potassium (K) and Phospurus (P) (Table 2) at the analytical laboratory Eurofins, Kristianstad, Sweden. It should be noted that the forest moraine soil (For-mor) contained larger stones (2 - 5 cm in diameter) than the agricultural soils. The soil pH was measured at a soil-to-water ratio of 1/2 (v/v).

Table 1. pH and soil Characteristics, including pH, humus, clay and sand content and the origin of the soil.

	Soil Characteristics					
	Ph	Humus (%)	Clay (%)	Silt (%)	Sand (%)	Soil origin
For-peat	3.7	34	3	5	58	Forest
For-mor	4.9	1.8	6	4	81	Forest
Agri-clay	7.0	6.1	48	33	19	Agricultural
Agri-sand	6.9	3.2	5	13	82	Agricultural

Note: the soils are; agricultural soils, a silty clay (Agri-clay) and a sandy soil (Agri-sand, and forest moraine soil (For-more and forest peat (For-peat).

Table 2. Soil conten of Nitrogen (N), Potasium (K) and Phosporus (P) Data shown are mg per kg soil.

	Soil content		
	Nitrogen (N)	Potasium (K)	Phosporus (P)
For-peat	18	34	73
For-mor	14	18	16
Agri-clay	160	320	260
Agri-sand	24	41	94

Note: the soils are; agricultural soils, a silty clay (Agri-clay) and a sandy soil (Agri-sand, and forest moraine soil (For-more and forest peat (For-peat).

2.3. Plant Material and Experimental Treatments

Populus trichocarpa clone SRF 93 (STT 7/3.26) was used as the plant material. This clone originates from British Colombia, Canada and was selected because it is robust and has a high rooting capacity. Dormant cuttings measuring 20 cm long and with a diameter of 12 - 15 mm were used in the experiment. Before planting, cuttings were soaked in water for 24 h. To avoid severe drought stress during the experiment, the lowest acceptable level of precipitation was set to 10 mm per week. If the natural precipitation was less than 10 mm per week the young plants were irrigated with water equivalent to 10 mm precipitation. Such irrigation was necessary three times during drought periods (Figure 1(a)). During the experimental period, the temperature varied between 13.4°C and 22.9°C (Figure 1(b)). Precipitation and air temperature were determined at a local weather station less than 1 km from the experimental site.

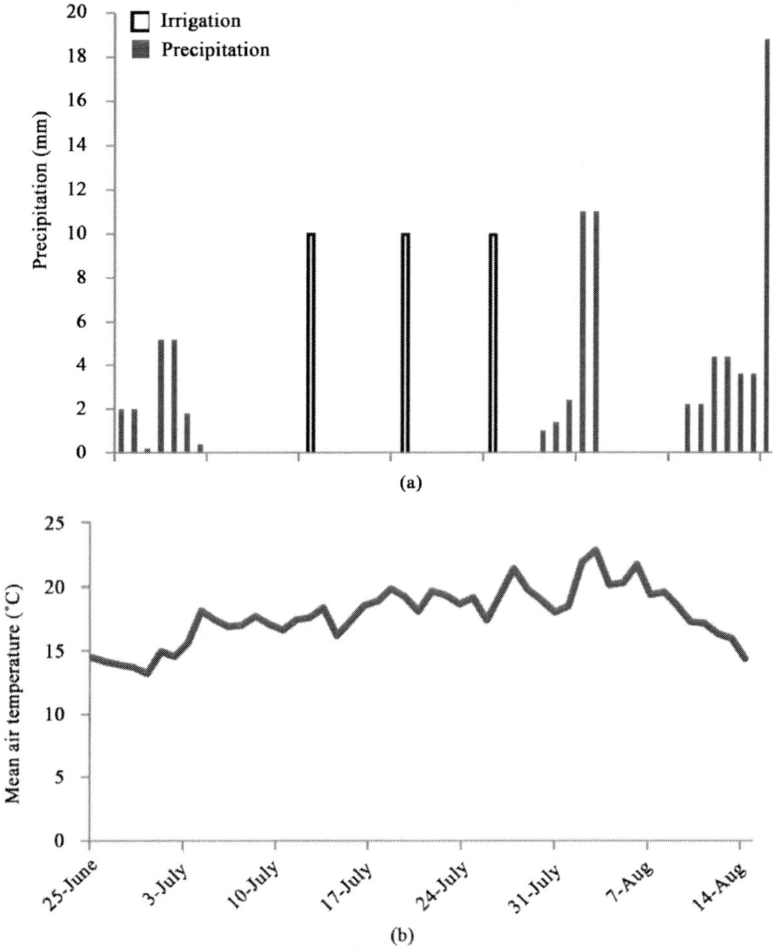

Figure 1. Temperature and precipitation at the experimental site. (a) precipitation and (b) air temperature at a local weather station.

2.4. Experimental Design and Plant Analysis

In each soil, eight rows each of seven cuttings were planted. Each week, seven plants from each soil were sampled according to a pattern were each of the planting positions were represented. The parameters analyzed were height, total root number, development of fine roots and mean root length of each cutting. Fine root development was recorded as follows: 0 = no fine roots, 1 = few fine roots, 2 = fine roots found on all roots, 3 = fine roots abundant on all roots. Mean and total leaf area were determined eight weeks after planting. Individual leaf area was determined from one fully expanded leaf of each plant eight weeks after planting. The total leaf area was calculated by multiplying the total number of leaves on each individual plant by the individual leaf area of the same plant. To determine the biomass eight weeks after planting, roots were washed and together with the leaves, the stems dried at 80°C for 48 hours. The nutrient analysis was performed on leaves and stem samples eight weeks after planting using material from four cuttings in each soil type, following Leco AN 203821-394, ISO 16634 (Nitrogen) and NMKL 161, 198, mod (minerals) at the analytical laboratory Eurofins, Kristianstad, Sweden.

3. RESULTS

3.1. Plant Survival, Height Growth and Root Development

All the cuttings survived and developed into plants in all soil types during the experiment (Figure 2). A gradual increase in height was observed throughout the experiment for plants growing in all soil types except For-peat, reaching heights of 47 cm in For- mor, 57 cm in Agri-clay and 69 cm in Agri-sand (Figure 2) For plants grown in For- peat, plants grow up to 10 cm at week 4 but after this no further height growth occurred.

After this week plants turned yellow and were dying. For this reasons plants grown in For-peat are not included in further analysis. The number of roots gradual increase for plants grown in For-mor and Agri-sand while those in Agri-clay remained quite constant throughout the experiment (Figure 3(a)). Mean root length were gradually increased during the experimental period in all soil types, reaching 8.2 cm and 10 cm for plants grown in For-mor and Agri-sand soils and 4.4 cm for plants grow in the Agri-clay soil (Figure 3(b)). On plants grown in For-mor and Agri-sand, fine roots were developed after three week while no fine roots were found in the Agri-clay soil (Figure 3(c)).

3.2. Biomass Allocation, Production and Leaf Area

Root biomass of plants grown in Agri-clay, Agri-sand and For-more were 0.4, 0.3 and 0.3 g (Figure 4(b)). Stem biomass production varied form 1.9 g for seedling in For- mor to 6.1g for plants in Agri-clay and 4.3 g in Agri-sand. Leaf biomass of plants grown in For-more reached 2.3 g while plants

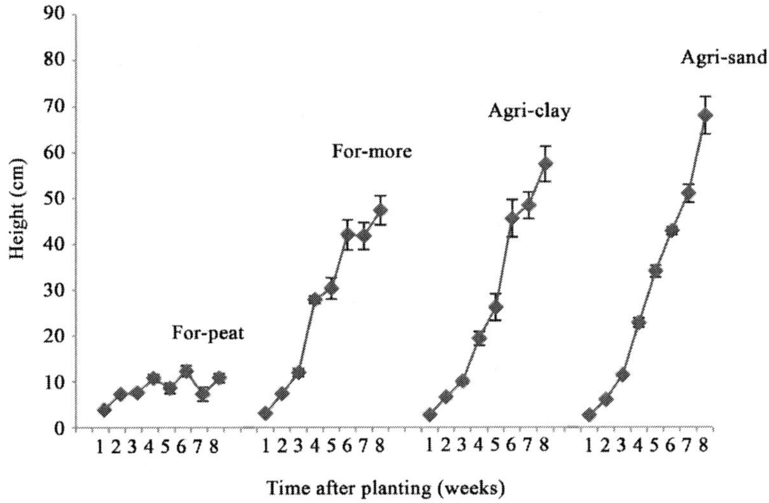

Figure 2. Height growth. *P. trichocarpa* cuttings were planted in forest peat (For-peat), forest moraine soil (For-mor), agricultural silty clay soil (Agri-clay) and agricultural sandy soil (Agri-sand). Data shown are mean values (n = 7). Error bars represent ± SE.

grown in Agri-clay and Agri-sand reached 7.3 and 5.0 g. Root fraction varied for plants with 3.6% For-more, 3.6% Agri-clay and 6.5% in Agri-sand Figure 4(a). Stem biomass allocation changed from 42% in seedlings growing in For-more to 40% - 38% in plants growing in Agri-clay and Agri-sand. For plants grown in For-more leaf biomass were 52%, Agri-clay 55% and Agri-sand 58%. Mean leaf area reached 17 cm^2 and total leaf area 335 cm^2 for plant grown in For-more (Figure 4(c)). In the agricultural soils, Agri-clay and Agri-sand, individual leaf area reached 30 to 28 cm^2 and the total leaf area was determined to 645 cm and 628 cm (Figure 4(d)).

3.3. Leaf and Stem Nutrient Content

In the stem and leaf, the N concentrations reached 0.75% and 1.5% when plants were grown in For-mor (Figure 5(a)). Plants grown in Agri-sand and Agri-clay, N concen- tration in stem reached 0.95% to 0.85% and in leave 2.3% to 2.0%. P concentration reached 0.11% in stem and 0.12% in leaves of plants grown in For-more. For plants grown in agricultural soils, Agri-clay and Agri-sand, P concentration in stem were 0.11 and 0.13 respectively and in leafs 0.15 and 2.0. K concentration was 0.96% in leaf and 0.8% in stem of plants grown in For-more. In Agri-clay and Agri-sand, K concentration reached 0.6% and 0.71% (leaf) and 0.45% and 0.71% (stem)

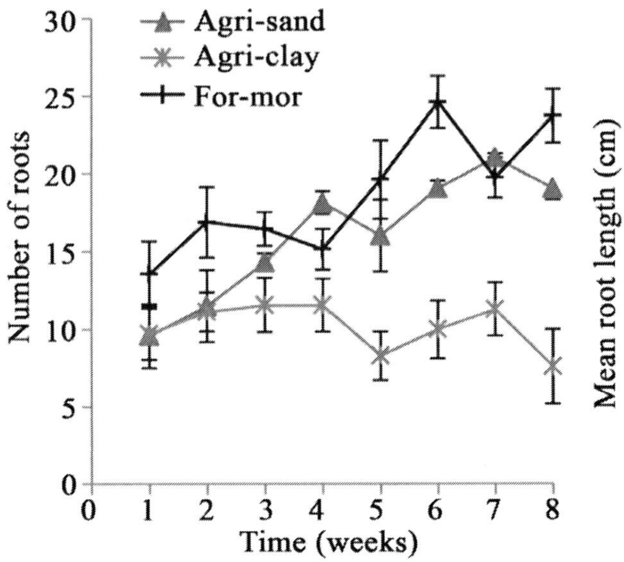
Agri-sand
Agri-clay
For-mor
Number of roots
Mean root length (cm)
Time (weeks)
0
5
10
15
20
25
30
0
1
2
3
4
5
6
7
8

(a)

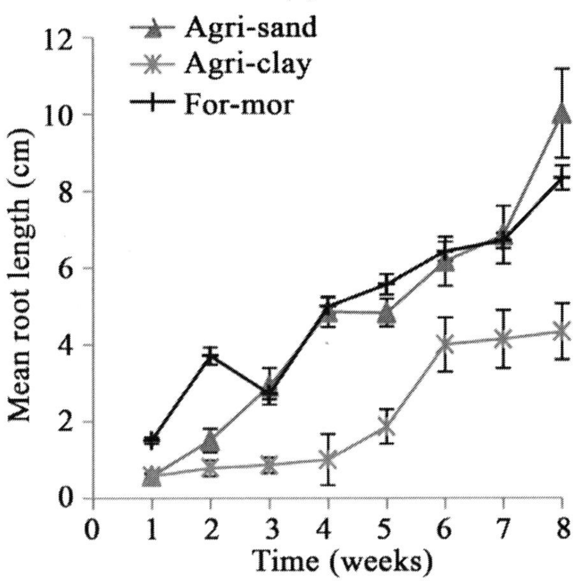
Agri-sand
Agri-clay
For-mor
Mean root length (cm)
Time (weeks)
0
2
4
6
8
10
12
0
1
2
3
4
5
6
7
8

(b)

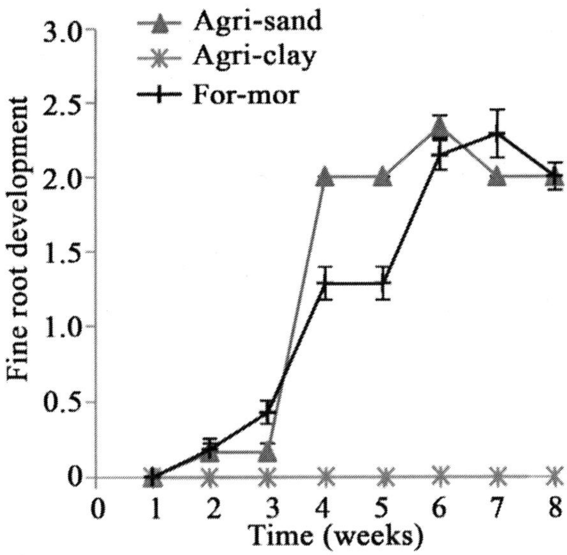

(c)

Figure 3.Root development. P. trichocarpa cuttings were planted in forest moraine soil (For-mor), agricultural silty clay soil (Agri-clay) and agricultural sandy soil (Agri-sand): the total number of roots per cutting (a); mean root length (b) and development of fine roots (c) were analyzed. Fine root values shown are means of the four root development categories 0 = no fine roots, 1 = few fine roots, 2 = fine roots on all roots and 3 = fine roots abundant on all roots. Data shown are mean values (n = 7), error bars represent ±SE.

(Figure 5(c)). Total N content for plants grown in For-mor reached 0.005g in leaves and 0.0025g in stem (Figure 5(d)) and for plants grown in Agri-clay or Agri-sand total N content varied between 0.2 and 0.14 (leaves) and 0.075 to 0.05 g (stem).

In the analyzed from all the soils, P:N ratios were below 0.1 for leaf and stem of plants grown in For-mor and total for Agri-clay. The other samples were all close or above 0.1 (Figure 5).

Except leaf samples from plants grown in the Agri-clay all other samples (leaves, stem and combination of Leaf and stem) K:N ratio that exceeded 0.35.

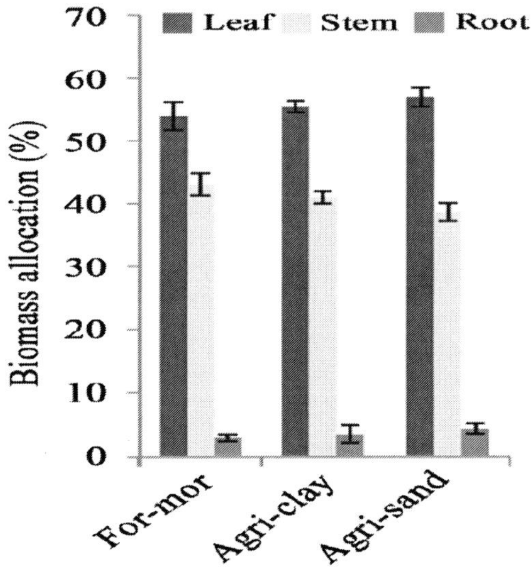
Leaf
Stem
Root
Biomass allocation (%)
70
60
50
40
30
20
10
0
For-mor
Agri-clay
Agri-sand

(a)

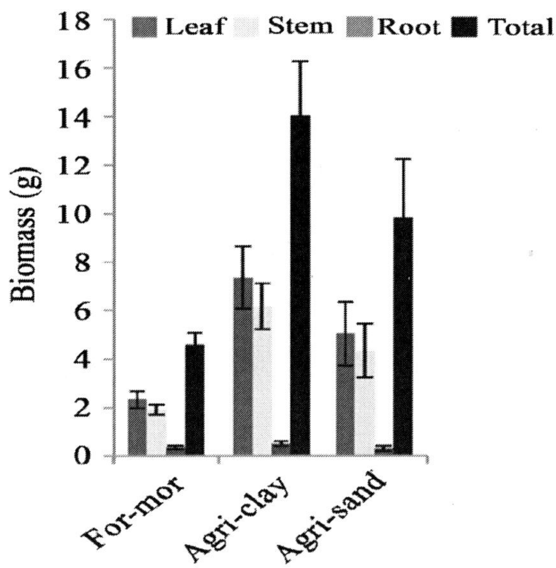
Leaf
Stem
Root
Total
Biomass (g)
18
16
14
12
10
8
6
4
2
0
For-mor
Agri-clay
Agri-sand

(b)

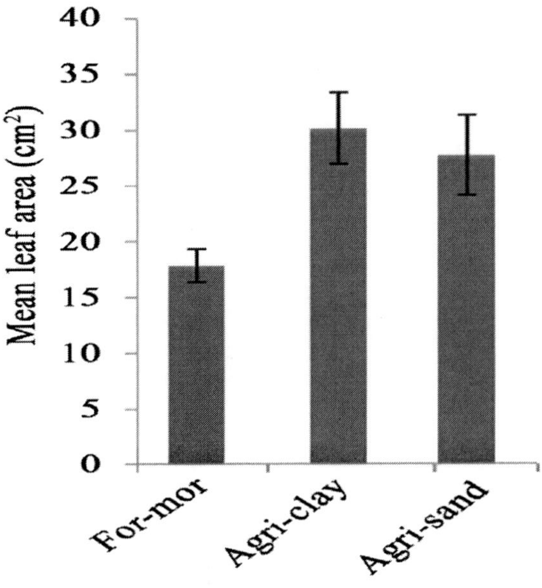

(c)

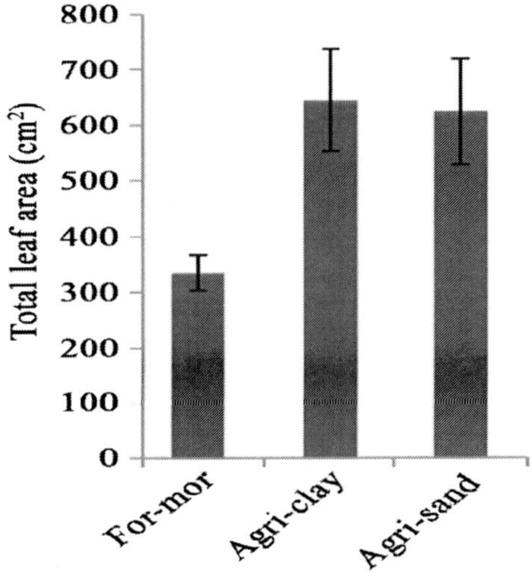

(d)

Figure 4. Biomass growth, allocation, individual and total leaf area. Biomass allocation (a) and leaf, stem, root and total biomass production (b); individual leaf area (c); total leaf area (d) of *P. trichocrapa* plants eight weeks after planting in forest moraine (For-mor), agricultural silty clay (Agri-clay) and agricultural sandy soil (Agri-sand). Data shown are mean values (n = 7). Error bars represent ±SE.

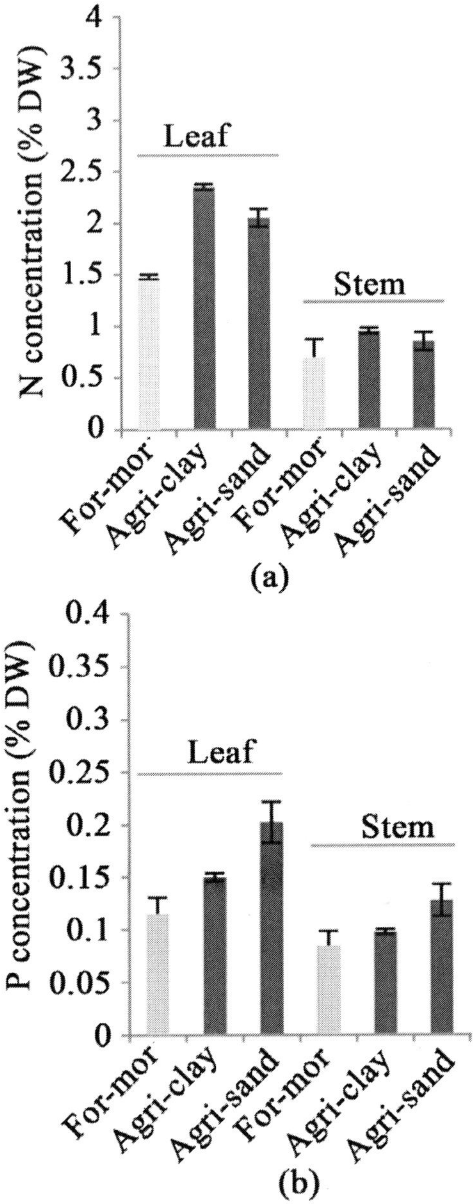
N concentration (% DW)
4
3.5
3
2.5
2
1.5
1
0.5
0
Leaf
Stem
For-mor
Agri-clay
Agri-sand
For-mor
Agri-clay
Agri-sand
(a)
P concentration (% DW)
0.4
0.35
0.3
0.25
0.2
0.15
0.1
0.05
0
Leaf
Stem
For-mor
Agri-clay
Agri-sand
For-mor
Agri-clay
Agri-sand
(b)

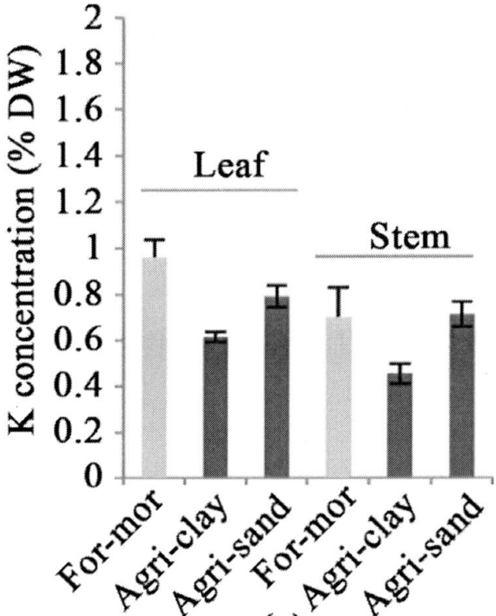
2
1.8
1.6
1.4
1.2
1
0.8
0.6
0.4
0.2
0
K concentration (% DW)
Leaf
Stem
For-mor
Agri-clay
Agri-sand
For-mor
Agri-clay
Agri-sand

(c)

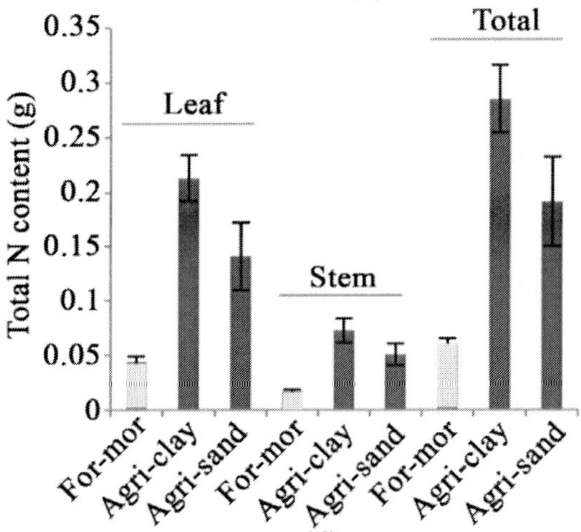
0.35
0.3
0.25
0.2
0.15
0.1
0.05
0
Total N content (g)
Leaf
Stem
Total
For-mor
Agri-clay
Agri-sand
For-mor
Agri-clay
Agri-sand
For-mor
Agri-clay
Agri-sand

(d)

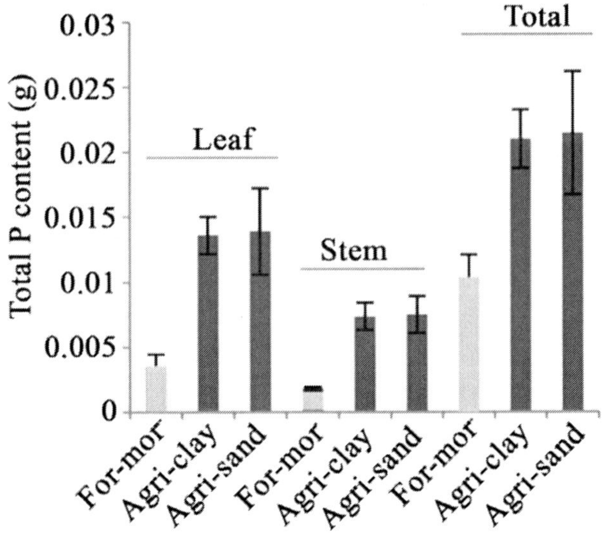
Total P content (g)
0
0.005
0.01
0.015
0.02
0.025
0.03
Leaf
Stem
Total
For-mor
Agri-clay
Agri-sand
For-mor
Agri-clay
Agri-sand
For-mor
Agri-clay
Agri-sand

(e)

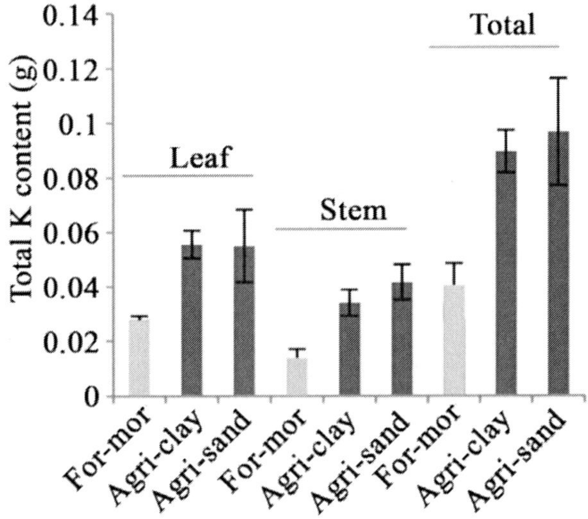
Total K content (g)
0
0.02
0.04
0.06
0.08
0.1
0.12
0.14
Leaf
Stem
Total
For-mor
Agri-clay
Agri-sand
For-mor
Agri-clay
Agri-sand
For-mor
Agri-clay
Agri-sand

(f)

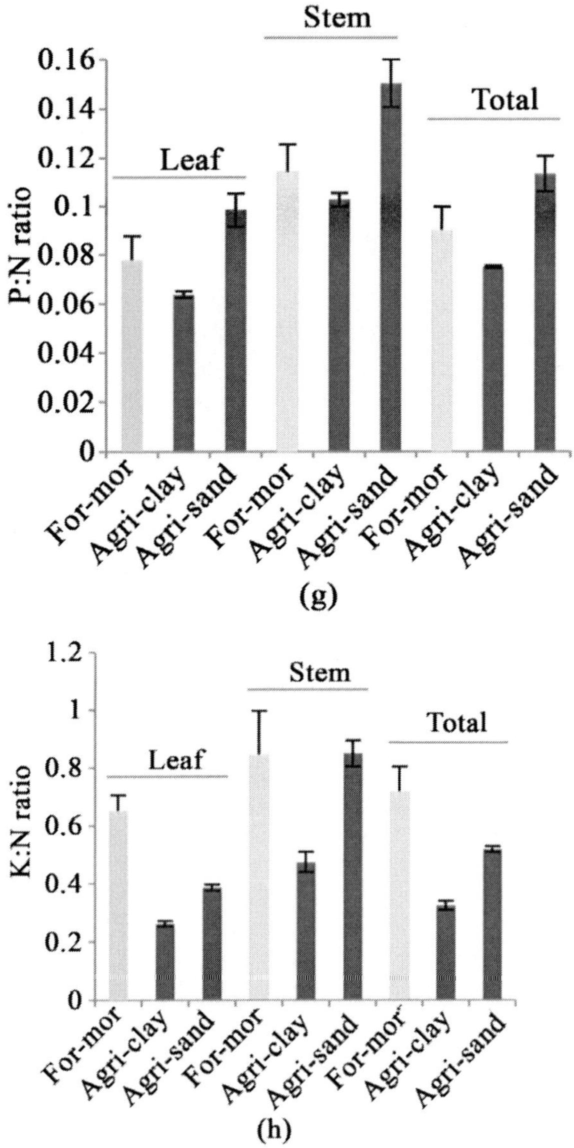

Figure 5. Concentration of macronutrients and total content. Concentrations of (a) Nitrogen; (b) Phosphorus and (c) Potassium; total content of (d) Nitrogen; (e) P:N ratio and (f) K:N ratio The analyzed soils were forest moraine (For-mor), agricultural silty clay soil (Agri-clay) and agricultural sandy soil (Agri-sand). Data shown are mean values (n = 4). Error bars represent ±SE.

4. DISCUSSION

We investigated how soils with different textures and origins influenced establishment with dormant *P. trichocarpa* cuttings and how development of plants proceeded during the first 8 weeks. Our results revealed that survival were not influenced by soil texture or origin. Cutting planted in the agricultural soils, Agri-sand and Agri-clay or in forest moraine (For-mor) all developed in to plants with no obvious differences. In contrast to height, soil texture influenced root development with more and longer roots with fine roots present in the soils with sandy texture. Our intention were not to identify the optimal soil for establishing poplars but to describe how transplanted poplar cuttings develops in to plants in different soils. Therefore results from this study can give additional information how poplar establishment is influenced by different soils. If proper growth comparisons between different soils should be possible, soils need to be translocated to one experimental site. It could very well be that this translocation will influence the soil as much as an intensive site preparation would but in our experiment the soils was left to settle for eight years. This time period would probably remove some of the issues concerning altered growth conditions due to soil translocation. An alternative to our approach would be to grow the plant sin pots. Using this method would probably have additional effects. For example, the possibilities that the soil structure is altered during transportation and that growth volume will be restricted are likely to be increased. The connection between the soil layer were cutting are planted and the lower soil layers are also lost. These problems might create soil conditions that are altered to a higher extent than if a whole soil structure to a depth of 1m are translocated and left to settle for eight years, as in our experiment. One limitation to our case study is that only one soil of each type was tested. Therefore, extra care should be taken before general conclusion is made on where poplars should be planted. Pinno and Bélanger, 2009 found that soil texture was the best predictor of poplar growth: soils with more silt and clay produced better growth. Another report (Pinno et al. (2009)found that maximum poplar tree growth occurred at a sand content between 55% and 70%. For willow, Bouma and Bryla (2000) found that leaf and stem biomasses increased with increasing proportions of silty clay loam, while Schaff (2003) reported that willow growth was higher in sandy sediments than in silt clay sediments. Bouma and Bryla (2000) , reported that growth of *Citrus volkameriana* was only slightly affected by soil texture. Our results suggest that poplar establishment is not dependent on a specific soil texture. We found that plants could develop and reach a height of 64 cm for Agri-clay and 68 cm for Agri-sand with stem biomass production of 7.3 g and 5 g form transplanted cutting (Figure 2 and Figure 4), soils which differed specifically with respect to texture (Table 1). Our results also indicate that establishment of poplar in For-more is possible but that plants only reached a height of 47 cm with a stem biomass of 1.9 g (Figure 5 and Figure 4), 17 cm shorter and 3.1 g lighter stems than plants grown in Agri-sand. This could indicate that the origin of the soils (agricultural or forest) are important for early growth and plant development. However, at other sites (similar soils as our case study), planting of other *Populus g*enotype could result in growth differences. If differences in available nutrients between forest soils and agricultural soils are the reason for growth differences, plant macronutrient concentrations and total content could give an indication. Indeed, we found that total N was 0.06g in plants grown in For-mor and exceeded 0.2 g for plants grown in Agri-clay and Agri-sand (Figure 5). We did observe, however, that if K concentration was reduced, N concentration is increased; this is consistent with other research reporting that soil K concentration and content is decreased as nitrogen levels increase (Ring et al., 2011) . Although N, P and K were low (1.5% (N), 0.125 (P) and K 0.96%) in plants grown in For-mor, total nutrient ratios were close to optimal (K:N 0.35 and P:N 0.1) (Aronsson & Elowson, 1980; Ingestad, 1979; Linder, 1995) , suggesting that the plants where not suffering from nutrient imbalance. We observed that in plants grown in Agri-clay, N concentrations were high and P and K concentrations relatively low and, as a consequence, total K:N ratios were below their optimal levels, suggesting a nutrient imbalance despite the high growth. Differences in nutrient concentrations and contents in plants grown the agricultural soils and forest soil could be the result of a combination of differences in the soil water holding capacity (Zhou et al., 2014) and/or nutrient availability that together facilitate nutrient uptake. In addition, we found longer mean root length and more fine root development on cuttings grown in Agri-sand and For-mor. This root development indicates that plants grown in these soils need more roots to increase water uptake and/or nutrient uptake. In contrast, the plants grown in Agri-clay produced less root growth, but their nutrient content was high, suggesting that cuttings were grown in a soil environment where water and nutrients were abundant and high root growth was not required to allow the plant to access sufficient nutrients and water. These results agree with others research, reporting reduced root growth with increased N availability (Burton et al., 2000; Gower et al., 1992; Kern et al., 2004; Keyes & Grier, 1981) . It should be noted that the results presented herein were obtained after a growth period of eight weeks. Height growth and biomass production influenced by soil texture may be more visible after a longer time period, as reported by others (Bouma & Bryla, 2000; Pinno et al., 2009; Schaff et al., 2003) . In a real poplar plantation a slow start could be problematic as early plant development is important for stand development and perhaps one of the reasons why heavy soils such as Agri-clay are considered less suitable for poplars (Stanturf et al. 2001) Our results suggest that, under similar climatic conditions, survival and early plant growth are not dependent on the soil texture and that any difference in growth between soils (Bouma & Bryla, 2000; Pinno et al., 2009; Schaff et al., 2003) probably occurs in older plants.

5. CONCLUSION

This study sheds some light on our understanding on early poplar plant establishment in different soil types. The results obtained indicate that establishment of poplar cuttings in agricultural soils is independent of soil texture. Also, on forest land, where moraine is the most common soil type, poplars can be established but soil nutrient concentration and/or availability and differences in uptake capacity are presumably some of the factors limiting growth. For the practical work in establishing poplar plantations on forest sites, increased growth during the first few weeks after planting could be important for successful establishment. To further elucidate how poplar plantations can best be established on forest sites, we need to increase our knowledge concerning plant- forest soil interaction as well as our understanding of the type of mechanical soil preparation required for regeneration of poplars on such sites.

ACKNOWLEDGEMENTS

The authors wish to thank Alexandra Nikolic for technical assistance and the staff and students at the Alnarp garden laboratory for sample collection and for providing the experimental site. The clone used was a kind gift from SweTree technology (STT), Sweden Umeå. The authors also thank Prof Urban Nilsson and Prof Björn Sundberg for early comments of the manuscript. This project was funded by the Swedish research program, Trees and Crops for the Future (TC4F), a Vinnova project Bio4energy—a Strategic Research Environment supported by the Swedish government and the Swedish research council Formas.

REFERENCES

1. Albaugh, T. J., Allen, H. L., Dougherty, P. M., Kress, L.W., & King, J. S. (1998). Leaf Area and Above- and Belowground Growth Responses of Loblolly Pine to Nutrient and Water Additions. Forest Science, 44, 317-328
2. Aronsson, A., & Elowson, S. (1980). Effects of Irrigation and Fertilization on Mineral Nutrients in Scots Pine Needles. Ecological Bulletins, 32, 219-228

3. Baker, J. B., & Broadfoot, W. M. (1979). A Practical Field Method of Site Evaluation for Commercially Important Hardwoods. General Technical Report SO-36, New Orleans, LA: USDA Forest Service Southern Forest and Range Experimental Station
4. Beets, P. N., & Whitehead, D. (1996). Carbon Partitioning in Pinus radiata Stands in Relation to Foliage Nitrogen Status. Tree Physiology, 16, 131-138. http://dx.doi.org/10.1093/treephys/16.1-2.131
5. Bergante, S., Facciotto, G., & Minotta, G. (2010). Identification of the Main Site Factors and Management Intensity Affecting the Establishment of Short-Rotation-Coppices (SRC) in Northern Italy through Stepwise Regression Analysis. Central European Journal of Biology, 5, 522-530
6. Bouma, T., & Bryla, D. (2000). On the Assessment of Root and Soil Respiration for Soils of Different Textures: Interactions with Soil Moisture Contents and Soil CO2 Concentrations. Plant Soil, 227, 215-221. http://dx.doi.org/10. 1023 /A:1026502414977
7. Burton, A. J., Pregitzer, K. S., & Hendrick, R. L. (2000). Relationships between Fine Root Dynamics and Nitrogen Availability in Michigan Northern Hardwood Forests. Oecologia, 125, 389-399
8. Christersson, L. (2008). Poplar Plantations for Paper and Energy in the South of Sweden. Biomass and Bioenergy, 32, 997-1000. http://dx.doi.org/ 10.1016/j.biombioe.2007.12.018
9. Christersson, L. (2010). Wood Production Potential in Poplar Plantations in Sweden. Biomass and Bioenergy, 34, 1289-1299. http://dx.doi.org/10.1016 /j.biombioe .2010.03.021
10. Coleman, M. D., Friend, A. L., & Kern, C. C. (2004). Carbon Allocation and Nitrogen Acquisition in a Developing Populus deltoides Plantation. Tree Physiology, 24, 1347-1357. http://dx.doi.org/10.1093/treephys/24.12.1347
11. DeBell, D. S., & Harrington, C. A. (1997). Productivity of Populus in Monoclonal and Polyclonal Blocks at Three Spacings. Canadian Journal of Forest Research, 27, 978-985. http://dx.doi.org/10.1139/x97-059
12. Gower, S. T., Vogt, K. A., & Grier, C. C. (1992). Carbon Dynamics of Rocky Mountain Douglas-Fir: Influence of Water and Nutrient Availability. Ecological Monographs, 62, 43-65. http://dx.doi.org/10.2307/2937170
13. Hartmann, H. T., & Kester, D. E. (1975). Plant Propagation: Principles and Practices (p. 609). Englewood Cliffs: Prentice-Hall
14. Hofmann-Schielle, C., Jug, A., Makeschin, F., & Rehfuess, K. E. (1999). Short-Rotation Plantations of Balsam Poplars, Aspen and Willows on Former Arable Land in the Federal Republic of Germany. I. Site-Growth Relationships. Forest Ecology and Management, 121, 41-55. http://dx.doi.org/10.1016/S0378-1127(98)00555-6
15. Ingestad, T. (1979). Mineral Nutrient Requirements of Pinus sylvestris and Picea abies Plants. Physiologia Plantarum, 45, 373-380. http://dx.doi.org/ 10.1111 /j.1399-3054.1979.tb02599.x
16. Johansson, T., & Karacic, A. (2011). Increment and Biomass in Hybrid Poplar and Some Practical Implications. Biomass and Bioenergy, 35, 1925-1934. http://dx.doi.org/10.1016/j.biombioe.2011.01.040
17. Kern, C. C., Friend, A. L., Johnson, J. M.-F., & Coleman, M. D. (2004). Fine Root Dynamics in a Developing Populus deltoides Plantation. Tree Physiology, 24, 651-660. http://dx.doi.org/10.1093/treephys/24.6.651
18. Keyes, M. R., & Grier, C. C. (1981) Above- and Below-Ground Net Production in 40-Year-Old Douglas-Fir Stands on Low and High Productivity Sites. Canadian Journal of Forest Research, 11, 599-605. http://dx.doi.org/10.1139/x81-082
19. Knops, J. M., & Kurt, R. (2000). Specific Leaf Area along a Nitrogen Fertilization Gradient. American Midland Naturalist, 144, 265-272. http://dx. doi.org/10.1674/0003-0031(2000)144[0265:SLAAAN]2.0.CO;2
20. Linder, S. (1995). Foliar Analysis for Detecting and Correcting Nutrient Imbalances in Norway Spruce. Ecological Bulletins, 178-190

21. Pinno, B. D., & Bélanger, N. (2009). Competition Control in Juvenile Hybrid Poplar Plantations across a Range of Site Productivities in Central Saskatchewan, Canada. New Forest, 37, 213-225. http://dx.doi.org/10.1007/s11056-008-9118-3

22. Pinno, B. D., Thomas, B. R., & Bélanger, N. (2009). Predicting the Productivity of a Young Hybrid Poplar Clone under Intensive Plantation Management in Northern Alberta, Canada Using Soil and Site Characteristics. New Forest, 39, 89-103. http://dx.doi.org/10.1007/s11056-009-9157-4

23. Ring, E., Jacobson, S., & Högbom, L. (2011). Long-Term Effects of Nitrogen Fertilization on Soil Chemistry in Three Scots Pine Stands in Sweden. Canadian Journal of Forest Research, 41, 279-288. http://dx.doi.org/10.1139/X10-208

24. Schaff, S. D., Pezeshki, S. R., & Shields Jr., F. D. (2003). Effects of Soil Conditions on Survival and Growth of Black Willow Cuttings. Environmental Management, 31, 748-763. http://dx.doi.org/10.1007/s00267-002-2909-y

25. Scotti, C., Facciotto, G., & Canestrale, R. (2010). Soil Map of Pedological Limitations to the Growth of Poplar Clones for Biomass Production in Emilia-Romagna Region (Italy). Proceedings of the 18th European Biomass Conference and Exhibition (pp. 115-122). Florence: ETA-Florence Renewable Energies

26. Stanturf, J. A., von Oosten, C., Netzer, D. A., Colman, M. D., & Prtwood, C. J. (2001). Ecology and Silviculture of Poplar Plantations. In D. I. Dickman, J. E. Eckenwald, & J. Richardson (Eds.), Poplar Culture in North America (pp. 152-206). Ottawa: National Council of Canada Research Press

27. Tullus, A., Rytter, L., Tullus, T., Weih, M., & Tullus, H. (2011). Short-Rotation Forestry with Hybrid Aspen (Populus tremula L. × P. tremuloides Michx.) in Northern Europe. Scandinavian Journal of Forest Research, 27, 10-29. http:// dx.doi. org /10.1080/02827581.2011.628949

28. Zhou, H., Ding, L., Fan, T., Ding, J., Zhang, D., & Guo, Q. (2014). Leaf-Inspired Hierarchical Porous CdS/Au/N-TiO2 Heterostructures for Visible Light Photocatalytic Hydrogen Evolution. Applied Catalysis B: Environmental, 147, 221-228. http://dx.doi.org/10.1016/j.apcatb.2013.08.025

CHAPTER 7

Agroforestry Practices Promote Biodiversity and Natural Resource Diversity in Atlantic Nicaragua

Seeta A. Sistla[1,2*], Adam B. Roddy[3], Nicholas E. Williams[4,5], Daniel B. Kramer[6], Kara Stevens[6], Steven D. Allison[2,7]

[1]*School of Natural Science, Hampshire College, Amherst, Massachusetts, 01002, United States of America,*

[2]*Department of Ecology and Evolutionary Biology, University of California Irvine, Irvine, California, 92697, United States of America,*

[3]*School of Forestry & Environmental Studies, Yale University, New Haven, Connecticut, 06511, United States of America,*

[4]*Department of Anthropology, University of California Santa Barbara, Santa Barbara, California, 93106, United States of America,*

[5]*Department of Environmental Studies, University of Colorado Boulder, Boulder, Colorado, 80309, United States of America,*

[6]*Department of Fisheries and Wildlife, Michigan State University, East Lansing, Michigan, 48824, United States of America,*

[7]*Department of Earth System Science, University of California Irvine, Irvine, California, 92697, United States of America*

ABSTRACT

Tropical forest conversion to pasture, which drives greenhouse gas emissions, soil degradation, and biodiversity loss, remains a pressing socio-ecological challenge. This problem has spurred increased interest in the potential of small-scale agroforestry systems to couple sustainable agriculture with biodiversity conservation, particularly in rapidly developing areas of the tropics. In addition to providing natural resources (i.e. food, medicine, lumber), agroforestry systems have the potential to maintain higher levels of biodiversity and greater biomass than lower diversity crop or pasture systems. Greater plant diversity may also enhance soil quality, further supporting agricultural productivity in nutrient-limited tropical systems. Yet, the nature of these relationships remains equivocal. To better understand how different land use strategies impact ecosystem services, we characterized the relationships between plant diversity (including species richness, phylogenetic diversity, and natural resource diversity), and soil quality within pasture, agroforests, and secondary forests, three common land use types maintained by small-scale farmers in the Pearl Lagoon Basin, Nicaragua. The area is undergoing accelerated globalization following the 2007 completion of the region's first major road; a change which is expected to increase forest conversion for agriculture. However, farmer agrobiodiversity maintenance in the Basin was previously found to be positively correlated with affiliation to local agricultural NGOs through the maintenance of agroforestry systems, despite these farmers residing in the communities closest to the new road, highlighting the potential for maintaining diverse agroforestry agricultural strategies despite heightened globalization pressures. We found that agroforestry sites tended to have higher surface soil %C, %N, and pH relative to neighboring to secondary forest, while maintaining comparable plant diversity. In contrast, pasture reduced species richness, phylogenetic diversity, and natural resource diversity. No significant relationships were found between plant diversity and the soil properties assessed; however higher species richness and phylodiversity was positively correlated with natural resource diversity. These finding suggest that small, diversified agroforestry systems may be a viable strategy for promoting both social and ecological functions in eastern Nicaragua and other rapidly developing areas of the tropics.

INTRODUCTION

Together, crop- and pasture-lands comprise one of the largest biomes on earth, representing ~ 40% of global terrestrial area [1]. While agricultural innovations have greatly increased food production, they have also caused extensive environmental degradation; nearly half of global croplands are impacted by soil erosion, declines in soil fertility, reduced biodiversity, and other socio-ecological concerns [2–5]. In the tropics, deforestation for agricultural expansion accounts for 8% of anthropogenic carbon dioxide (CO_2) emissions–nearly all global land-use change emissions–and is the primary cause of species extinctions worldwide [6–8]. Coupling sustainable agriculture to biodiversity conservation through small, diversified farms, such as those typified by traditional tropical agroforestry ecosystems, may be a viable complementary land use strategy in rapidly developing areas of the tropics [1,3,4,9–12]. Despite the potential socio-ecological benefits of agroforestry systems and the possibility for them to support conservation efforts in ecologically fragile areas of high biodiversity, tropical conservation policy remains dominated by efforts to reduce intact forest conversion and promote natural reforestation in lieu of supporting more socio-ecologically integrative practices [3,13–15]. Increasing our understanding of the ecosystem service benefits of traditional agroecosystems could help encourage policy that supports a broader range of conservation objectives [16].

Agroforestry ecosystems that incorporate perennial trees into agriculture, such as those typified by smallholder farmer-dominated areas in the tropics (typically 0.01 to 5 ha), can be a fundamental component of both biodiversity conservation and socio-ecological resilience [10,12,17–20]. In addition to provisioning natural resources (i.e. food, medicine, and building materials), agroforestry ecosystems have the potential to maintain higher levels of biodiversity and greater biomass than conventional agriculture. Relative to monoculture or pasture techniques, agroforestry ecosystems may also enhance soil quality, including C content and nutrient status, by increasing litter inputs and soil organic matter accumulation [2,3,17,21–23]. In many tropical systems, agricultural productivity is constrained by low nutrient availability due to highly leached, acidic soils; this problem is amplified because many subsistence farmers in these areas often cannot access mineral fertilizers [24].

Positive relationships between plant diversity and ecosystem functions such as C sequestration—which can be driven by either niche complementarity or the greater likelihood of including functionally-important species in more diverse assemblages—have been identified in a number of model ecosystems [25–27], although these relationships are complex and positive species/functional diversity relationships are not always observed [21,28]. Both plant community functional diversity and phylogenetic diversity, rather than simply the number of taxonomic units (e.g. species or functional group richness), appear to underlie observed biodiversity–ecosystem-service relationships [29,30]. However, the nature of these relationships remains equivocal in human-managed systems, such as smallholder agroecosystems typical in developing areas of the tropics [9,21,31–33].

We characterized the relationships between common smallholder land use strategies, plant diversity, and soil quality in the Pearl Lagoon Basin, Nicaragua (Fig 1) to better understand how these strategies impact a variety of ecosystem functions and properties. The Basin, which is situated within the Mesoamerica Biodiversity Hotspot, is an area historically characterized by a matrix of agroforestry systems and unmanaged secondary forest [34] and is on the eastern edge of a front of rapid pasture expansion [35]. The area is also undergoing accelerated globalization following the 2007 completion of the region's first major road, which connects the Basin with the rest of Nicaragua [36].These changes are expected to increase forest conversion for agriculture, particularly pasture [34]. Intriguingly, a recent study in the Basin found that farmer agrobiodiversity maintenance is positively correlated with affiliation to local agricultural NGOs through the maintenance of agroforestry systems, despite these farmers residing in the communities closest to the new road [37]. These findings suggest that despite heightened globalization pressures, the potential for supporting more ecologically diverse agricultural strategies can be maintained with culturally-specific land use norms [38].

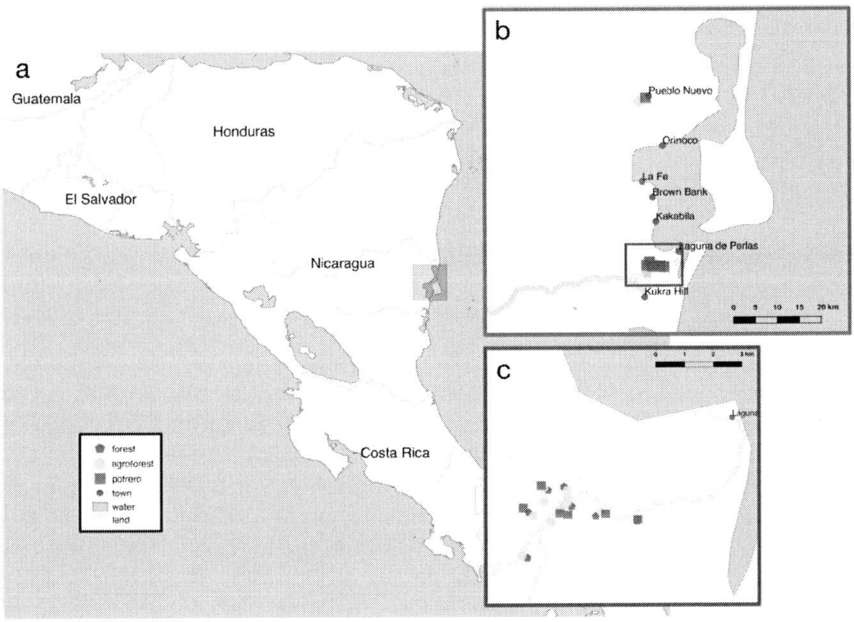

Figure 1. Maps of the study region. Nicaragua (a) with an inset (b) defining the Pearl Lagoon Basin. Sampled plots (c) are shown by points colored by land use type: secondary forest (green), agroforest (yellow), pasture (red).

Limited understanding of the interrelated effects of land use strategy (agroforestry vs. pasture vs. unmanaged secondary forest) on soil quality and plant diversity in this system hinders our ability to project how accelerated expansion of agroforestry versus pastureland will impact the region's terrestrial ecosystem services. Further, characterizing how land use affects natural resource availability is fundamental to assessing whether traditional agroforestry practices are comparable to natural reforestation within this system. To help fill these knowledge gaps, we characterized how three land-use types common in Atlantic Nicargua (agroforestry, pasture, secondary forest, Fig 2) affect total (wild and cultivated) plant species diversity and a suite of soil properties. We also estimated potential natural resource diversity provided by plants in each land-use type within four categories (food, medicine, lumber, and other uses, collectively termed 'natural resources'). We hypothesized that: 1) secondary forests would be the most species-rich and phylogenetically-diverse system, while pasture would support the lowest diversity; 2) soil nutrient status would positively correlate with species-richness and phylogenetic diversity; and 3) agroforestry ecosystems would support the highest functional diversity of ecosystem services, while pasture would have the lowest functional diversity. Identifying how land use affects biodiversity and ecosystem services from both a social and ecological perspective is fundamental to assessing whether traditional agroforestry practices might serve as a tractable alternative to unmanaged reforestation in Atlantic Nicaragua and similar systems.

Figure 2. Images of landscapes sampled. Secondary forest (a), agroforest (b), pasture (c) in the Pearl Lagoon Basin, Nicaragua.

METHODS

Site description

The Territory of the 12 Indigenous & Afro-descendant Communities of the Pearl Lagoon Basin, Región Autónoma de la Costa Caribe Sur (RACCS) encompasses 5200 km^2 in Atlantic Nicaragua. The Basin is located within the Mesoamerican biodiversity hotspot, which includes all subtropical and tropical ecosystems from central Mexico to the Panama Canal [36]. Our focal study area was characterized by Central American Atlantic moist forest, which is the dominant forest ecoregion along the Nicaraguan Atlantic Coast. The Pearl Lagoon region moist forest is characterized by a history of disturbances including the Class IV hurricane Joan (1989) and a patchwork of previous agroforestry sites [36]. Atlantic Nicaragua has a tropical wet climate and is characterized by a wet season with an average rainfall (MAP) exceeding 500 cm with distinct wet (May—November) and dry (December—April) seasons and an average temperature (MAT) of 26°C [39]. The region is dominated by highly weathered, iron- and aluminum-oxide rich ultisols [39].

The Pearl Lagoon Basin is predominantly inhabited by indigenous and afro-descendant populations who utilize marine and aquatic fishery resources in conjunctions with agricultural and forest resources to meet their subsistence needs. Basin indigenous and afro-descendant farmers commonly maintain a wide variety of ethnobotanically-important cultivated and wild species for food, medicinal, timber, fuel, ornamental, and other uses [37]. Agroforestry systems in the Pearl Lagoon Basin are typified by small-scale subsistence farms characterized by a diversity of annual crops interspersed with perennial fruit trees. Common cultivated plants include: citrus varietals, coconuts (*Cocos nucifera*), supa (*Bactris gasipaes*), cacao (*Theobroma cacao*), breadfruit (*Artocarpus altilis*), mango (*Mangifera indica*), avocado (*Persea americana*), varieties of plantain and banana (*Musa sp.*); pineapples (*Ananas comosus*), and starchy roots, including cassava (*Manihot esculenta*) and dasheen (*Colocasia esculenta*). In addition to agroforestry systems, farmers in the region also commonly maintain homegardens for daily use as well as small cattle pastures for personal consumption or local market. Farm land is interspersed with secondary forest held in communal tenure [37].

This type of agroforestry system has been used by Basin afro-descendant and indigenous residents for at least a century [37,40,41]. Mestizo farmers who are part of a wave migration from Nicaragua's highland regions rely on market-oriented cattle farming leading to the conversion of forest to pastureland and accelerating forest fragmentation. Following the end of the

Nicaraguan civil war (1979–1987), cultural diffusion has increased the prevalence of cattle ranching among afro-indigenous farmers [36,37]. However, local agricultural NGOs and Nicaraguan governmental organizations are working to mitigate land conversion by promoting local agroforestry practices through interaction with individual farmers, which are correlated with greater agrobiodiversity maintenance at the farm-level [37].

Farm identification

We identified thirteen farms in the Pearl Lagoon Basin that included agroforest (n = 17 sites), adjacent secondary forest (n = 8 sites), and pasture sites (n = 7 sites). Sample farms were identified by conducting interviews with farmers to determine their land use practices (method approved by UCSB Human Subjects Committee, submission ID: 13–0324). Literacy rates are low in the region, as such, verbal consent to sample each farmer's land was obtained from participants. Farmer consent was digitally recorded by the researchers (S. Sistla, N. Williams). Farmers were asked if they had cleared forest on their site, when their sites were established, and if their pasture was actively grazed. Areas within any of the three land use types that had been burned within the past decade, were subject to external fertilizer and/or pesticide application, or had been planted less than a decade ago were excluded from the survey. This time period was chosen because previous studies in the region found that a 5-year interval following disturbance from hurricane or agricultural abandonment was sufficient to allow pioneer species to establish [42,43].

All sample farms included at least one agroforestry site, and no farm had more than one pasture area. Specific agroforestry land use strategies (coconut-dominated, cacao-dominated, plantain-dominated, mixed fruit) were identified through the farmer survey and sampled as separate agroforestry sites within each farm. All agroforestry sites in a given farm that met the sampling criteria were included in the plant and soil surveys.

Plant surveys

Plant surveys and soil sampling were conducted during the dry season from February through April, 2013. At each agroforestry, pasture, and secondary forest area, four haphazardly selected 3x3 m plots, whose location was recorded with GPS, were sampled within each land use type. Understory species were sampled from 1-m^2 quadrats centered within the 3x3 m plots [44,45]. Within each 1-m^2 quadrat, all identifiable plant species (including sub-adults) were recorded by common name in Spanish or Creole with assistance from a community member identified through a local agricultural NGO as an expert on local plants. Digital photographs were taken of each quadrat, and unknown plant species were individually photographed for post-sampling identification to the family level; unidentifiable species were recorded as unknown morphotypes. Species abundance was not considered in these analyses due to the dense growth patterns of many observed species, particularly grasses and herbaceous species.

Natural resource diversity

Potential human uses for known species (i.e. natural resource diversity) were identified through ethnobotanical literature focused on eastern Nicaragua. Plant uses were cross-validated between the ethnobotanical literature and ethnobotanical surveys conducted in tandem with our study [37]. Natural resource uses were defined as: "timber", "medicinal", "food", and "other" (which included dyes and tannins, fuel, animal forage, ornamental, and household utensils) [40,46–51]. Each identified species was categorized as either serving or not serving each of those 4 categories; a species' 'total functionality' under these criteria could therefore range from 0 (no use) to 4 (all uses).

Potential natural resource diversity for each quadrat was estimated in three ways. First, the total number of functions was summed across species in a quadrat, which we termed F_T. For example, a community with two species that both have two functions would have a F_T equal to four, regardless of whether the two functions were the same or different. The F_T metric allows for functionally redundant communities that may nonetheless be functionally limited (i.e. many species serving the same few functions) to be equivalent to communities that have low redundancy but high functional diversity (i.e. many species that do not have common functions). Second, we divided F_T by the number of species in each quadrat to calculate the average functional value of a species in that quadrat, termed here $\bar{F}$. This accounts for differences in species richness that can inflate F_T such that quadrats with a high F_T would be composed of species that each serve many functions. Third, we calculated an analog of Simpson's Index of Diversity for potential natural resource diversity as:

$$D = 1 - \sum \left(\frac{F_i}{F_T} \right)^2$$

where F_i is the total number of functions of a particular species and F_T is the total number of all functions of all species, as defined above. Similar to other applications of Simpson's Index of Diversity, D is scaled between 0 and 1, with a larger D value equivalent to high functional diversity. This functional analog to the traditional Simpson's Index of Diversity metric measures the probability that any two species selected from a community will have the same function. These three metrics, therefore, have different and complementary properties that allow for discrimination between functional diversity, functionally redundancy, and the average functional value of different communities. We chose to focus on species richness rather than abundance to avoid confounding non-productive (i.e. non-fruit bearing or immature lumber species) with productive individuals of a given species for a given resource.

Plant phylodiversity

A regional species pool was assembled from (ethno)botanical assessments of the region, sociological surveys of plants growing on farms, and supplemented by our plant surveys. Plants identified in our surveys but not present in either the botanical or sociological assessments were identified to the family level and scored as distinct morphotypes. From this combined list, we generated a family-level phylogeny using the megatree in Phylomatic [52] and manually added basal land plants, based on current phylogenetic understanding, that were not in the Phylomatic megatree but that appeared in our plant surveys.

This master tree, which was used for all subsequent phylogenetic analyses, constituted 468 total species, of which 136 occurred in the plant surveys in the present study. We did not scale branch lengths proportional to time and instead calculated branch lengths proportional to diversity [53]. Incorporating more accurate branch length estimates does not necessarily improve estimates of phylodiversity [54]. We calculated three metrics of phylodiversity using the R package *picante* (v. 1.6–2) [55]: Faith's Index (hereafter 'phylodiversity'), the Net Relatedness Index (NRI), and the Nearest Taxon Index (NTI). Faith's Index describes the phylogenetic diversity of each community based on the total phylogenetic branch lengths in each community and is correlated with species richness while NRI characterizes the average phylogenetic distance between species in a community and NTI characterizes the average phylogenetic distance to the most closely related species in the community. NRI and NTI showed similar patterns and so, for brevity, we focus only on NRI. Because our master tree included many more species than were observed in the plant surveys, estimates of NRI are deflated and will resemble more phylogenetic overdispersion (less clustering) than if we had sampled from a phylogeny composed solely of the 136 species observed in the plant surveys. Although this biases our estimates of NRI towards overdispersion (as compared to other studies), the absolute differences among our plots are still comparable to each other.

To account for species turnover among land use types, we used the aggregated species lists for each land use type to further explore how management strategy impacted biodiversity and species dispersal. We contrasted plant species composition among land-use types using a Jaccard similarity coefficient, which compares species presence/absence data for two sample sets (e.g. land-use types). The Jaccard similarity coefficient is calculated as:

$$J = \frac{A \cap B}{A \cup B}$$

where A and B are the species sets in each of two land use types.

Soil sampling and analyses

Two 10-cm deep, 5-cm diameter soil cores within the 1-m^2 quadrat were collected and pooled for all subsequent analyses. Wet weight of the total plot-level pooled duplicate cores and a sub-sample from the pooled cores was recorded the day of sampling. Subsamples were then air dried and sealed. Due to the remote conditions of the study site, subsequent analyses were carried out at the Iowa State University Soil and Plant Analysis Laboratory. The soil subsamples were dried to a constant mass, and total organic carbon (C) and total nitrogen (N) concentrations were determined by combustion using a C/N analyzer. Available soil phosphorus (P) was estimated using the Bray and Kurtz P-1 test [56]. Soil pH was measured using a 1:1 soil/slurry. Bulk density was determined by calculating the total dry weight of the cores and dividing by the total core volume.

Statistical analyses

The impact of land use type on diversity was analyzed using nested mixed models in R (*lmer*package) [57], with land use type (secondary forest, agroforest, pasture) coded as a fixed factor and specific land use strategy (agroforestry type) nested within farmer, which was coded as a random effect. No significant difference in species richness ($p = 0.7$), phylodiversity ($p = 0.7$), or NRI ($p = 0.9$) were identified between specific agroforestry land use strategies, so these strategies were pooled under the land use type 'agroforest' for all subsequent analyses to simplify interpretation and the potential for application. Post hoc comparisons of means were performed with Tukey's honest significant difference (HSD) test. The relationships between diversity metrics (species richness and phylodiversity) and soil qualities (%C, %N, P, pH, and bulk density) and plant natural resource diversity were tested using a linear mixed effects model (LME).

RESULTS

The influence of land use on plant species diversity

Across land use types, we identified a total of 136 species to the family level or below. Between 0 and 4 unidentifiable morphospecies were found across sample quadrats with no difference in average unknown species among land use types; unidentified species were not included in the phylodiversity metrics. The total number of morphospecies per quadrat (m^{-2}) ranged from 17 morphospecies (in the secondary forest) to 1 (in the pasture). We note that pasture systems tended to be both species-poor and extremely small statured (typically < 5 cm height). Secondary forests and agroforests shared the greatest number of species (38 species with Jaccard similarity (J) = 0.32), while secondary forest and pasture shared 11% of their species (15 species with J = 0.17). Agroforests and pasture shared 20.5% of their species (28 species with J = 0.24; Fig 3).

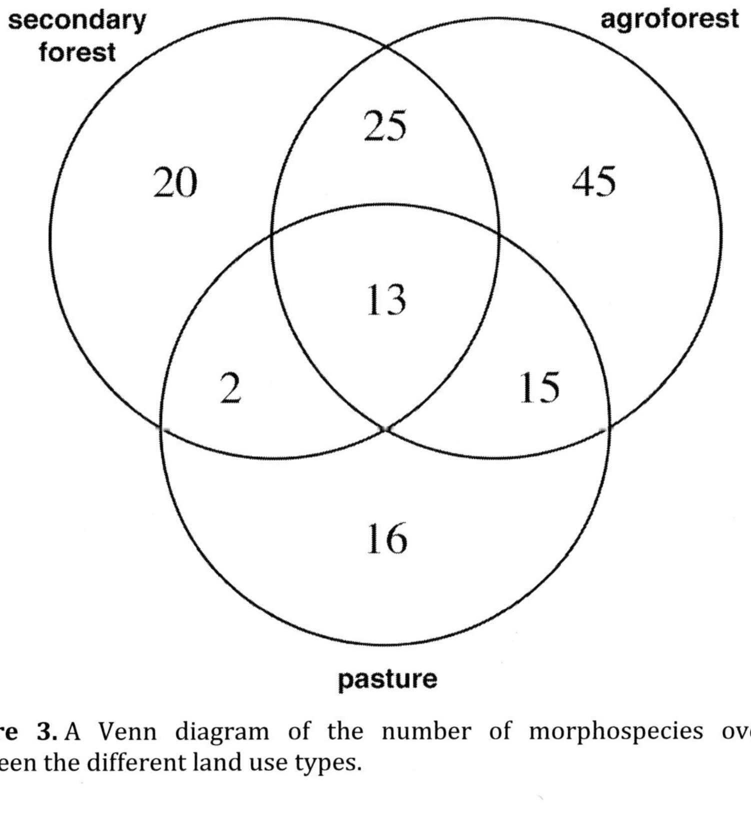

Figure 3. A Venn diagram of the number of morphospecies overlap between the different land use types.

Land use type significantly affected plant diversity across all metrics tested (Fig 4), including species richness ($p = 0.02$; this effect was equivalent when unidentified species were included), Faith's phylodiversity ($p = 0.009$), and net relatedness index (NRI, $p = 0.01$). Secondary forest and agroforest species richness per m^2 was 8.14 ± 0.15 and 7.97 ± 0.15, respectively, while pasture species richness was 6.18 ± 0.6. Tukey HSD post hoc comparisons indicated that secondary forest and agroforest were significantly more diverse in terms of species richness and Faith's phylodiversity (1248.7 ± 13.2, 1095.1 ± 12.4, respectively) than pasture (911.6 ± 25.7), but did not significantly differ from each other. In contrast, secondary forest had a significantly lower NRI (-0.97 ± 0.03) than either agroforest or pasture (-0.50 ± 0.03, -0.22 ± 0.03, respectively), which did not significantly differ from each other.

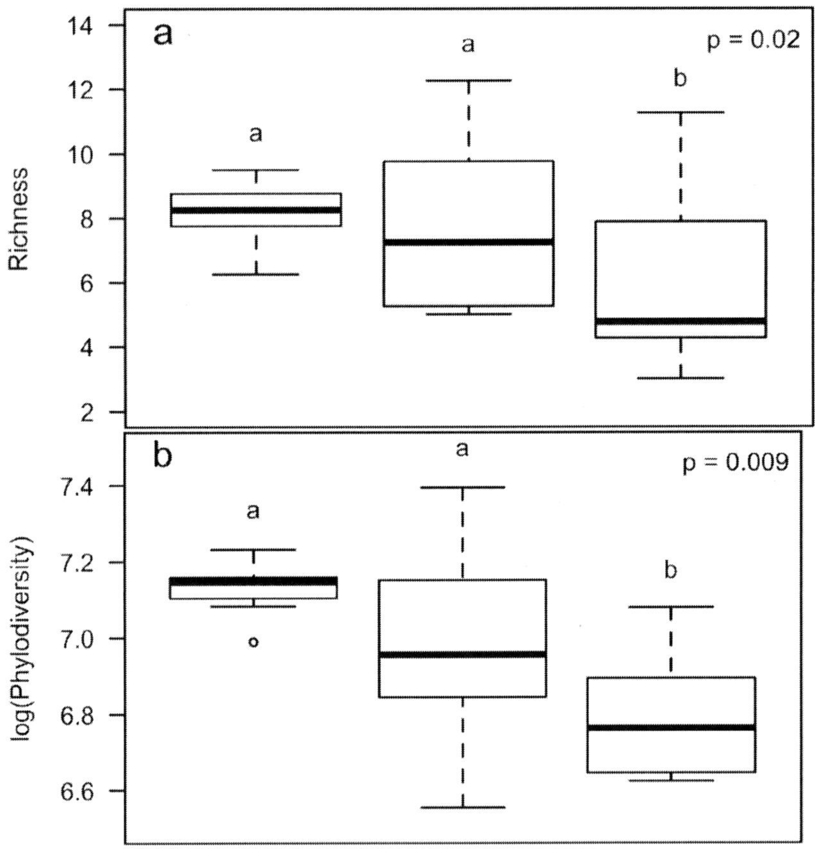

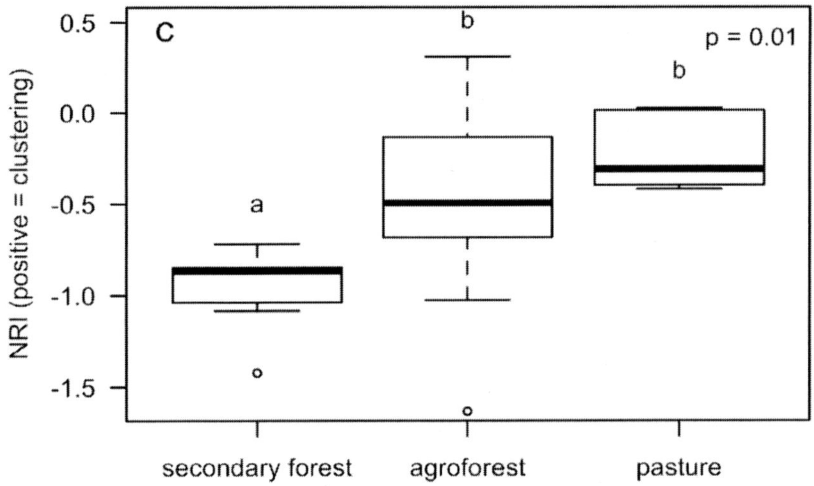

Figure 4. Land use effects on plant species diversity. Effect of land use type (secondary forest (N = 8), agroforest (N = 17) and pasture (N = 7)) on three metrics of diversity: species richness (a), phylodiversity (b), and Net Relatedness Index (C). Non-overlapping letters signify significant difference at $\alpha \leq 0.05$.

The influence of land use on plant-based natural resource diversity

Land use type also significantly affected total potential natural resource diversity (F_T; p = 0.0003), average functional value of each species ($\bar{F}$; p = 0.003), and Simpson's Index of Functional Diversity (D; p = 0.0004; Fig 5). Tukey HSD post hoc comparisons indicated that secondary forest and agroforest had a significantly greater F_T, $\bar{F}$, and D than pasture, but did not significantly differ from each other.

Paralleling this pattern, for three of the four specific functions assessed, land uses differed significantly in the average potential functional value per species: food ($p < 0.0001$), medicine (p = 0.03), and lumber (p = 0.006; Fig 6). Tukey HSD post hoc comparisons indicated that secondary forest and agroforest had higher average food and medicinal value than pastureland, but did not significantly differ from each other in either function. Secondary forest was more functionally valuable for lumber than both pasture and agroforest, which did not differ from each other. The proportion of species in a given community that had one or more potential natural resource functions did not vary for agroforest, whereas secondary forest and pasture were predominantly composed of one- or two-function species with almost no three-function species (S1 Fig). Across land use types, more biodiverse communities were also more functionally diverse (Fig 7). Species richness and phylodiversity had significant positive effects on F_T (both $p < 0.0001$; Fig 7A and 7B) and on D ($p = 0.002$ and $p < 0.0001$, respectively; Fig 7E and 7F), but no effect on $\bar{F}$ (Fig 7C and 7D).

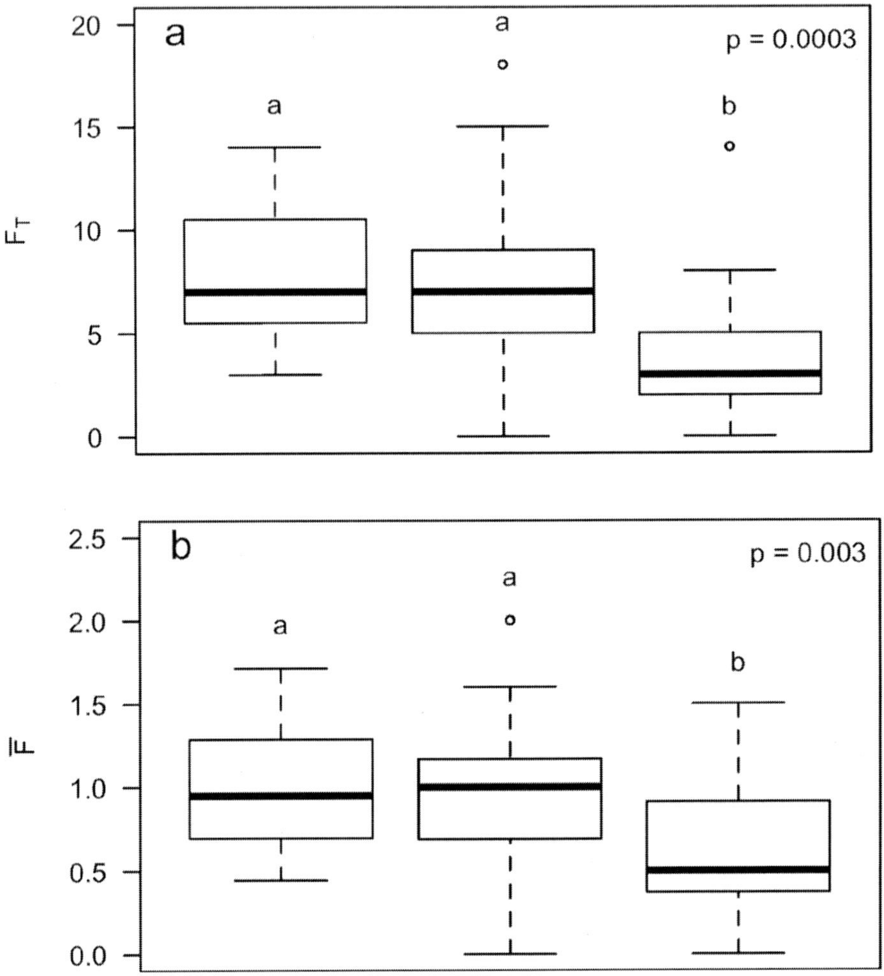

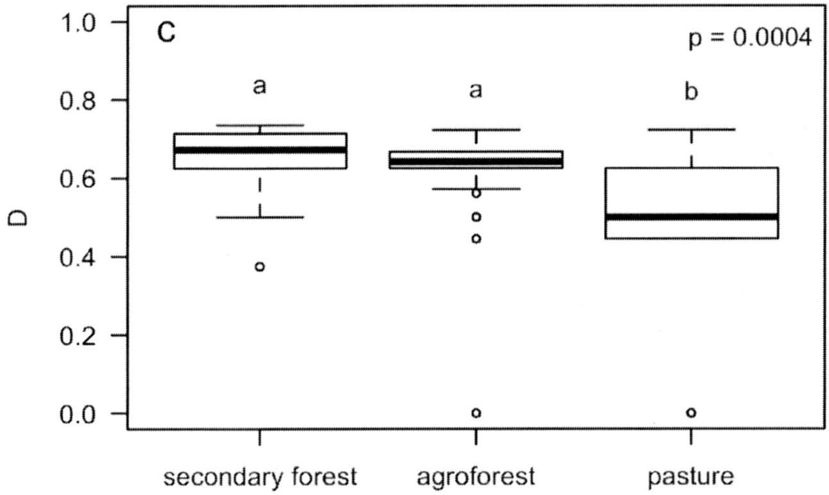

Figure 5. Land use effects on natural resource diversity. Effect of land use type (secondary forest (N = 8), agroforest (N = 17) and pasture (N = 7)) on total natural resource diversity (F_T) (a), average natural resource diversity $\bar{F}$ (b), and Simpson's Index of Natural Resource Diversity (D) (c). Non-overlapping letters signify significant difference at $\alpha \leq 0.05$.

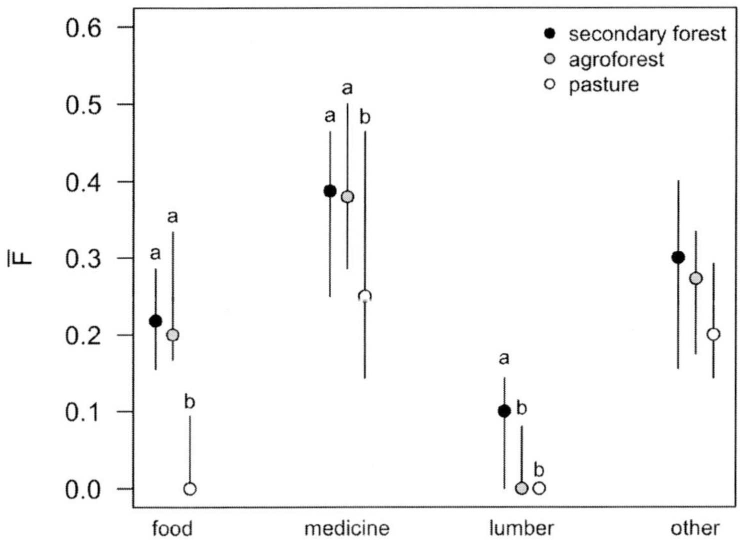

Figure 6. The proportion of plant species used for various social functions in each land use type. The proportion of species in each land use type (secondary forest (N = 8), agroforest (N = 17) and pasture (N = 7)) that are used for: food, medicine, lumber, and other. Non-overlapping letters signify significant difference at $\alpha \leq 0.05$ for each functional group.

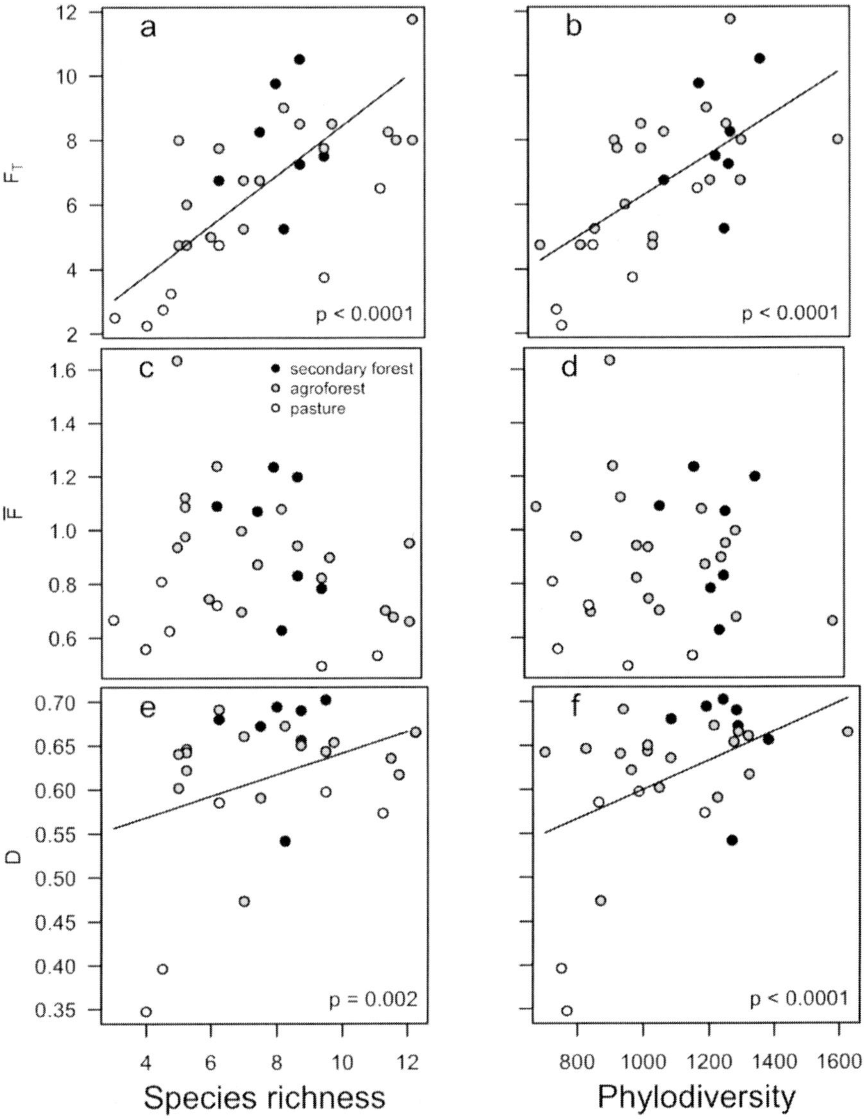

Figure 7. Relationships between plant diversity and natural resource diversity. Effect of species richness and phylodiversity on F_T (a,b), $\bar{F}$ (c,d), and D (e,f). Non-overlapping letters signify significant difference at $\alpha \leq 0.05$.

Relationships between land use and soil qualities

Soil %C had a statistically significant relationship with land use type ($p = 0.05$) and soil %N was marginally affected by land use type ($p = 0.08$), whereas there was no significant relationship was detected between land use and available P (Fig 8). Tukey HSD post hoc comparisons indicated that agroforest had significantly greater %C and %N than secondary forest in the top 10 cm of the soil, but pasture did not significantly differ from either agroforest or secondary forest (Fig 8A–8C). Land use also was correlated with soil pH ($p = 0.03$); Tukey HSD post hoc comparisons indicated that agroforest soil was less acidic than both secondary forest and pasture, which did not significantly differ from each other (Fig 8D). Soil bulk density was correlated with land use ($p = 0.002$). Tukey HSD post hoc comparisons indicated that pasture had a significantly greater bulk density than either secondary forest or agroforest, which did not differ from each other (Fig 8E). When data for soil nutrients and morphospecies richness from sites in all land-use types were pooled, we observed no effect of morphospecies richness or phylodiversity on any soil property.

DISCUSSION

We found that agroforestry sites may be comparable to unmanaged secondary forest across a variety of ecosystem functions and properties. Because tropical agricultural conversion causes considerable primary habitat and species loss [58], conservation efforts generally focus on closed-habitat species and their frontier forest ecosystems [59]. However, our results suggest that well-established agroforests composed of perennial tree and shrub crops intermixed with non-crop trees have the potential to be comparable in both species richness and phylodiversity to uncultivated secondary forests (Fig 4A and 4B). Managed agroforests were sometimes more speciose at the plot scale than secondary forests, but these species largely differed among land use types. While comparable to some studies [60], our pasture systems had a lower average species diversity than has been observed in other neotropical sites [45], which may reflect differences in pasture maintenance and landscape heterogeneity. Notably, pasture species richness and phylodiversity were significantly lower than secondary forest and agroforest, while the NRI of agroforest and pasture was greater than secondary forest (but did not differ from each other, Fig 4C). This pattern suggests that despite evidence that agroforests may support a similar species richness and phylodiversity to unmanaged secondary forests, active human management of agroforests appears to select for more closely related species.

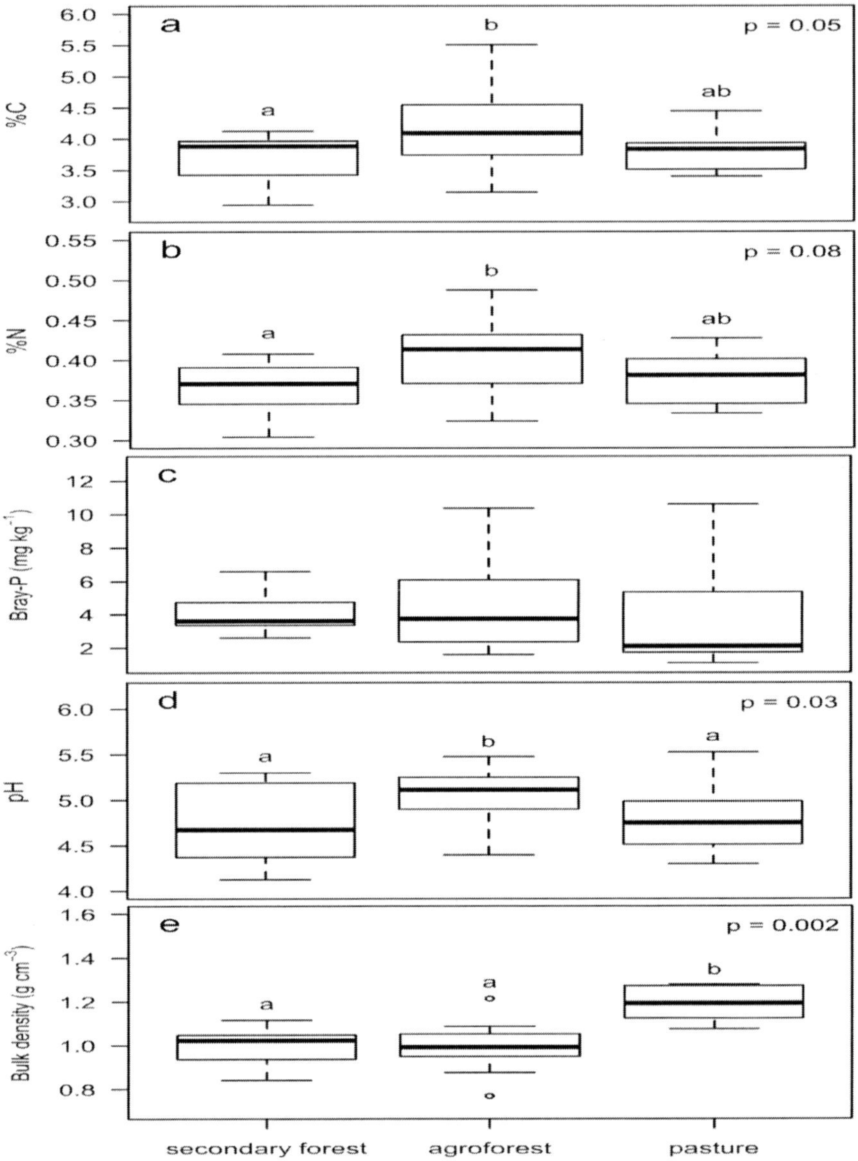

Figure 8. Relationship between land use type and surface soil qualities. Relationship between land use type (secondary forest (N = 8), agroforest (N = 17) and pasture (N = 7)) on soil %C (a), %N (b), Bray-P (c), pH (d), and bulk density (e). Non-overlapping letters signify significant difference at Non-overlapping letters signify significant difference at $\alpha \leq 0.05$ for %C and ≤ 0.1 for %N.

Despite minimal active management in secondary forests, potential natural resource diversity tended to be comparable between the secondary forest and agroforest, while it was lowest in the pasture (Fig 5). Agroforests had an even mixture of single-, double-, and triple-function species, as compared to the secondary forest, which primarily had one- and two-function species (S1 Fig). The relatively high number of species with such high functional utility in actively managed agroforest explains why this system's overall functional value was comparable to secondary forests despite having slightly lower species richness and phylodiversity, both of which positively correlate with functional diversity (Fig 7). Nonetheless, unmanaged secondary forests remain critical, as they are home to the species predominantly used for lumber, in addition to having comparable potential functional value of both food and medicine as agroforests (Fig 6). However, our results point to the utility of maintaining a matrix of accessible secondary forests alongside agroforests to promote both plant biodiversity conservation and socioeconomic development.

Complementing our assessment of aboveground natural resource diversity, we documented the influence of land use type on soil factors which are critical for regulating the productive potential and ecological functions of agroforestry systems [61]. The traditional practice of adding husks and other plant residues back to agroforestry sites–in addition to slash and burn land clearing, which transiently raises soil pH [61]–may increase soil pH while also enriching soil nutrient status [62]. Although the top 10 cm of soil in the well-developed agroforestry sites surveyed had a higher pH than secondary forest or pasture, no influence of land use on plant-available P was detected. However, agroforest soil was more C- and N-enriched than neighboring secondary forest, and comparable to pastureland (Fig 8). Paralleling past studies, pasture also had a higher bulk density relative to agroforest (and secondary forest), which can reduce soil aeration, porosity, and permeability, thereby negatively affecting fertility [60]. Therefore, established agroforestry systems tend to maintain soil structure relative to pastureland, which can increase nutrient cycling efficiency, and minimize leaching losses, a primary cause of soil fertility declines in annual cropping and pasture systems [61]. No significant relationships were identified between plant diversity and soil properties when pooled across land use types suggesting that land use decisions (such as adding back plant residues) may influence soil properties more strongly than plant species diversity.

Soil P limitation is a common phenomenon in old Neotropical forest soils; low available P-availability is a commonly noted constraint in tropical agriculture [63]. All land use types assessed had available P values significantly below proposed critical levels (the available P concentrations correlated with 80% maximum crop yield in tropical Oxisols, Ultisols, and

Alfisols) of 7–10 mg kg^{-1} [64]. Similar to other studies in P-poor tropical agroforests [61], land use choice appears to have little influence on available soil P in this lowland tropical system. The low available P concentrations observed across soil types suggest that P limits plant growth and agricultural yield in both unmanaged and managed land use types. Recommendations for stimulating productivity should therefore include the addition of P-rich organic matter.

Agroforestry systems, which are within the scope of the Kyoto Protocol sanctioned Clean Development Mechanism intended to increase C sequestration while improving land quality in deforested or degraded lands, are well-documented to have higher potential to sequester aboveground C than pastures and field crops [18,21,65–67]. Less emphasis, however, is typically placed on other functions that agroforestry systems may support. Our study suggests that agroforestry systems have the potential to be comparable to unmanaged secondary forests in their biological and natural resource diversity. This finding does not imply, however, that secondary forests are unimportant; secondary forests are home to many species not found in managed systems and can be useful for other socioeconomic functions that may not require direct management strategies.

Allowing agroforestry systems to develop alongside secondary forests and reserve areas may be a viable strategy to promote biodiversity conservation and socio-ecological stability in Atlantic Nicaragua and elsewhere [23,38]. Our results support the paradigm that highly impacted systems such as pasture may be too altered to meaningfully serve socio-ecological functions outside the designed purpose of grazing land [14]. Thus, the expansion of pasture has a much stronger impact on biodiversity and ecosystem functions than other management strategies while conferring few other socio-ecological benefits. By coupling a suite of ecological and social metrics, we were able to more comprehensively assess the impacts of common land use strategies in a rapidly developing area of Atlantic Nicaragua on terrestrial ecosystems. An integrative approach like the one employed here may strengthen future assessments of the tradeoffs associated with different land use strategies in this region and elsewhere.

ACKNOWLEDGMENTS

We thank Vanny Brown for expert field assistance, the Foundation for the Autonomy and Development of the Atlantic Coast of Nicaragua (FADCANIC) for field support, Iñigo Granzo de la Cerda for help with plant identification, and Anna Schürkmann for GIS support. This work was supported by the U.S. National Socio-Environmental Synthesis Center (SESYNC), the Helmholtz

Centre for Environmental Research (UFZ), and the German Center for Integrative Biodiversity Research (iDIV) Biodiversity and Ecosystem Services working group. SS was supported by the NOAA Climate and Global Change Postdoctoral Fellowship Program, administered by the University Corporation for Atmospheric Research.

REFERENCES

1. Foley JA, Defries R, Asner GP, Barford C, Bonan G, Carpenter SR, et al. Global consequences of land use. Science. 2005;309: 570–574. pmid:16040698
2. Wright HL, Lake IR, Dolman PM. Agriculture-a key element for conservation in the developing world. Conserv Lett. 2012;5: 11–19.
3. Tscharntke T, Clough Y, Wanger TC, Jackson L, Motzke I, Perfecto I, et al. Global food security, biodiversity conservation and the future of agricultural intensification. Biol Conserv. Elsevier Ltd; 2012;151: 53–59.
4. McNeely J a., Schroth G. Agroforestry and biodiversity conservation—Traditional practices, present dynamics, and lessons for the future. Biodivers Conserv. 2006;15: 549–554.
5. Tilman D, Fargione J, Wolff B, D'Antonio C, Dobson A, Howarth R, et al. Forecasting agriculturally driven global environmental change. Science. 2001;292: 281–4. pmid:11303102
6. Gullison RE, Frumhoff PC, Canadell JG, Field CB, Nepstad DC, Hayhoe K, et al. Tropical Forests and Climate Policy. Science. 2007;316: 985–986. pmid:17495140
7. Raupach MR, Field CB, Buitenhuis ET, Ciais P, Canadell JG, Que C Le, et al. Contributions to accelerating atmospheric CO2 growth from economic activity, carbon intensity, and efficiency of natural sinks. Proc Natl Acad Sci U S A. 2007;104: 18866–18870. pmid:17962418
8. Tubiello FN, Salvatore M, Ferrara AF, House J, Federici S, Rossi S, et al. The Contribution of Agriculture, Forestry and other Land Use activities to Global Warming, 1990–2012. Glob Chang Biol. 2015;21: 2655–2660.
9. Nair PKR, Nair VD, Kumar MB, Haile SG. Soil carbon sequestration in tropical agroforestry systems: a feasibility appraisal. Environ Sci Policy. 2009;12: 1099–1111.
10. Perfecto I, Vandermeer J. The agroecological matrix as alternative to the land-sparing/agriculture intensification model. Proc Natl Acad Sci U S A. 2010;107: 5786–91. pmid:20339080
11. Vandermeer J, Perfecto I. The agricultural matrix and a future paradigm for conservation. Conserv Biol. 2007;21: 274–7. pmid:17298536
12. Zimmerer KS. Understanding agrobiodiversity and the rise of resilience: analytic category, conceptual boundary object or meta-level transition? Resilience. 2015;3293: 1–16.
13. Lamb D. Restoration of Degraded Tropical Forest Landscapes. Science. 2005;310: 1628–1632. pmid:16339437
14. Knoke T, Calvas B, Aguirre N, Román-Cuesta RM, Günter S, Stimm B, et al. Can tropical farmers reconcile subsistence needs with forest conservation? Front Ecol Environ. 2009;7: 548–554.
15. Schelhas J, Pfeffer MJ. When global environmentalism meets local livelihoods: policy and management lessons. Conserv Lett. 2009;2: 278–285.
16. Norris K. Agriculture and biodiversity conservation: opportunity knocks. Conserv Lett. 2008;1: 2–11.

17. Nair PKR, Nair VD, Kumar BM, Haile SG. Soil carbon sequestration in tropical agroforestry systems: a feasibility appraisal. Environ Sci Policy. 2009;12: 1099–1111.
18. Montagnini F, Nair PKR. Carbon sequestration: An underexploited envionmental benefit of agroforestry systems. Agrofor Ecosyst. 2004;61: 281–295.
19. Albrecht A, Kandji ST. Carbon sequestration in tropical agroforestry systems. Agric Ecosyst Environ. 2003;99: 15–27.
20. Scales BR, Marsden S. Biodiversity in small-scale tropical agroforests: a review of species richness and abundance shifts and the factors influencing them. Environ Conserv. 2008;35: 160–172.
21. Kirby KR, Potvin C. Variation in carbon storage among tree species: Implications for the management of a small-scale carbon sink project. For Ecol Manage. 2007;246: 208–221.
22. Lal R. Soil carbon sequestration impacts on global climate change and food security. Science. 2004;304: 1623–1627. pmid:15192216
23. Jose S. Agroforestry for ecosystem services and environmental benefits: an overview. Agrofor Syst. 2009;76: 1–10.
24. Glaser B, Lehmann J, Zech W. Ameliorating physical and chemical properties of highly weathered soils in the tropics with charcoal—A review. Biol Fertil Soils. 2002;35: 219–230.
25. Lambers JHR, Harpole WS, Tilman D, Knops J, Reich PB. Mechanisms responsible for the positive diversity-productivity relationship in Minnesota grasslands. Ecol Lett. 2004;7: 661–668.
26. van Ruijven J, Berendse F. Diversity-productivity relationships: initial effects, long-term patterns, and underlying mechanisms. Proc Natl Acad Sci U S A. 2005;102: 695–700. pmid:15640357
27. Tilman D, Lehman CL, Thomson KT. Plant diversity and ecosystem productivity: Theoretical considerations. Proc Natl Acad Sci. 1997;94: 1857–1861. pmid:11038606
28. Cadotte MW, Carscadden K, Mirotchnick N. Beyond species: functional diversity and the maintenance of ecological processes and services. J Appl Ecol. Blackwell Publishing Ltd; 2011;48: 1079–1087.
29. Flynn DFB, Mirotchnick N, Jain M, Palmer MI, Naeem S. Functional and phylogenetic diversity as predictors of biodiversity–ecosystem-function relationships. Ecology. Ecological Society of America; 2011;92: 1573–1581. pmid:21905424
30. Milcu A, Allan E, Roscher C, Jenkins T, Meyer ST, Flynn D, et al. Functionally and phylogenetically diverse plant communities key to soil biota. Ecology. Ecological Society of America; 2013;94: 1878–1885. pmid:24015531
31. Paul KI, Polglase PJ, Nyakuengama JG, Khanna PK. Change in soil carbon following afforestation. For Ecol Manage. 2002;168: 241–257.
32. Johnson DW. Effects of Forest Management on Soil Carbon Storage: metaanalysis. Water Air Soil Pollut. 1992;64: 83–120.
33. Guo LB, Gifford RM. Soil carbon stocks and land use change: a meta analysis. Glob Chang Biol. 2002;8: 345–360.
34. Schmitt KM, Kramer DB. Road development and market access on Nicaragua's Atlantic coast: implications for household fishing and farming practices. Environ Conserv. 2010;36: 289–300.
35. Wassenaar T, Gerber P, Verburg PH, Rosales M, Ibrahim M, Steinfeld H. Projecting land use changes in the Neotropics: The geography of pasture expansion into forest. Glob Environ Chang. 2007;17: 86–104.
36. Stevens K, Campbell L, Urquhart G, Kramer D, Qi J. Examining complexities of forest cover change during armed conflict on Nicaragua's Atlantic Coast. Biodivers Conserv. 2011;20: 2597–2613.
37. Williams NE. A Political Ecology of "Ethnic" Agricultural Biodiversity Maintenance. J Polit Ecol. 2016;23: 223–245.

38. Harvey CA, Komar O, Chazdon R, Ferguson BG, Finegan B, Griffith DM, et al. Integrating Agricultural Landscapes with Biodiversity Conservation in the Mesoamerican Hotspot. Conserv Policy. 2008;22: 8–15.
39. McCloskey TA, Liu K. A sedimentary-based history of hurricane strikes on the southern Caribbean coast of Nicaragua. Quat Res. 2012;78: 454–464.
40. Coe F, Anderson G. Ethnobotany of the Miskitu of Eastern. Jounral Ethnobiol. 1997;17: 171–214.
41. Coe FG, Anderson GJ. Ethnobotany of the Garífuna of Eastern Nicaragua. Econ Bot. 1996;50: 71–107.
42. Guariguata MR, Ostertag R. Neotropical secondary forest succession: changes in structural and functional characteristics. For Ecol Manage. 2001;148: 185–206.
43. Boucher DH, Vandermeer JH, Yih K, Zamora N. Contrasting Hurricane Damage in Tropical Rain-Forest and Pine Forest. Ecology. 1990. pp. 2022–2024.
44. Katovai E, Burley AL, Mayfield MM. Understory plant species and functional diversity in the degraded wet tropical forests of Kolombangara Island, Solomon Islands. Biol Conserv. 2012;145: 214–224.
45. Mayfield MM, Daily GC. Countryside biogeography of Neoropical herbaceous and shrubby plants. Ecol Appl. 2005;15: 423–439.
46. Dennis PA. Herbal medicine among the miskito of Eastern Nicaragua. Econ Bot. 1988;42: 16–28.
47. Coe FG, Anderson GJ. Ethnobotany of the Garifuna of Eastern Nicaragua. Econ Bot. 1996;50: 71–107.
48. Coe FG, Anderson GJ. Snakebite ethnopharmacopoeia of eastern Nicaragua. J Ethnopharmacol. 2005;96: 303–323. pmid:15588683
49. Coe FG. Rama midwifery in eastern Nicaragua. J Ethnopharmacol. 2008;117: 136–157. pmid:18337033
50. Barrett B. Medicinal plants of Nicaragua's Atlantic Coast. Econ Bot. 1994;48: 8–20.
51. Coe FG, Anderson GJ, Coe FG, Anderson GJ. Ethnobotany of the Sumu (Ulwa) of Southeastern Nicaragua and comparisons with Miskitu plant lore. Econ Bot. 1999;53: 363–386.
52. Webb CO, Donoghue MJ. Phylomatic: tree assembly for applied phylogenetics. Mol Ecol Notes. 2005;5: 181–183.
53. Grafen A. The phylogenetic regression. Philos Trans R Soc London. 1989;326: 119–157.
54. Cadotte MW. Phylogenetic diversity-ecosystem function relationships are insensitive to phylogenetic edge lengths. Narwani A, editor. Funct Ecol. 2015;29: 718–723.
55. Kembel SW, Cowan PD, Helmus MR, Cornwell WK, Morlon H, Ackerly DD, et al. Picante: R tools for integrating phylogenies and ecology. Bioinformatics. 2010;26: 1463–4. pmid:20395285
56. Bray RH, Kurtz LT. Determination of total, organic, and available forms of phosphorus in soils. Soil Sci. 1945;59: 39–46.
57. R Core Team. R: A language and environment for statistical computing. R Foundation for Statistical Computing, Vienna, Austria. URL http://www.R-project.org/ [Internet]. 2013. Available: http://www.r-project.org
58. Lawrence D, Vandecar K. Effects of tropical deforestation on climate and agriculture. Nat Clim Chang. Nature Publishing Group, a division of Macmillan Publishers Limited. All Rights Reserved.; 2014;5: 27–36.
59. Bond WJ, Parr CL. Beyond the forest edge: Ecology, diversity and conservation of the grassy biomes. Biol Conserv. 2010;143: 2395–2404.
60. Reiners WA, Bouwman AF, Parsons WFJ, Keller M. Tropical Rain Forest Conversion to Pasture: Changes in Vegetation and Soil Properties. Ecol Appl. 1994;4: 363.
61. Alfaia SS, Ribeiro GA, Nobre AD, Luizão RC, Luizão FJ. Evaluation of soil fertility in smallholder agroforestry systems and pastures in western Amazonia. Agric Ecosyst Environ. 2004;102: 409–414.

62. Giardina C, Sanford R, Døckersmith IC. Changes in Soil Phosphorus and Nitrogen During Slash-and-Burn Clearing of a Dry Tropical Forest. Soil Sci Soc Am J. 2000;64: 399.
63. Townsend AR. Unexpected changes in soil phosphorus dynamics along pasture chronosequences in the humid tropics. J Geophys Res. 2002;107: 8067.
64. Schoenholtz SH, Miegroet H Van, Burger JA. A review of chemical and physical properties as indicators of forest soil quality: Challenges and opportunities. For Ecol Manage. 2000;138: 335–356.
65. Wara M. Is the global carbon market working? Nature. 2007;445: 595–596. pmid:17287794
66. Jandl R, Lindner M, Vesterdal L, Bauwens B, Baritz R, Hagedorn F, et al. How strongly can forest management influence soil carbon sequestration? Geoderma. 2007;137: 253–268.
67. García-Oliva F, Masera OR. Carbon Sequestration in Land-Use, Land-Use Change, and Forestry (LULUCF) Projects Under the Kyoto Protocol. Clim Change. 2004;65: 347–364.

CHAPTER 8

The Extent and Drivers of Deforestation and Forest Degradation in Masito-Ugalla Ecosystem, Kigoma Region, Tanzania

John E. Makunga[1*], Salome B. Misana[2]

[1]*Pasiansi Wildlife Training Institute, Mwanza, Tanzania*

[2]*Department of Geography, University of Dar es Salaam, Dar es Salaam, Tanzania*

ABSTRACT

Deforestation and forest degradation has been observed to be rampant in Masito-Ugalla ecosystem, Kigoma Region, western part of Tanzania. This paper therefore, intended to assess the extent of deforestation and forest degradation in the area, and to determine their causes. A total of 101 respondents were considered as the sample size for this study. The methods used for data collection were household questionnaire interviews, in-depth interviews, focus group discussions, analysis of satellite images and direct observation. The findings indicated that deforestation was occurring in the study area. Satellite data revealed diminished closed woodland, bushed grassland, forest and thickets between 1990 and 2014. On the contrary, settlement area, cultivated land and open woodland had increased during the same time frame. Proximate factors causing deforestation and forest degradation included agricultural expansion, wood extraction and expansion of settlement area. Underlying factors included population growth, poverty, poor levels of education, lack of employment, corruption and embezzlement of public funds by politicians and senior government officials; and high demand for fuel-wood. Biophysical drivers like incidences of unplanned wildfires and socio trigger events notably civil strife were also important. In order to minimize the problem and based on the factors augmenting deforestation and forest degradation in the Masito-Ugalla ecosystem and their coupled negative consequences, effective environmental conservation education, increased patrols, effective law enforcement and provision of alternative energy sources are necessary.

Keywords: Deforestation, Forest Degradation, Drivers of Deforestation, Masito-Ugalla Ecosystem

1. INTRODUCTION

Although forests cover around 30% of the Earth's land surface, they are getting lost at an alarming rate (FAO, 2011). Ecologically, forest ecosystems contribute to the protection of watersheds. Forests are key elements in the regulation of hydrological cycles, climate and the reduction of atmospheric pollution and they play an essential role in the global carbon cycle (FAO, 2011). After oceans, forests are the world's largest storehouses of carbon and they provide ecosystem services that are important to human wellbeing (FAO, 2010). These services include absorption of harmful greenhouse gasses that produce climate change. According to WWF (2016) , tropical forests alone store a quarter of a trillion tons of carbon in above and below ground biomass. They also provide clean water for drinking, bathing, and other household needs; protect watersheds and reduce or slow the amount of erosion and chemicals that reach waterways; provide food and medicine; act as a buffer in natural disasters like flood and rainfalls as well as provide habitat to more than half of the world's land-based species (WWF, 2016).

Regardless of such immense importance, forests have continued to be degraded and severely depleted. While deforestation is regarded by Hosonuma et al. (2012) as the complete removal of trees and conversion of forests into other land uses, forest degradation involves a change process that negatively affects the characteristics of a forest such that the value and production of its goods and services decline (FAO, 2009). This change process is caused by disturbance (although not all disturbance causes degradation), which may vary in extent, severity, quality, origin and frequency. Disturbance may be natural (e.g. that caused by fire, storm or drought), human-induced (e.g. through harvesting, road construction, shifting cultivation, hunting or grazing) or a combination of the two (FAO, 2009). A degraded forest is a secondary forest that had lost, through human activities, the structure, function, species composition or productivity normally associated with a natural forest type expected on that site UNEP, 2001).

High rates of deforestation have been experienced globally particularly in tropical areas. FAO (2016), for example, reports of a net annual loss of forest of about 7 million hectares between 2000 and 2010 and a decline of forest by 129 million hectares between 1990 and 2010. Such deforestation was reported by FAO (2010) to have been highest in South America, central West Africa, and South and Southeast Asia, with 13 million hectares of global forest lost each year. In contrast, Africa was facing a huge deforestation problem, with the rate of deforestation being higher than in any other continent (Johnson & Chenje, 2008). Accordingly, of the ten countries in the world with the largest annual net loss of forested area, six were in Africa. On the other hand,Pickering & Owen (1997) reported that in Europe, widespread deforestation took place from the 11th century as lands were cleared for use as fuel and settled agriculture, thus in 200 years the forests of central Europe were almost entirely depleted.

Empirical evidence reveals that a combination of factors and not a single factor are responsible for the high rates of deforestation that have been experienced globally particularly in developing countries (Hosonuma et al., 2012; Geist & Lambin, 2002; Sitarz, 1994). These include inappropriate land tenure systems, agricultural expansion, e infrastructure expansion and increasing demand for forest products. For example, felling of trees for domestic and agricultural fuel-wood, charcoal, building poles and indiscriminate bush clearing and bush fires for different reasons were reported by Reed (1996) and Misana & Nyaki (1993) to be causes of deforestation in parts of Africa. Charcoal making, a commercial activity, which in most countries is largely unregulated and outside government controls, has also been reported to be an important cause of deforestation, with its effects largely seen close to cities (Timberlake, 1991).

Forests in Tanzania were being lost at a very high rate of 412,000 hectares per year out of a total forest estate of around 35 million hectares (TFWG, 2009). Although forest loss and degradation were occurring inside Forest Reserves, deforestation was particularly rapid and widespread outside reserved forests (TFWG, 2009), including the Masito-Ugalla ecosystem (MUE). According to URT (2014) the Masito-Ugalla ecosystem is potential as a catchment area for Lake Tanganyika, which serves as an irreplaceable source of clean water, transportation and economic opportunities for an estimated 10 million people in its four riparian countries (Tanzania, Democratic Republic of Congo, Burundi and Zambia). The major challenge facing Lake Tanganyika (URT, 2014) is environmental degradation caused mainly by human induced activities, which pose serious threats to the biodiversity and sustainable use of natural resources in the Lake Tanganyika basin.

Analysis of the deforestation situation in Masito-Ugalla ecosystem was thus, vital to help improve appreciation of the role of rural communities as potential guardians in the management of forest resource to avoid deforestation and forest degradation. The study therefore, had two major objectives: to assess the extent of deforestation and forest degradation, and to determine the causes of such deforestation and forest degradation in the study area. The study examined the drivers of deforestation and forest degradation focusing on both the proximate (direct) and underlying (indirect) causes (Geist & Lambin, 2002; Hosonuma et al., 2012).

2. MATERIALS AND METHODS

2.1. The Study Area

This research was conducted in Ilagala, Karago and Kirando villages, located within the Masito-Ugalla Ecosystem in Uvinza District along the shoreline of Lake Tanganyika (Figure 1). Svoboda, and McNamara (2009) revealed that the MUE is located near the western edge of the East African Rift Valley and Lake Tanganyika, in Western Tanzania and lies between grid reference 138,000E and 278,000E and between 9,260,000N and 9,400,000N. The ecosystem is surrounded by nine wards, out of which, Ilagala, Sunuka, Sigunga, Mganza, Uvinza and Nguruka are in Uvinza District, Mishamo and Mpanda ndogo wards in Mpanda District and Usinge in Kaliua District, Tabora Region. MUE covers an extensive forested landscape of 10,827 km^2 under different management and ownership regimes (Deloitte, 2012 ; CIFOR, 2014). Two types of native forest predominate in the area; these are the miombo woodland dominated by Brachystegia species and the lowland rain forest in the valley bottoms (URT, 2014). In addition to woodlands, a large proportion of the region is covered by grasslands (URT, 2014).

2.2. Data Collection Methods

A triangulation of methods was used in this study to collect both quantitative and qualitative data. The methods used included remote sensing techniques and geographical information systems (GIS), household questionnaire interviews, in-depth interviews of key informants, focus group discussions and field observations. These methods are described below:

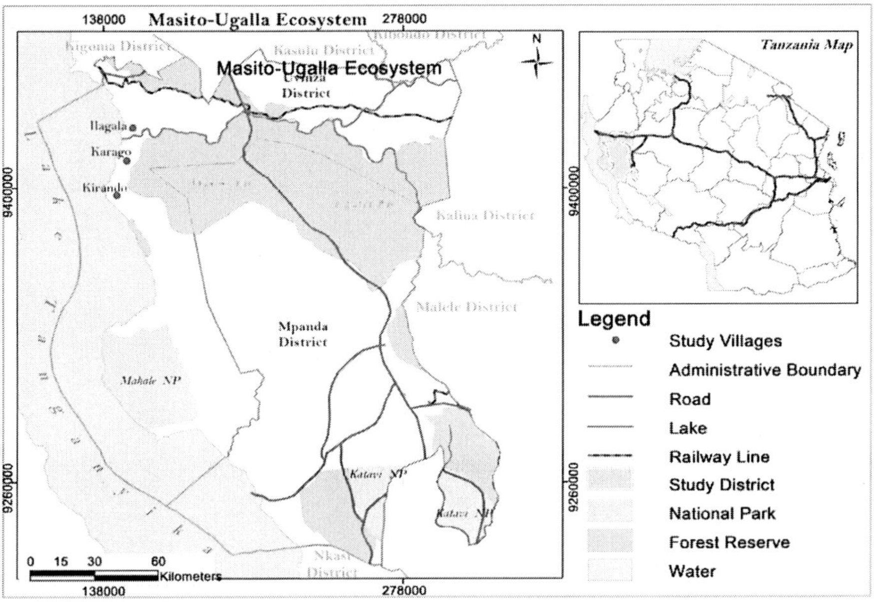

Source: IRA Lab, University of Dar es Salaam (2014)

Figure 1.Location of the study villages in Uvinza District.

2.2.1. Remote Sensing and GIS

Land cover and land use change detection in the forests of Masito and Ugalla, Uvinza and Mpanda Districts were analyzed based on the interpretation of satellite images. Landsat Thematic Mapper ™ for 1990 and Landsat Enhanced Thematic Mapper Plus (ETM+) for 2000 and 2014 were used to extract spatial and temporal land cover and use data. Basically, the study intended to detect land cover changes for every 10 years, from 1990 to 2010. However, it was found interesting to include four extra years from 2010 in order to investigate the effectiveness of the REDD+ project, which lasted for three years from January 2010 to December 2013. It was therefore, logical to detect such change using the 2014 Landsat image, since it was the same year in which this study was conducted.

Before analysis of the data, image processing, classification and interpretation were done. Image processing (Vicente-Serrano, 2008) was done by undertaking radiometric and geometric correction. Radiometric correction was employed to make the analyzed images to be recognizable and also avoid the obtained results being invalid (i.e. to reduce the influence of inconsistencies in image brightness values that may limit an ability to interpret). Rocchini & Rita (2005) found that since all metrics calculated on a landscape (from area coverage by vegetation classes to patches shape and size) are sensible to geometric distortions, thus, the geometric correction of aerial photos is due to allow a realistic multi-temporal study. The process enhances and improves the quality of images for better data extraction. Geometric correction was also employed to remove internal and external distortions in order to identify the location of each pixel on the earth by overlaying with other reference data such as topographical maps (Baboo & Devi, 2011). The processing activities were done using ERDAS IMAGINE V11. Before image classification, image interpretation was done by identifying all the land cover and use types based on pattern, colour, shape, size, texture and association (Caelli & Reye, 1993). Image classification was done (Table 1) using supervised classification method.

Table 1. Land cover types obtained after image classification.

Id	Class Name	Description
1	Closed Woodland	Land dominated by closed trees (>65%), shrubs and herbaceous plants may be present but are closed to absent (>65%)
2	Cultivated Land	An area characterized by herbaceous vegetation that has been planted or is intensively managed for the production of food, feed, or fiber; or is maintained in developed settings for specific purposes. Herbaceous vegetation accounts for 75 - 100 percent of the cover.
3	Dense Bushland	The land in which trees and herbaceous plants may be present, but are closed to absent, while shrubs are dominant and closed
4	Bushed Grassland	An area characterized by shrubs and grasses. Trees and herbaceous may be present, but are closed to absent
5	Forest	The plant life form dominated by tree cover, generally greater than 6 meters tall); tree canopy accounts for 25 - 100 percent of the cover. Shrubs and trees may be present
6	Open Grassland	An area in which grasses are the dominant vegetation and are open (65% - 15%), other vegetation such as trees, shrubs and herbaceous may be present but are sparse to absent (15% - 1%)
7	Open Bushland	Trees and herbaceous plants are closed to absent, shrubs usually form the main layer and are open
8	Open Land	Any open piece of land that is undeveloped (has no buildings or other built structures)
9	Open Woodland	The land in which the dominant vegetation cover is represented by open trees, with herbaceous plants and shrubs closed to absent.
10	Settlement Area	An area characterized by community of people
11	Swamp	A swamp is a wetland that is dominated by woody plants (i.e. shrubs and trees, defined at the height of less than 5 m or above 5 m respectively)
12	Thicket	A plant growth form characterized by shrubs, herbaceous plants and trees may be present, but with sparse cover (15% - 1%)
13	Water	All areas of open water, generally with less than 25% cover of vegetation/land cover

Source: EPA (2007) & FAO (2007).

2.2.2. Household Questionnaire Interviews

Household interviews were conducted using structured questionnaires consisting of both open and closed-ended questions. A pretest was necessary to determine how the questionnaire could be improved to minimize response errors, such as a respondent misinterpreting a question (Bolton, 1993). A total of 101 respondents, one from each household were interviewed. The types of data, which were collected using this method, included demographic data, education, size of households and income.

Other data were the extent of deforestation and forest degradation, causes as well as impacts of deforestation.

2.2.3. In-Depth Interviews

In-depth interviews were used to gather qualitative data from key informants, who comprised of the Village Chairpersons, Village Executive Officers, Ward Executive Officer; the REDD+ CBO leader, forest monitors, fire breakers as well as the District Natural Resources Officer (DNRO). There were eight key informants from whom information was obtained. Key informants were asked pre-determined questions and then allowed to respond. Their responses were guided by the researcher who was the moderator. Types of data collected using this method included socio-economic activities of villagers; awareness of causes of deforestation and forest degradation.

2.2.4. Focus Group Discussions (FGDs)

Focus group discussions were also used to collect qualitative data. The FGDs involved two processes, which were brainstorming and discussions. According to Pripathy & Pripathy (2017) , in a focus group discussion, a small number of individuals (e.g. 8 - 12) are brought together to talk about some topic of interest to a focus group moderator. There were four focus groups comprising 10 individuals each. The first group included farmers; the second one represented women as the major forest resource users; the third group consisted of businessmen; and the last focus group represented honey gatherers. The four focus groups were identified as the main forest resource users in the study area. The same types of data as those collected using semi-structured interviews such as socio-economic activities of villagers; awareness of causes of deforestation and forest degradation were obtained.

2.2.5. Field observation

In this study, unstructured observation was used to gather information on socio-economic activities such as logging/lumbering. Visits were made to different locations in the study area, such as village forest lands, Masito forest and within villages in the study area to observe socio-economic activities and development (Mulhall, 2003). Photographs were taken to illustrate the status of deforestation and forest degradation.

2.3. Data Analysis

The vector data obtained from the satellite imageries after classification of land cover and use types were used for analysis so as to identify changes using ArcGIS V10.1. Overlaying analysis was done by comparing land cover of 1990, 2000 and 2014 to identify the changes in land cover and use whereby the area of each land cover was computed to know the increase or decrease of a certain land cover. A land cover change detection matrix, as well as graphs using Microsoft Excel was generated for easy identification of changes.

Qualitative data, which involved those obtained using FGDs and semi-structured interviews, were analyzed using the content analysis approach (GAO, 1989; Stewart & Shamdasani, 2017). Under the content analysis, materials to be analyzed were first determined. The next step involved categorization of the recorded units into several themes falling under the study objectives. Qualitative data collected using questionnaires were intended to generate outputs in terms of frequencies and percentages, which revealed perceptions of the respondents on particular matters. Quantitative data gathered using household questionnaires were analyzed using IBM Statistical Package for Social Sciences (SPSS), version 20.0. The first step in the analysis of quantitative data was preparation of variables followed by coding the responses in a code book and analysis of different variables (Stewart & Shamdasani, 2017).

3. RESULTS AND DISCUSSION

3.1. Extent of Deforestation and Forest Degradation

Majority of the respondents (79.2%) agreed that deforestation and forest degradation had happened in the area; only few (20.8%) reported that there was no deforestation or forest degradation. These findings are supported by results from satellite imagery interpretation (Figures 2-4; Table 2 & Table 3), which show a decrease in closed woodland, dense bushland and bushed grassland and an increase in open woodland and cultivated land between 2000 and 2014. There is clear evidence of woodland degradation and deforestation, illustrated by the opening up of dense woodlands, and replacement of woodlands, bush lands, thickets and bushed grasslands by cultivated lands, respectively.

The patterns of land use and cover for 1990 (Figure 2 & Table 2) reveal that during that time the dominant cover types in the study area were open woodland, closed woodland and open grassland. Cultivated land only covered a small area (2.23%) while settlement area and forest had the least coverage. By 2000, closed woodland, open grassland and forest had decreased (Table 2 & Figure 2). Closed woodland decreased by 11.1% (Table 3) indicating degradation of the woodland. Open grassland decreased to 11.88%. Other cover categories that showed a decrease were dense bushland, bushed grassland and forest while cultivated land increased during the two time periods. On the other hand, open woodland and cultivated land increased during the same time period (Table 2 & Figure 2). Open woodland had increased by 10% between 1990 and 2000 while cultivated land increased by 7.5% (Table 3).

The same pattern was observed in 2014 (Table 2 & Figure 4). In actual fact the dominant cover types during this period had changed to open woodland and cultivated land. Closed woodland had diminished to only 5.6% from 23.2% in 1990 (Table 2 & Table 3) while cultivation increased to 17.12 in 2014 from 2.23% in 1990. This gives a clear indication of the extent of deforestation and woodland degradation, which may have resulted from anthropogenic activities. Figure 5 visualizes how the different land cover and use types had increased or decreased between 1990 and 2014.

These findings were also supported by Plate 1 & Plate 2, which illustrate tree felling in the study area and a typical deforested area, respectively. One of the key informants revealed that deforestation was an act, which did not necessarily require capital; it was just someone's effort and time. This means that everyone was free to engage in acts that would lead to deforestation and degradation.

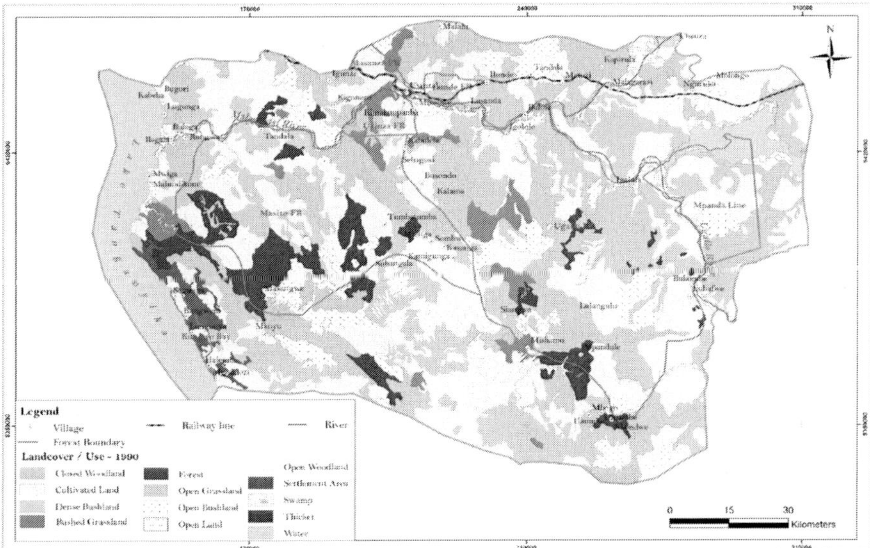

Figure 2. Land cover in Masito-Ugalla Ecosystem, 1990.

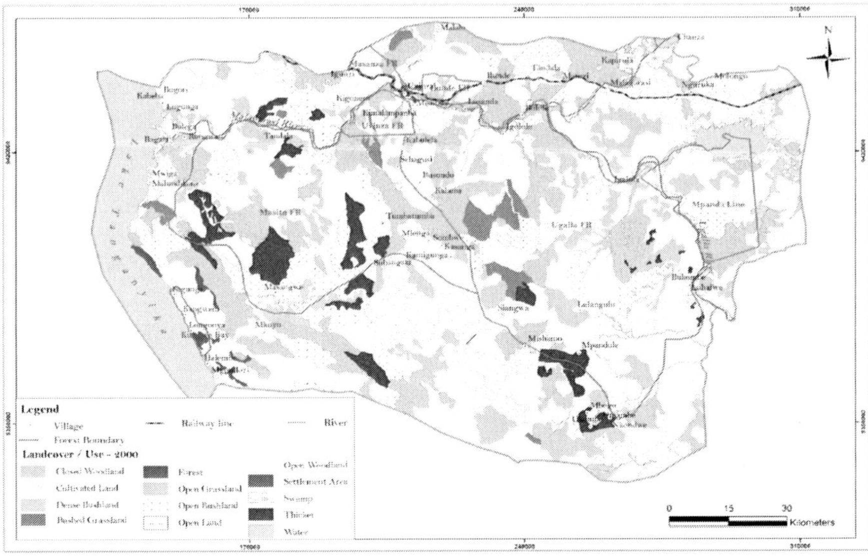

Figure 3. Land cover in Masito-Ugalla Ecosystem, 2000.

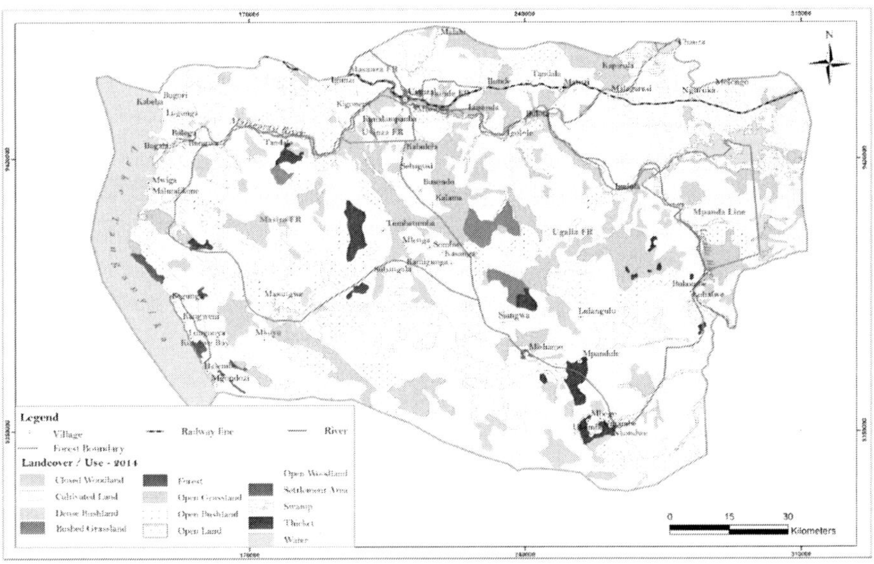

Figure 4. Land cover in Masito-Ugalla Ecosystem, 2014.

Table 2. Area coverage of land cover/use categories in Masito-Ugalla Ecosystem for 1990- 2014.

Land use/cover classes	Area coverage of land use/cover classes					
	1990		2000		2014	
	Ha	%	Ha	%	Ha	%
Bushed grassland	43,039	2.72	24,459	1.55	14,488	0.92
Closed Woodland	366,710	23.17	190,479	12.03	88,285	5.58
Cultivated Land	35,343	2.23	154,113	9.74	271,020	17.12
Dense Bushland	19,170	1.21	6443	0.41	3066	0.19
Forest	16,311	1.03	6165	0.39	4143	0.26
Open Bushland	124,203	7.85	132,831	8.39	127,770	8.07
Open Grassland	226,897	14.34	188,078	11.88	152,627	9.64
Open Land	1386	0.09	1164	0.07	991	0.06
Open Woodland	489,896	30.95	648,858	41.00	721,863	45.61
Settlement Area	272	0.03	1247	0.08	1831	0.12
Swamp	47,681	3.01	51,240	3.24	59,483	3.76
Thicket	84,105	5.31	54,082	3.42	20,643	1.30
Water	127,542	8.06	123,596	7.81	116,545	7.36
Total	1,582,755	100	1,582,755	100	1,582,755	100

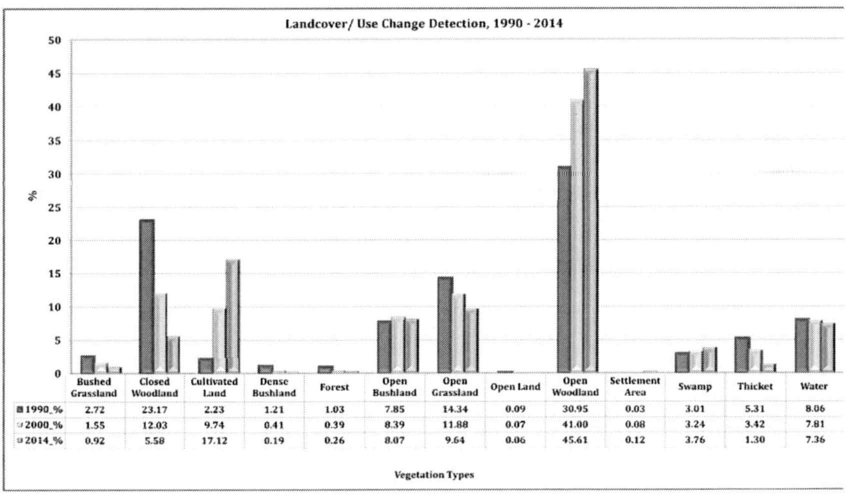

Figure 5. Visualization of land cover/use change in Masito-Ugalla Ecosystem for 1990-2014.

Table 3. The rate of change (in hectares and percent) of land use/cover classes for 1990- 2014.

Land use/cover classes	Net cover change					
	1990-2000		2000-2014		1990-2014	
	Ha	%	Ha	%	Ha	%
Bushed grassland	−18,580	−1.17	−9971	−0.63	−28,551	−1.8
Closed Woodland	−17,6231	−11.13	−102,194	−6.46	−278,425	−17.59
Cultivated Land	118,770	7.50	116,707	7.39	235,677	14.89
Dense Bushland	−12,727	−0.80	−3377	−0.21	−16,104	−1.02
Forest	−10,146	−0.64	−2022	−0.13	−12,168	0.77
Open Bushland	8628	0.55	−5061	−0.32	3567	0.22
Open Grassland	−38,819	−2.55	−35,451	−2.24	−74,270	−4.7
Open Land	−222	−0.01	−173	−0.01	−395	−0.03
Open Woodland	158,962	10.04	73,005	4.61	231,967	14.66
Settlement Area	775	0.05	−584	0.04	1359	0.09
Swamp	3559	0.22	8243	0.52	11,802	0.75
Thicket	−30,023	−1.90	−33,439	−2.11	−63,462	−4.01
Water	−3946	−0.25	−7051	−0.45	−10,997	−0.7

The above findings supported the findings of other studies, which also reported the problem of deforestation in other parts of the world including some REDD+ project areas as explained below. For example, deforestation was reported by Su et al. (2011) to be a common phenomenon in Quezon Province, Philippines (The presence of deforestation at a rate of 2.4% in the Sofala Community Carbon Project, in Mozambique, between 1999 and 2007 was observed (Mori, 2013). Other REDD+ project areas that were reported to have been experiencing deforestation were the Rimba Raya Biodiversity Reserve REDD+ Project, in Indonesia, where 4.29% of the forest had been lost within a period of six years, from 2003 to 2008, and Bugondo-Bugoma Landscape REDD+ Project, in Hoina, Kibaale and Kyenjojo Districts, in Uganda, where 3.5% of the forest, equal to 945 ha were lost due to deforestation during the first year of the project implementation (Mori, 2013).

Plate 1. Tree felling in Masito forest, which is likely to degrade the forest.

Plate 2. A typical deforested area at Ilagala village, Uvinza district.

Findings by Kashaigili et al. (2013) in Tanzania revealed that during the period 1980-1995 and 1995-2010, closed forest decreased by 635.5 ha (−11.9%) and 1674.9 ha (−31.3%), respectively. Similarly, in the period 1980-1995, settlement and other land uses increased by 85 ha (1.6%) and 439.2 ha (8.2%) between 1995 and 2010. During the period 1980-1995 closed forest decreased by 109.2 ha (−4.5%) and 611.1 ha (−25.3%) for the period 1995-2010. On the contrary, in the period 1980-1995, settlement and other land uses increased by 75 ha (3.1%) and 100.2 ha (4.1%) between 1995 and 2010. According to them, while closed forest decreased between 1980 and 2010, bushland, grassland, settlement and other land uses, and Open Forest increased.

Hansen & De Fries (2004), who examined the long term change of global forest cover using Advanced Very High Resolution Radiometer (AVHRR), observed that the tree cover of the world had decreased between 1984 and 1997 and the annual deforestation rate in tropical Africa was about 0.09%. In his study to identify and map deforestation using Landsat images and Global Inventory Modeling and Mapping Studies (GIMMS) data from 1980s to 2000s, Wu (2011) found that Kenya still remained at a deforestation rate of about 0.3% per year. Large loss of forest area was indicated for Tanzania and Uganda, about 1.2% and 2.7% per year, respectively.

3.2. Causes of Deforestation and Forest Degradation

Factors that enhance deforestation and forest degradation fall into three categories namely: proximate causes, underlying causes and other factors. Proximate factors include agricultural expansion, wood extraction, and infrastructure extension while the underlying causes are demographic, economic, technological, policy and institutional as well as cultural or socio-political factors (Madeira, 2008; Abdullah & Mahmud, 2005; Geist & Lambin, 2002). Other factors include pre-disposing environmental factors, such as land characteristics like soil quality and topography; biophysical factors such as wildfires and pest; and socio trigger events such as wars/ civil strife and sudden displacement. In this study, the causes that were identified could also be grouped into proximate, underlying and other factors (Madeira, 2008; Abdullah & Mahmud, 2005).

3.2.1. Proximate Causes

The proximate causes of deforestation and woodland degradation are presented in Table 4. The demand for forest products and land for cultivation and settlement, among other needs, were the major factors accelerating the loss of forest ecosystems in the study area. Uncontrolled farming was reported by the majority of the respondents as the dominant factor for deforestation and forest degradation. This is in line with the findings presented in Table 2 & Table 3, which show a tremendous increase of cultivated land between 1990 and 2014. Other factors mentioned were uncontrolled encroachment, illegal logging/lumbering and illegal honey gathering.

Table 4. Causes of deforestation and forest degradation in Masito-Ugalla Ecosystem.

Factors	Frequency of responses (N = 101)	Percent of Respondents
Uncontrolled Encroachment	22	21.8
Uncontrolled Farming	31	30.7
Illegal Logging/Lumbering	16	15.8
Illegal Honey Gathering	12	11.9
Poor Institutional Framework	19	18.8
No response	1	1.0
Total	101	100.0

Plate 3. Traditional processing of oil from palm kernels at Ilagala village.

It was reported during the focus group discussions and key informant interviews that there were incidences of encroachment in natural forests, especially in village forests, made purposely for clearing and expanding farm lands as well as obtaining fuel wood. Field observations showed that oil processing from palm kernels (Plate 3) and burnt bricks making were among the activities, which demanded large amounts of fuel-wood. It was also observed that traditionally, drying fish in villages surrounding Lake

Tanganyika was done using fuel-wood. Therefore, the expansion of markets for fish within and beyond Kigoma region would thus require huge amounts of firewood to meet the demands. Rural areas within Masito-Ugalla ecosystem are used as exit points for charcoal, which is transported using Sigunga-Ilagala road to Kigoma urban.

The above findings were in line with findings from other studies such as Hogqvist et al. (1998);Ahrends (2005); Malugu (2007); Svoboda & McNamara (2009); Su et al. (2011). For example,Svoboda & McNamara (2009) reported that deforestation in the Masito-Ugalla ecosystem was due to the increase in human practices such as farming, cutting down trees for charcoal burning and fuel- wood (). While public forests and village lands were commonly used for fuel wood, lack of reforestation activities were leaving open areas forcing some people to illegally obtain resources from nearby forest reserves. Random repeated burning was also reported to be the biggest impediment to forest maintenance and regeneration (Svoboda & McNamara, 2009).

Some of the human activities that contributed to the degradation of the forest reserves in Tanzania include charcoal burning, logging, encroachment for agriculture and pole cutting and expansion of farms within the forest reserves (Hogqvist et al., 1998; Ahrends, 2005; Malugu, 2007). Similarly, Su et al. (2011) revealed that in Quezon Province, in Philippines, the problem of deforestation emanated from illegal logging, excessive faming and conversion of forests to agricultural lands as well as conversion of forests to urban settlements ().

3.2.2. Underlying Causes

This subsection discusses a number of underlying causes of deforestation in the study area. They include: demographic factors, socio-economic factors, political and institutional factors and technological and biophysical factors.

1) Demographic factors

Population increase was found to be among the underlying factors augmenting deforestation in the study area. Gauging from the high population growth rates in other districts, there is no doubt that Uvinza district has experienced a similar pattern of population increase, with negative impacts on forests. It was observed in the study area that, majority of the villagers were living a low life standard, which made them dependent on forest resource as a means of attaining livelihood, thereby, affecting the resource. Because of rapid population growth (Kumar et al., 2010) among other reasons, in Kodagu District, Southern India, more than 200 hectares of forest in the region had been encroached since 1990 onwards.

The size of households has a bearing on deforestation and forest degradation. The larger the household size the higher the demand for land for cultivation to meet the food needs as well as the demand for forest products. Findings from the household interviews revealed a big number of household members in the study area. About half of the households (50.5%) had household size ranging from 6 to above 9 members. Out of these, 27.7% had 6 - 9 members while 22.8% had above 9 members. An additional 43.6% of the households had 2-5 members, the mean and standard deviation being 2.67 and .89, respectively, and only very few (5.9%) had a single member. This study examined reasons for the high household size in the study area. It was noted that habitually, many adult males in the study area preferred polygamous marriages, which have higher chances of increasing the number of children in a single household.

2) Socio-economic factors

Local social and economic factors such as poverty, poor education and lack of employment were found to impinge on efforts to conserve forest resources in the study area. Respondents' income was found to be very low. Majority of the respondent (68.3%) had a monthly income level below TZS. 40,000 ≠ and very few (5.9%) earned between TZS. 201,000 ≠ and 300,000 ≠. The highest income group consisted of government employees and those engaged in small businesses. Generally, females were found to have less income than males. However, there was no statistically significant difference in the monthly income between males and females: $t(99) = 1.056$, $p = 0.293$). The magnitude of the differences in the means was very small, with males having a slightly higher income than females (the eta squared statistics was 0.011).

It was observed that majority of the villagers were farmers who depended on selling their agricultural products. They also lacked reliable income generating activities, which would supplement their monthly income. This situation suggests extreme income poverty in the study area, a factor which indicates over- exploitation of forest resources in the absence of alternative livelihood options. Some key informants mentioned higher rates of poverty and limited employment opportunities in the study area as potential causes of deforestation. One of the key informants said that poverty increased the rate of deforestation and forest degradation in the study area due to increased demand for wood-fuel for palm oil processing and other domestic uses.

The above findings are supported by Klein (2002) and Ngalason (2009) who found that people in Uvinza District had the lowest per capita income, as low as $150 per year, which was among the factors contributing to the excessive pressure on the utilization of natural resources. Kigoma was among the regions in the country with low rates of literacy and school completion, thus impinging on majority of the people to access employment, all of which contribute to the extreme pressure on natural resources (Ngalason, 2009).

3) Political and institutional factors

Poor natural resources governance was among the hurdles limiting the success in conserving forests and their associated biological diversity in the study area. Factors like corruption and embezzlement of funds impinged the performance of the REDD+ project. These factors had detrimental consequences, accelerating the violation of natural resource conservation laws among others. It was also reported by majority of the respondents that they had no title deeds for the land they owned and only 2% agreed that they had a title deed. When asked about the reasons for not owning land, fifty percent of the respondents reported that they were not owning land due to lack of knowledge in interpreting the Village Land Act, while the other 50% of the respondents said that accessing land in their area was expensive.

The close proximity of the refugee camps to the forest reserves has also been a key contributing factor to environmental degradation and deforestation. Both refugees and locals continued to encroach upon the forest reserves, primarily to collect firewood or wood for construction, or to clear areas for cultivation (UNEP & CBD, 2001; Relief, 2003). Deforestation and depletion of forest resources (Relief, 2003) were the most often-cited environmental problem in Western Tanzania, the rate of deforestation being reportedly higher than the national average. Deforestation was most apparent when the areas outside of the refugee camps were contrasted with those within the camp boundaries, because unlike the village and forestlands, the lands within the camp boundaries were fairly being well patrolled. The major causes of refugees' influx were political and institutional factors (TRAFFIC, 2006).

The above findings support the findings of other scholars such as (Madeira, 2008; Svoboda & McNamara 2009; Burgess et al., 2012; NORAD, 2013). Weak governance institutions and corruption were associated with illegal logging in parts of Asia and with agricultural expansion in Latin America (Madeira, 2008). In the Democratic Republic of Congo (DRC), a number of politicians, including those in the government circles and Parliament, as well as high-ranking civil servants allegedly engaged in industrial and artisanal logging (NORAD, 2013). It has also been well established that military operations caused deforestation in Vietnam during the Vietnam War and elsewhere (Sands, 2005). The role of military in deforestation has also been documented in Southeast Asia and South America (Mather, 1991; Sands, 2005).

4) Technological factors

Technological factors have contributed to acceleration of forest degradation and deforestation in the Masito-Ugalla ecosystem. This is due to industrial activities that demand fuel-wood, thus causing negative consequences on forest conservation. A good example is the Uvinza Salt Mine, which was allegedly utilizing 20,000 cubic meters of fuel-wood per day for drying ovens (Svoboda & McNamara, 2009).

5) Biophysical drivers

Unplanned wildfires are among the environmental problems facing the study area. During the study several fire incidences were observed in Masito forest (Plate 4). One key informant explained that wildfires in Masito-Ugalla ecosystem were largely contributed by unsustainable farming activities. Fires were being used to help clear land for farming, re-juvenate rangeland grasses, and were sometimes due to cultural norms. Sometimes hunters and honey gatherers also used fire to attract animals during growth of new pastures and as a traditional honey collection method, respectively, all of which posed negative impacts on forest resources.

Plate 4. Unplanned wildfire in Masito forest, such fire is likely to cause tremendous negative impacts to the ecosystem.

According to research findings (Svoboda & McNamara, 2009) recent and active ires were observed throughout the Masito-Ugalla ecosystem. Accordingly, the combined effects of fire and drought destroyed 25,500 km^2 of primary and secondary forest and a further 7500 km^2 of settlement areas in previous years before the project was initiated. The misuse and over-use of fire as a cheap method of land clearing had also caused massive changes to the vegetation of large areas in Indonesia (Murdiyarso & Adiningsih, 2003). It has also been found that fire poses the greatest threat to Amazon forests (Cochrane & Laurance, 2002).

Apart from fire, there is an interrelationship between climate warming, drought and loss of vegetation cover. Drought can be characterized as a deviation from normal conditions in the physical system such as climate and hydrology (Lanen et al., 2007). As a result of drought many forests are threatened by secondary effects of climate change, such as a potential increase in the frequency and possibly intensity of fires (Lanen et al., 2007).

Although it has been difficult to attribute changes in forests and woodlands to climate change because of the interactions between natural climate variability, anthropogenic climate change and human induced land use change (IPCC, 2014) there are evidences that associate such changes to climate change. In the Amazon forest, for example, a link between forest fires, habitat fragmentation, climate change, and extreme El Niño events has been well established (Cochrane & Laurance, 2002). Sanderson et al. (2010), also observed that in all forest types, a warming climate could promote the frequency and severity of pest outbreaks, which will damage the health of the forest and may reduce the numbers of certain species. They further observed that reduced precipitation could negatively affect agricultural yields and other economic activities which majority of the people depend up on, thus forcing them to go to forests where they might lead to deforestation and forest degradation for the purpose of earning a living. Long term declines in density of trees and shrubs in the Sahel zone of Senegal have also been attributed to drought stress induced by decreased precipitation (IPCC, 2014). All in all, the IPCC (2014) reports of the general consensus that future changes in terrestrial ecosystems throughout Africa will likely be driven by continued changes in precipitation and carbon dioxide associated with climate change.

3.3. Measures for Minimizing Deforestation and Forest Degradation

When respondents were asked about measures that needed to be taken to minimize deforestation and degradation, 45.5% recommended environmental conservation education as the best measure. Effective law enforcement was reported by 18.8% and 15.5% of the respondents mentioned the provision of land tenure rights as the appropriate measure. Only very few (8.9%) mentioned capacity building on non-farm activities and controlling corruption in the forestry sector (10.9%). Environmental conservation education was seen as a reliable measure because it has been commonly used to deliver conservation message among local communities.

A Chi-square test of respondents' education level and measures to minimize deforestation and forest degradation revealed that the relationship between the two variables was statistically significant [$\chi^2(12, N = 101) = 23.24$, $p > 0.026$]. This meant that respondents' knowledge and skills in natural resources management were of paramount importance in understanding suitable measures needed for minimizing deforestation and forest degradation in the study area.

4. CONCLUSION

From the preceding discussion, it may be concluded that deforestation and forest degradation was a big problem within forest reserves as well as village lands in the Masito-Ugalla ecosystem. Findings generated through analysis of satellite data indicated a diminished closed woodland, open grassland, bushed grassland, forest and thickets between 1990 and 2014. On the contrary, settlement area, cultivated land and open woodland had increased in the same time frame. Major factors causing deforestation and forest degradation were uncontrolled farming, uncontrolled encroachment for farm lands and fuel-wood and illegal logging. Providing effective environmental conservation education and alternative sources of energy would to a large extent minimize the problem. Further, effective law enforcement and increased patrols will ultimately reduce chances of people to encroach in the forests, thereby, increasing effectiveness in conserving the Masito-Ugalla ecosystem.

REFERENCES

1. Abdullah, S. M. S., & Mahmud, M. (2005). Tropical Deforestation and Its Impacts on Environment and Quality of Life. http://www.apn-gcr.org/images/programmes AndActivities/HyogoActivities/Hyogo5/Hyogo5_Mastura_Mahmud.pdf
2. Ahrends, A. (2005). Patterns of Degradation in Lowland Coastal Forests in Coast Region, Tanzania. MSc Thesis, Greifswald: Greifswald University.
3. Baboo, S., & Devi, M. R. (2011). Geometric Correction in Recent High Resolution Satellite Imagery: A Case Study in Coimbatore, Tamil Nadu. International Journal of Computer Applications, 14, 32-37. https://doi.org/10.5120/1808-2324

4. Bolton, R. N. (1993). Pretesting Questionnaires: Content Analyses of Respondents' Concurrent Verbal Protocols. Marketing Science, 12, 280-303. https://doi.org/10.1287/mksc.12.3.280
5. Burgess, R., Hansen, M., Olken, B., Potapov, P., & Sieber, S. (2012). Corruption and REDD + Identifying Risks amid Complexity.
6. Caelli, T., & Reye, D. (1993). On the Classification of Image Regions by Color, Texture and Shape. Pattern Recognition, 26, 461-470.
7. CIFOR (2014). Chapter 12: Building REDD Readiness in the Masito Ugalla Ecosystem Pilot Area in Support of Tanzania's National REDD Strategy. http://www.cifor.org/redd-case-book/case-reports/tanzania/
8. Cochrane, M. A., & Laurance, W. F. (2002). Fire as a Large-Scale Edge Effect in Amazonian Forests. Journal of Tropical Ecology, 18, 311-325. https://doi.org/10.1017/S0266467402002237
9. Deloitte (2012). Mid-Term Review Report of Nine NGO REDD + Pilot Projects in Tanzania. http://www.norway.go.tz/
10. FAO (2009). Towards Defining Degradation by Markku Simula (p. 11). FRA Working Paper 154, Rome: FAO.
11. FAO (2010). Global Forest Resources Assessment 2010—Main Report. Forestry Paper No. 163, Rome: FAO.
12. FAO (2011). Economic and Social Significance of Forests for Africa's Sustainable Development. Nature and Faune, 25.
13. FAO (2016). State of the World Forests. Rome: FAO. http://www.fao.org/publications/sofo/2016/en/
14. GAO (General Accounting Office) (1989). Content Analysis: A Methodology for Structuring and Analyzing Written Material. Washington DC: GAO. http://archive.gao.gov/d48t13/138426.pdf
15. Geist, H. J., & Lambin, E. F. (2002). Proximate Causes and Underlying Driving Forces of Tropical Deforestation. BioScience, 52, 143-150. https://doi.org/10.1641/0006-3568(2002)052[0143:PCAUDF]2.0.CO;2
16. Hansen, M. C., & DeFries, R. S. (2004). Detecting Long-Term Global Forest Change using Continuous Fields of Tree-Cover Maps from 8-km Advanced Very High Resolution Radiometer (AVHRR) Data for the Years 1982-99. Ecosystems, 7, 695-716. https://doi.org/10.1007/s10021-004-0243-3
17. Hogqvist, M., Nummelin, S., Stahl, M., & Miljoventen-skap, B. (1998). The View of Nature of Villagers in Kisarawe District. Karlstad: The Department of Nature and Environment, Karlstad University.
18. Hosonuma, N., Herold, M., De Sy, V., De Fries, R. S., Brockhaus, M., Verchot, L., Angelsen, A., & Romijn, E. (2012). An Assessment of Deforestation and Forest Degradation Drivers in Developing Countries. Environmental Research Letters, 7, 4009. https://doi.org/10.1088/1748-9326/7/4/044009
19. IPCC (2014). Climate Change 2014 Impacts, Adaptation, and Vulnerability: Summaries, Frequently Asked Questions, and Cross-Chapter Boxes. A Working Group II Contribution to the Fifth Assessment Report of the Intergovernmental Panel on Climate Change. https://www.ipcc.ch/pdf/assessment-report/ar5/wg2/WGIIAR5-IntegrationBrochure_FINAL.pdf
20. Johnson, R. L., & Chenje, M. (2008). Africa: Atlas of Our Changing Environment. Nairobi: United Nations Environment Programme (UNEP).
21. Kashaigili, J. J., Mdemu, M. V., Nduganda, A. R., & Mbilinyi, B. P. (2013). Integrated Assessment of Forest Cover Change and Above-Ground Carbon Stock in Pugu and Kazimzumbwi Forest Reserves, Tanzania. Advances in Remote Sensing, 2, 1-9. https://doi.org/10.4236/ars.2013.21001
22. Klein, J. (2002). Deforestation in the Madagascar Highlands-Established "Truth" and Scientific Uncertainty. Geo Journal, 56, 191-199. https://doi.org/10.1023/A:1025187422687
23. Kumar, G. R. P., Hemanjali, A. M., Ravikumar, P., Somashekar, R. K., & Nagaraja, B. C. (2010). Assessing the Historical Forest Encroachment of Kodagu Region of Western

Ghats, South India using Remote Sensing and GIS. https://www.nrsc.gov.in/sites/all/pdf/pram.pdf

24. Lanen, H. A. J., Tallaksen, L. M., & Rees, G., (2007). Droughts and Climate Change. http://www.geo.uio.no/edc/downloads/droughts_and_climate_change_2007.pdf
25. Madeira, C. E. (2008). Policies to Reduce Emissions from Deforestation and Degradation (REDD) in Developing Countries: An Examination of the Issues Facing the Incorporation of REDD into Market-based Climate Policies. http://www.rff.org/files/sharepoint/WorkImages/Download/RFF-Rpt-REDD_final.2.20.09.pdf
26. Malugu, I. O. E. (2007). Resources-Use Conflicts and Management Challenges for Pugu and Kazimzumbwi Forest Re-Serves in Kisarawe and Ilala Districts, Tanzania. Discovery and Innovation, 19, 149-174. https://doi.org/10.4314/dai.v19i1-2.15781
27. Mather, A. S. (1991). Global Forest Resources. Dehra Dun: International Book Distributors.
28. Misana, S. B., & Nyaki, F. R. (1993). Population and Deforestation in Tanzania. In D. Salaam, & United Nations, Population, Environment and Development in Tanzania (pp. 41-60). New York, NY: United Nations.
29. Mori, H. (2013). Review of Selected REDD + Project Designs. https://pub.iges.or.jp/system/files/publication_documents/pub/policyreport/3496/Booklet_09-04-2013.pdf
30. Mulhall, A. (2003). Methodological Issues in Nursing Research in the Field: Notes on Observation in Qualitative Research. Journal of Advanced Nursing, 41, 306-313. https://doi.org/10.1046/j.1365-2648.2003.02514.x
31. Murdiyarso, D., & Adiningsih, E. S. (2003). Causes of Indonesian Vegetation/Land Fires and Terrestrial Carbon Emissions. http://www.fire.uni-freiburg.de/summit-2003/3-IWFC/Papers/3-IWFC-109-Murdivarso.pdf
32. Ngalason, P. (2009). Building REDD Readiness in the Masito-Ugalla Ecosystem Pilot Area in Support of Tanzania's National REDD Strategy. http://www.norway.go.tz/
33. NORAD (2013). Implementing REDD+ in the Democratic Republic of Congo How to Manage the Risk of Corruption. A Report Commissioned by NORAD Prepared by PricewaterhouseCoopers (PwC) in Collaboration with NORAD and UNDP. http://www.redd-monitor.org/wp-content/uploads/2012/01/NORAD-PwC-Implementing-REDD-in-DRC-Corruption-risks-NORAD-Final-Version-E.pdf
34. Pickering, U., & Owen, G. (1997). Impact of Deforestation on Environment. New York, NY: Herper and Row Publications.
35. Pripathy, P., & Pripathy, K. P. (2017). Fundamentals of Research: Dissective View. https://books.google.com/books?isbn=3954894173
36. Reed, D. (1996). Structural Adjustment, the Environment, and Sustainable Development. London: Earthscan.
37. Relief (2003). Report on Buyungu Forest Reserve. Kibondo: REDESO Kibondo Office.
38. Rocchini, D., & Rita, A. D. (2005). Relief Effects on Aerial Photos Geometric Correction. Applied Geography, 25, 159-168.
39. Sanderson, M., Santini, M., Valentini, R., & Pope, E. (2010). Relationships between Forests and Weather. EC Directorate General of the Environment. http://ec.europa.eu/environment/forests/pdf/EU_Forests_annex1.pdf
40. Sands, R. (2005). Forestry in a Global Context. Wallingford: CABI Publishing.
41. Sitarz, D. (1994). Agenda 21: The Earth Summit Strategy to Save Our Planet. Boulder: Earthpress.
42. Stewart, D. W., & Shamdasani, P. (2017). Online Focus Groups. Journal of Advertising, 46, 48-60. https://doi.org/10.1080/00913367.2016.1252288
43. Su, G. S., Macawile, J., Villarino, A., Agapito, J., & Gomez, N. (2011). Recognizing Local People's Perceptions towards Deforestation in Quezon Province, Philippines. Environmental Research Journal, 5, 131-135. https://doi.org/10.3923/erj.2011.131.135
44. Svoboda, D., & McNamara, M. (2009). USDA Forest Service Technical Assistance Trip Watershed Assessment of the Masito-Ugalla Landscape. https://rmportal.

net/library/content/usda-forest-service/africa/usfs-watershed-assessment-of-the-masito-ugalla-landscape-1/at_download/file

45. TFWG (2009). Recommendations for the National Strategy on Reduced Emissions from Deforestation and Forest Degradation (REDD). http://www.tfcg.org /pdf/Brief%201%20Recommendations%20on%20REDD%20Strategy_final.pdf
46. Timberlake, L. (1991). Africa in Crisis: The Causes, the Cures of Environmental Bankruptcy (2nd ed.). London: Earthscan.
47. TRAFFIC (2006). Anti-Poaching Conference, 17 October 2006.
48. UNEP and CBD (2001). Main Theme: Forest Biological Diversity. https:// www.cbd.int /doc/meetings/sbstta/sbstta-07/information/sbstta-07-inf-03-en.pdf
49. URT (2014). Lake Tanganyika Integrated Regional Development Programme (PRODAP): Exit Strategy (Draft). http://www.vpo.go.tz/
50. Vicente-Serrano, S. M., Pérez-Cabello, F., & Lasanta, T. (2008). Assessment of Radiometric Correction Techniques in Analyzing Vegetation Variability and Change Using Time Series of Landsat Images. Remote Sensing of Environment, 112, 3916-3934.
51. Wu, Y. (2011). Investigation of Deforestation in East Africa on Regional Scales. Master's Thesis Geography, 30 HECs, Stockholm: Stockholm University.
52. WWF (2016). Importance of Forests. http://wwf.panda.org/ about_ our_ earth/deforestation/importance_forests/

CHAPTER 9

Species Diversity and Structure of an Intact Freshwater Swamp Forest in the Niger Delta

Nwabueze I. Igu

Department of Geography and Meteorology, Nnamdi Azikiwe University, Awka, Nigeria

ABSTRACT

Tropical forest ecosystems are generally seen as diverse ecosystems with rich biodiversity. However, how this applies to the freshwater swamp forests (especially in West Africa) is largely unknown. To verify how diverse this ecosystem is and how its structure varies in an intact forest landscape, one hectare forest plots were set up at 8 different points. This was used to collect information on the diversity and structure of the ecosystem. Consistent with the findings from other freshwater swamp forests, the ecosystem was seen to have a low species occurrence which ranged from 4 to 19 (mean value = 11 species) across the forest plots. Its diversity was equally low (mean = 1.66), unlike other tropical forest ecosystems. Stem heights varied as in mature tropical forest ecosystems; with the middle stratum recording the highest proportion of trees (54.63%) and the emergent layer having the least (0.83%). Its basal area, biomass and relative density were similar with other tropical ecosystems and equally had its highest species contribution from Leguminosae as in some other ecosystems as well. The ecosystem was seen to have features and characteristics that were common and similar with other tropical forest ecosystems, apart from its low diversity. Ensuring that effective and appropriate forest-tree species conservation measures are enhanced across the landscape are vital steps to securing the already existing (few) species and preventing species extinction across the ecosystem.

Keywords: Biodiversity, Conservation, Forest Disturbance, Forest Management, Species Richness, Swamp Forests

1. INTRODUCTION

Tree species in tropical forests are varied at different spatial scales and characterized with high diversity across the Tropics and Neotropics. This could be broadly estimated to be more than 100 species per hectare (Gentry, 1990 ; Huang et al., 2003), or specifically seen to range between 56 and 283 (>10 cm DBH) in mature tropical forests (Philips & Gentry, 1994) and 300 or more species of trees (≥10 cm DBH) per hectare in the Neotropics (Gentry, 1988 ; Valencia et al., 1994 ; Huang et al., 2003 ; Kelloff, 2008). The dominant species have been reported to vary from place to place; from Leguminosae in Neotropical lowland forests and African lowland forests, to Dipterocarps in Southeast Asia (Gentry, 1988 ; Whitmore, 1998 ; Huang et al., 2003). Though the African forests are composed of large tree biomass, its diversity has been reported to be relatively poor compared to higher diversities in Asian and American forests (Parmentier et al., 2007 ; Chuyong et al., 2011 ; Malhi et al., 2013). While this could be attributed to climate (mainly annual rainfall and rainfall seasonality) (Francis & Currie, 2003 ; Field et al., 2005 ; Parmentier et al., 2007), a good baseline is yet to be established for its diversity as only few systematic studies have been conducted in the region regarding the basic attributes such as biomass, species diversity and structure (Parmentier et al., 2007 ; Chuyong et al., 2011 ; Gourlet-Fleury et al., 2013 ; Malhi et al., 2013 ; Peh et al., 2014). While some studies in Nigeria have enumerated the diversity and species composition of different forest sites (Ihenyen et al., 2009 ; Ihuma et al., 2011 ; Adekunle et al., 2013 ; Ogbemudia et al., 2013 ; Udofia et al., 2014), studies that systematically showed the species diversity and structural baselines with particular reference to the freshwater swamp ecosystem are still much needed.

The freshwater swamp forests are moist forests found in freshwater environments. Though they are found in tropical rainforest regions, its species diversity is varied from that of the lowland forests (which supports a higher diversity of plant species) on account of constraints in dispersal, germination and establishment, due to its seasonal extremes (Lopez & Kursar, 2007 ; Corlett & Primack, 2011). This ecosystem is instead characterized by lower diversity and strong dominance of few tree species (Bacon, 1990 ; Richards, 1996 ; Koponen et al., 2004). Within the ecosystem, the flora is equally varied according to site conditions, inundation and floral history (Fickert & Grüninger, 2010), disturbance regimes, land use and water quality. Since the ecosystem offer veritable uses like groundwater recharge, water quality improvement, floral and faunal habitat and varieties of ecosystem services ranging from provisioning to regulatory services, understanding its diversity at different scales are important. With increased forest degradation, especially by selective logging and large scale human-induced changes that are altering and modifying the forest landscapes of the Niger Delta region, there is need to understand the diversity and structure of the freshwater swamp forest ecosystem. Forest degradation in Nigeria has accelerated due to recent development activities, population increase and agricultural expansion. Such degradations have not only fragmented the ecosystem and left it as forest mosaics and islands, but equally reduced its capacity to provide vital ecosystem functions. With a continuous trend in forest degradation continuing across the ecosystem, there is need to document its biodiversity and patterns. This work is aimed at understanding the species diversity of the ecosystem across intact forest landscapes, as well as its structural patterns. Understanding these patterns are important and useful for providing baseline information on community coexistence, stability and ecosystem functions that will be vital to underpin future management, use and abstraction for the ecosystem.

2. MATERIALS AND METHODS

2.1. Study Site

The study was carried out in Otuwe forest (Figure 1)—an intact forest location in the flood forest freshwater swamp forests of the Niger Delta. This forest land is owned by the Akarai-Obodo community and accessible mainly during the dry season due to poor accessibility (no access roads or waterway connecting it to the community or elsewhere). The community as well as the forest is located in Ndokwa East Local Government Area of Delta state, Nigeria (Figure 1). The forest which is named after a lake, is found in the Niger basin of Nigeria and is made up of marine sediments of the upper and lower cretaceous age. Being categorized broadly as hydromorphic soils (Areola, 1982), the soils are characterized by both the seasonally

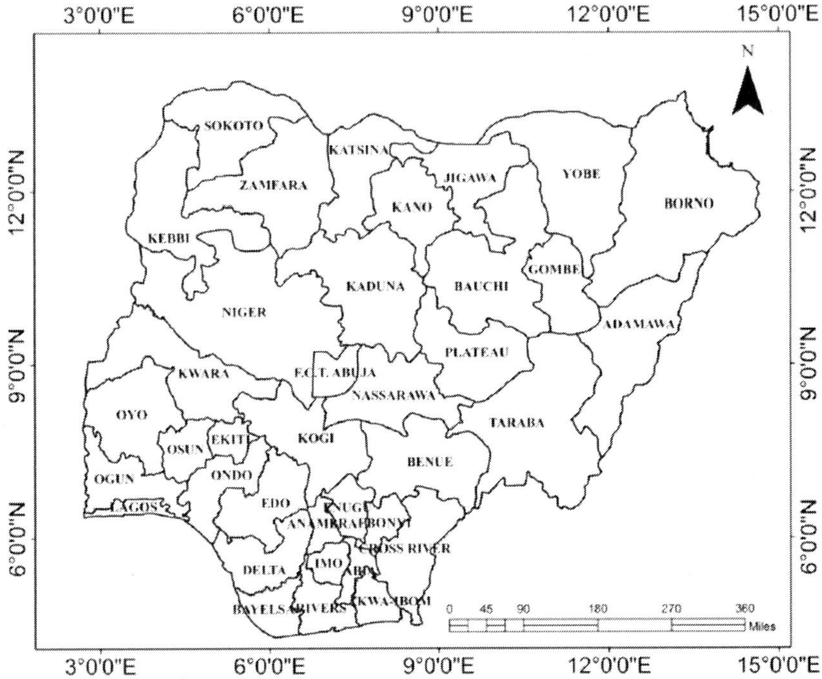
3°0'0"E
6°0'0"E
9°0'0"E
12°0'0"E
15°0'0"E
12°0'0"N
9°0'0"N
6°0'0"N
N
SOKOTO
KATSINA
ZAMFARA
JIGAWA
YOBE
KANO
BORNO
KEBBI
GOMBE
KADUNA
BAUCHI
NIGER
ADAMAWA
PLATEAU
KWARA
F.C.T. ABUJA
NASSARAWA
OYO
TARABA
OSUN
EKITI
KOGI
BENUE
OGUN
ONDO
LAGOS
EDO
CROSS RIVER
DELTA
IMO
RIVERS
0
45
90
180
270
360
Miles

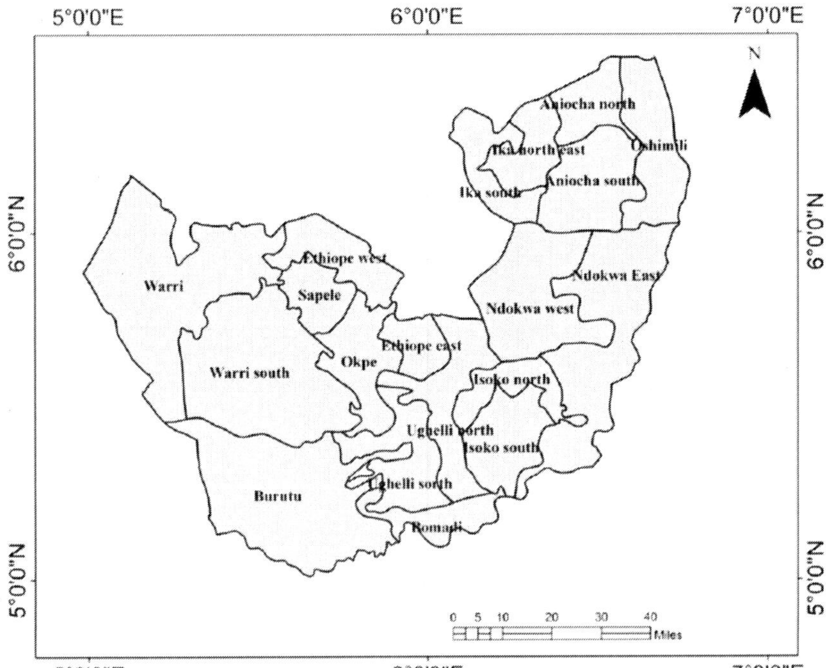
5°0'0"E
6°0'0"E
7°0'0"E
6°0'0"N
5°0'0"N
N
Aniocha north
Oshimili
Ika north east
Aniocha south
Ika south
Ethiope west
Warri
Ndokwa East
Sapele
Ndokwa west
Ethiope east
Okpe
Warri south
Isoko north
Isoko south
Burutu
0
5
10
20
30
40
Miles

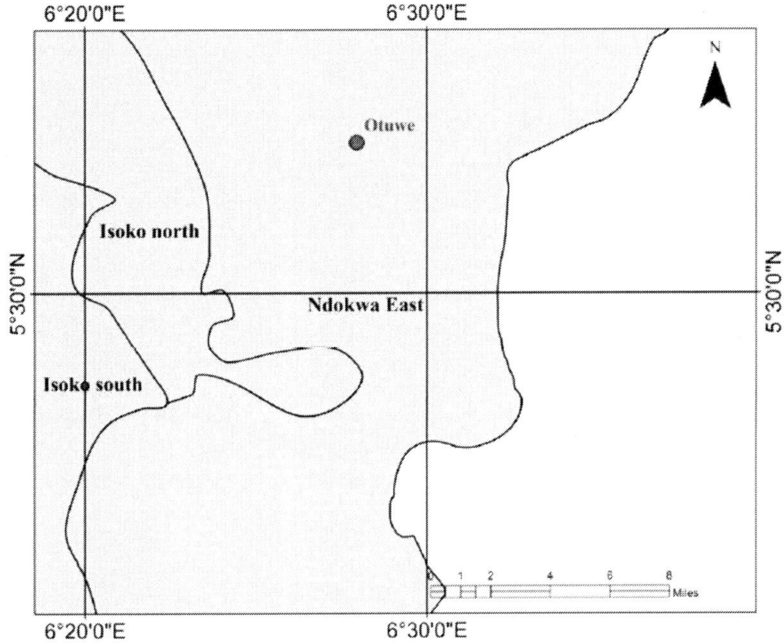

Figure 1. Map of the study site.

and permanently waterlogged soils; which are uniformly alluvial. The study region is a vast sedimentary region and flat area that is criss-crossed by a large number of meandering streams and creeks. It has a tropical climate with long rainy season, relative humidity which rarely dips below 60% and an average monthly maximum and minimum temperatures varying between 28°C to 33°C and 21°C to 23°C, respectively. The flood regime of the region begins toward the end of the rainy season in August, peaks in October, and tapers off in December.

The freshwater swamp forest of the Niger Delta is the most extensive and largest across Nigeria and West Africa (NDES, 1997 ; Spalding et al., 2010). It is divided into the seasonally flooded and permanently flooded zones. The ecosystem has transition zones with the lowland rainforest and mangrove ecosystems in the northern and southern parts of the ecosystem respectively. The region has been inhabited since about 5000 Years before present (YBP) (Nzewunwa, 1985) and much of the forest ecosystem have been exploited for timber production, industrial activities and domestic functions, mainly as sources of firewood and charcoal. Much of the forest in this region has been converted into plantation farms and overtaken by urbanization and industrial activities—mainly crude oil exploration. The ecosystem is threatened by salt water intrusion from the canalization and crude oil prospecting activities in the region, dredging and land reclamation, and population pressure.

2.2. Vegetation Data Collection

Standard plot-based method, which involves a one-time census of all stems ≥ 10 cm in diameter using a 1 ha square shaped (Phillips et al. 2003; Newton, 2007) was employed for this work between December 2013 and April 2014. Each of the 1 hectare square plots were established on transects that were 1 km wide apart, and contained plots of 100 × 100 m each at the interval of 500 m between each of the plots. The interval between the plots and transects were to ensure that there is a sizeable distance to enable floristic variation, yet maintaining consistency. All stems ≥10 cm in diameter at breast height (130 cm) were analysed and smaller ones omitted in order to ensure that species identification had minimal bias. Stem heights was determined using a laser rangefinder and where the tree height was less than 10 meters, graduated poles was used in measuring the height. Tree species were identified in the field by a trained taxonomist from the Forestry Research Institute, Ibadan (FRIN), while the specimen for the unidentified ones were collected and verified at FRIN herbarium. Specie identification followed the taxonomy of Nigerian plants (Keay, 1989).

2.3. Data Analyses

2.3.1. Forest Structure Analyses

In order to characterize the forest sites, the importance values of the species (species importance values) and the families (family importance values) were calculated after Cottam and Curtis (1956) , Mori et al. (1983) and Husch et al. (2003) as follows:

$$\text{Relative Density} = 100 \times \frac{\text{Number of stems of a species}}{\text{Total number of stems}} \quad (1)$$

$$\text{Relative frequency} = 100 \times \frac{\text{Frequency of a species}}{\text{Sum of all species}} \quad (2)$$

$$\text{Relative dominance} = 100 \times \frac{\text{Total basal area of a species}}{\text{Total basal area of all species}} \quad (3)$$

$$\text{Species importance values (SIV)} = \text{Relative density} + \text{Relative frequency} + \text{Relative dominance} \quad (4)$$

$$\text{Relative diversity} = 100 \times \frac{\text{Number of species in a family}}{\text{Total number of species}} \quad (5)$$

$$\text{Family importance values (FIV)} = \text{Relative density} + \text{Relative dominance} + \text{Relative diversity} \quad (6)$$

Dominance was defined as species or families with ecological values ≥ 10 for this study.

2.3.2. Basal Area

The basal area was calculated as follows:

$$BA = \left(\frac{dbh}{2}\right)^2 \times \pi \quad (7)$$

where *BA* is the basal area (m^2); dbh is the diameter at breast height (cm) and π as pie (3.142).

The stem sizes were categorized into the following forest structural analysis: small (10 - 20 cm dbh); medium (21 - 50); large (51 - 100) and largest (>100) (Adekunle et al., 2013). Heights of the trees were categorized into classes using an interval of 5 (≤5, 6 - 10, 11 - 15, ...) and used to plot a graph to depict the vertical structural pattern of the forest.

2.3.3. Species Diversity Analyses

The diversity indices were calculated following Kent and Coker (1992) , Magurran (1998) andMagurran (2004) as follows:

Shannon-Wiener index:

$$H' = -\sum_{i=1}^{s} pi \ln pi \quad (8)$$

where H^l is the Shannon-Weiner index, s is the total number of species, *pi* is the proportion of individuals in the ith species, and ln is the natural logarithm.

Pielou's evenness index:

$$E = \frac{\sum_{i=1}^{s} pi \ln pi}{\ln(S)} \quad (9)$$

Margalef's species richness index:

$$M = \frac{(S-1)}{\ln N} \tag{10}$$

where S is the total number of species in a community, N is the number of individuals and ln is the natural logarithm.

Pearson correlation was used to show the relationship that existed among the biodiversity indices using IBM SPSS software version 20.

3. RESULTS

3.1. Floristic Composition, Abundance and Dominance

This forest ecosystem is composed of 35 species within 28 genera and 18 families. The number of species found across the 8 plots in the forest site ranged from its lowest of 4 species found in plot 3 to the highest number (19) found in plot 4. The average abundance of the species across the forest was concentrated on four species: Celtis zenkeri Engl, Diospyros mespiliformis Hochst, Sterculia oblonga Mast and Sterculia rhinopetala K. Schum; which contributed 92.89% (Table 1) of the total species occurrence, while the remaining 7.10% were from the other 31 species in the forest. Among the 35 species present in the forest, the dominant species across the forest site were: Diospyros mespiliformis, Sterculia oblonga, Sterculia rhinopetala, Celtis zenkeri Engl and Erythrophleum ivorense A. Chev, with species importance values of 83.61, 66.29, 35.51, 26.32 and 13.31 respectively. The dominant families (family importance value) across the forest site were: Malvaceae, Ebenaceae, Leguminosae and Cannabaceae, with importance values of 106.93, 72.36, 40.51 and 20.75 respectively.

3.2. Species Diversity and Rarity

The species found in the ecosystem had a total of 2043 (stems) occurrences across the 8 plots. The pattern of species occurrence varied from the species that had > 100 stems (14.28%), >1 < 100 stems (62.85%), and those with only a single stem occurrence (22.85%). Leguminosae had the highest number (9) of species and was followed by Malvaceae (5); Moraceae (3); Annonaceae, Apocynaceae, Ebenaceae and Olacaceae (2 species each), while the remaining families were represented by one species each (Table 2).

The species diversity across the site was low: with the highest species diversity (2.13) seen in plot 1 and the least diversity of 0.98 in plot 3; and a mean value of 1.66 (Table 3). Pielou's evenness index ranged from 0.83 to

0.55 (Table 3), while the mean value across the site was 0.72. Margalef's species richness index ranged from 0.52 to 3.26; with a mean value of 1.85. Mean number of species recorded across the plots was 11 (which ranged from 4 to 19 species) (Table 3).

Table 1. Summary of the abundance and characterizing species across the sites.

Species	Ave. Abundance	Ave. Similarity	% Contribution	Cumulative %
Diospyros mesipiliformis	94.75	22.1753	45.3885	45.3885
Sterculia oblonga	36.25	10.6965	21.8937	67.2822
Sterculia rhinopetala	34.125	8.62638	17.6565	84.9387
Celtis zenkeri	39.5	3.8888	7.95961	92.8983

Table 2. Number of species per family across the forest site.

Family	Number of species
Annonaceae	2
Apocynaceae	2
Arecaceae	1
Cannabaceae	1
Chrysobalanaceae	1
Combretaceae	1
Ebenaceae	2
Gentianaceae	1
Lecythidaceae	1
Leguminosae	9
Malvaceae	5
Meliaceae	1
Moraceae	3
Myristicaceae	1
Olacaceae	2
Rhizophoraceae	1
Rubiaceae	1
Sapindaceae	1

Table 3. Biodiversity indices across the forest site.

Plot number	Margalef index	No of stems	Diversity	Evenness	Singleton	No. of species
1	2.47	128	2.13	0.83	3	13
2	1.43	264	1.78	0.81	2	9
3	0.52	304	0.98	0.7	0	4
4	3.26	250	1.88	0.64	7	19
5	2.16	409	1.44	0.55	4	14
6	1.34	183	1.47	0.71	2	8
7	1.91	310	1.85	0.75	1	12
8	1.71	193	1.76	0.77	2	10

The pattern of species rarity across the forest showed the highest value of 7 single (rare) species and a no occurrence value (0) as its least. On the average, the pattern of rarity across the forest site was low (2.62).

The results of the Pearson correlation coefficient (significant at 0.05 levels) showed that species richness correlated with species diversity and rarity. However, the highest correlation (r value) of 0.78 was between species richness and species rarity, an average correlation of 0.48 between species richness and diversity, and no significance between species richness and evenness.

3.3. Forest Structure

The highest total basal area of 30,995.62 m^2 was contributed by Sterculia oblonga, followed by Diospyros mespiliformis (29,214.11 m^2) and Sterculia rhinopetala (10,397.52 m^2), respectively. The least basal area per hectare (0.0086 m^2) was contributed by Elaeis guineensis, while the highest basal area (stem size) across the forest was 448 cm (as seen in Celtis zenkeri).

Ebenaceae had the highest relative density (38.17) across the forest and was followed by Malvaceae (29.98). However, Malvaceae had the highest total basal area of 79,107.98 m^2 and was followed by Ebenaceae with 34,419.65 m^2. Annonaceae had the highest mean dbh of 118.9 cm, followed by Malvaceae (55.79 cm), while Arecaceae had the least mean dbh (10.5 cm). Sterculia oblonga had the highest dbh (dominant dbh) of 19,870.8 cm, while Elaeis gunieensis and Millettia thonningii had the least dbh (10.5 cm). Diospyros mespiliformis had the highest mean height of 837.81 m, the tallest single tree in the forest was 82.5 m (Cleistopholis patens Engl. & Diels) and the mean height for the entire forest was 15.61 m.

The stem height intervals had variations in the number of stems per class, as expected in mature forests. While stem heights between 11 - 15 cm recorded the highest number of stems (662), the least value of 2 was found in the highest stem interval (≥46 cm) (Figure 2). The highest proportion (54.63%) of the trees in the forest occupied the middle stratum (11 - 25 cm), followed by the lower stratum (≤10 cm) (30.76%) and upper stratum (26 - 40 cm) (13.76%). The emergent layer (>40 cm) occupied the least value (0.83%) (Figure 2).

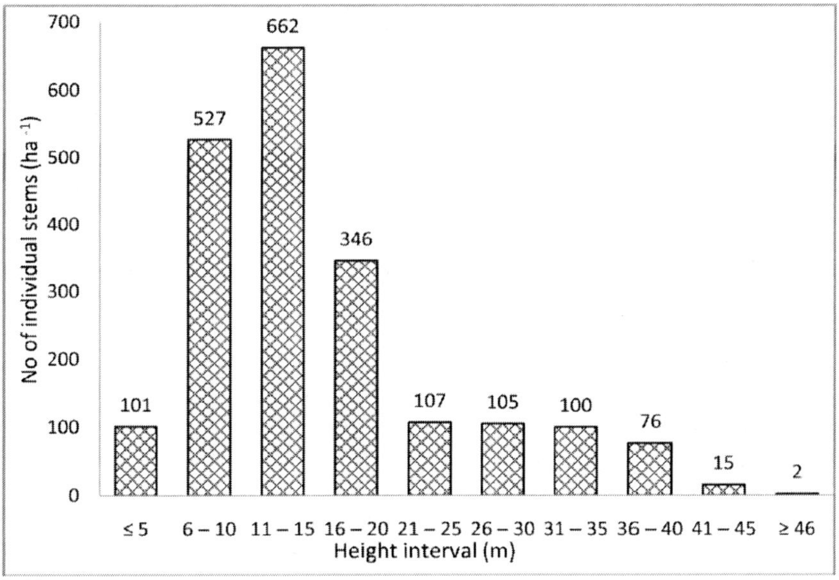

Figure 2. Height intervals of individual stems per hectare.

The pattern of distribution of the basal area of the forest had the highest values (dbh) from the lower class to the higher ones. The highest numbers of stems (1279) in the forest were found in dbh class of 10 - 20 cm, and reduced as the dbh classes increased consistently until the 91 - 100 cm dbh class (Figure 3). It however, experienced another increase in the number of stems for the > 100 cm dbh class (Figure 3). The number of stems in the forest was inversely proportional to the diameter sizes, shown by the reverse J (curve) pattern of the dbh interval distribution (Figure 3). This showed that the forest had a balanced species regeneration and recruitment.

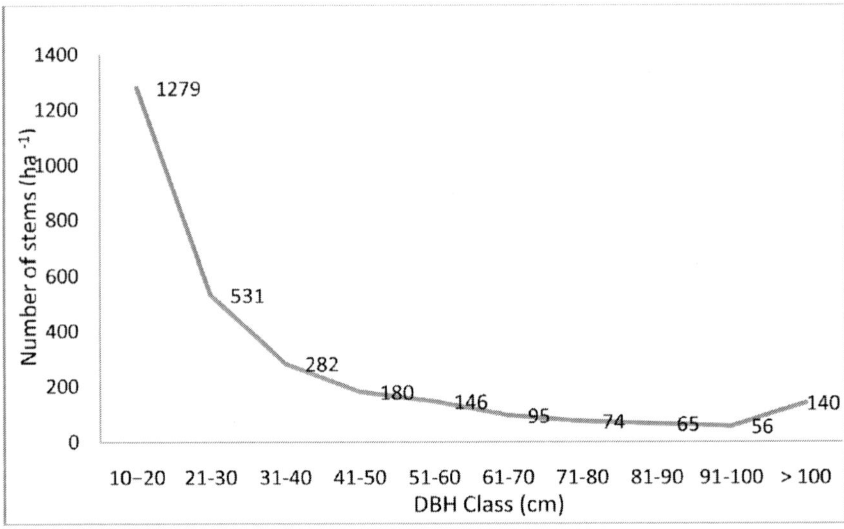

Figure 3. Diameter class of trees in the forest.

4. DISCUSSION

The freshwater swamp forest ecosystem in the Niger Delta is an ecologically important biome with unique biodiversity of regional and local importance. However, unlike other tropical forest ecosystems with higher species richness such as the Neotropics which records as much as 300 species per hectare (Gentry, 1988) and other African forests where the species richness is even as low as 60 species per hectare (Bernhard-reversat et al., 1978); it is much lower. Hence, while the forest recorded as much as 35 species across the entire forest plots, its abundance and dominance is concentrated only on four species. Though its mean species richness is far below other forest ecosystems both in Africa and across the Amazonia (Gentry, 1988 ; Phillips & Gentry, 1994 ;Richards, 1996 ; Ndah et al., 2013), it is similar with other freshwater swamp forests which are taken to be species poor as well (Scarano et al., 1997 ; Kurtz et al., 2013). While this could be generally attributed to the variations that exist in their abiotic environment, history and biogeography (Orians et al., 1996), it is equally as a result of the environmental constraints associated with the ecosystem. On the other hand, the diversity of the forest ecosystem was equally low, with a mean value that is far lesser than other tropical forest ecosystems (Knight 1975 ; Duran et al., 2006). This low diversity across the ecosystem, as in other swamp forests across the tropics, is not only due to the oxygen stress that plants in such flooded terrains go through, but also due to the fragmented nature of such landscapes (van Andel, 2003). Even though such disturbances and fragmentation of the ecosystem are expected to enable the gap dynamics that would facilitate a higher diversity across the landscape, the rare species that eventually thrive in the ecosystem are limited and quite few. This understanding that new species recruitment and establishment is constrained by the environment is not only beneficial for establishing guidelines and restraints targeted at protecting and conserving rare species, but also useful in promoting the functional diversity of the ecosystem.

The forest structure of the ecosystem was mainly influenced by the dominant species which had the higher basal area and relative density. With a high evenness, though with a low diversity, the forest's community structure was able to attain a mature and stable status; evidenced by the inverse J-shape. The forest had representative tree stands for each of the structural intervals or categories and a healthy juvenile/middle class tree population size (Figure 3), which is suitable for good recruitment and replacement for the mature and emergent populations. Such mature forests' with large trees are veritable assets in carbon sequestration for the region (Igu & Marchant, 2016) and useful in providing necessary ecosystem services that support the livelihood of its owners. Tree stems which ranged from 128 - 409 stems per hectare (with a mean of 255) across the forest plots were within the range of 245 - 467 stems per hectare, as reported for other tropical forests (Campbell et al., 1992). Though the mean number of tree stems per hectare was less than the average numbers of 323 trees per hectare (Aigbe et al., 2014), 387 stems per hectare (Adekunle et al., 2013) in tropical forest reserves in Nigeria, it was seen to exhibit an adequate population structure that had both a regenerating and climax population. Its lower tree stem numbers however, is attributed to the disturbances associated with the ecosystem (such as flooding) as well as the biogeography (shallow soil depth) of the region; which promotes tree falls and uprooting of tree stems across the forest ecosystem. Relative density varied across the ecosystem as would in mature forest ecosystems, according to the different families represented in the forest. Ebenaceae had the highest relative density among the families in the forest through its single genus Diospyros (and two species—mespiliformis and crassiflora). With other families such as Malvaceae and Leguminosae with five and nine species respectively, it shows that higher species number does not necessarily translate to a higher relative density in forest ecosystems. On the other hand, while it is likely that a higher frequency of occurrence of species, genus and family across an area (relative density) would translate to a higher total basal area, however, basal area of individual large trees which may be a bit fewer in number (as in the case of Malvaceae) are more likely to record (as much as double the size) a higher cumulative basal area.

5. CONCLUSION

Species diversity across the ecosystem was found to be low as in other freshwater ecosystems across the globe. As this is mainly as a result of the environmental constraints associated with the swamp (oxygen stress) which limits reproduction and adaptability of most of the plants, caution needs to be applied as it concerns logging in the ecosystem, so that the ecosystem will not become bare of trees or turned to shrubs. Ensuring that only mature trees are logged will help to guarantee a balanced regeneration and recruitment process across the ecosystem, the environmental constraints and low diversity notwithstanding. Such measures are equally needed across other tropical forest ecosystems which are lost at astronomical rates, so as to ensure that such landscapes are preserved and that its species do not become extinct.

REFERENCES

1. Adekunle, V. A. J., Olagoke, A. O., & Akindele, S. O. (2013). Tree Species Diversity and Structure of a Nigerian Strict Nature Reserve. Tropical Ecology, 54, 275-289.
2. Aigbe, H. I., Akindele, S. O., & Onyekwelu, J. C. (2014). Tree Species Diversity and Density Pattern in Afi River Forest Reserve, Nigeria. International Journal of Scientific and Technology Research, 3, 178-185.
3. Areola, O. (1982). Soils. In K. M. Barbour, J. S. Oguntoyinbo, J. O. C. Onyemelukwe, & J. C. Nwafor (Eds.), Nigeria in Maps. USA: Africana Publishing Company.
4. Bacon, P. R. (1990). Ecology and Management of Swamp Forests in the Guianas and Caribbean Region. In A. Lugo, M. Brinson, & Brown, S. (Eds.), Ecosystems of the world 15. Forested wetlands (pp. 213-250) Amsterdam: Elsevier.
5. Bernhard-reversat, F., Huttel, C., & Lemee, G. (1978). Structure and Functioning of Evergreen Rain Forest Ecosystem of the Ivory Coast. In Tropical Forest Ecosystems: A State-of-Knowledge Report (pp. 557-574). Paris: UNESCO.
6. Campbell, D. G., Stone, J. L., & Stone, A. (1992). A Comparison of Phytosociology and Dynamics of Three Floodplain (Várzea) Forests of Known Ages, Rio Juruá, Western Brazilian Amazon. Botanical Journal of the Linnaean Society, 108, 213-237. https://doi.org/10.1111/j.1095-8339.1992.tb00240.x
7. Chuyong, G. B., Kenfack, D., Harms, K. E., Thomas, D. W., Condit, R., & Comita, L. S. (2011). Habiat Specificity and Diversity of Tree Species in an African Wet Tropical Forest. Plant Ecology, 212, 1363-1374. https://doi.org/10.1007/s11258-011-9912-4
8. Corlett, R. T., & Primack, R. B. (2011). Tropical Rain Forests: An Ecological and Biogeographical Comparison (2nd ed.). Hoboken, NJ: Wiley-Blackwell. https://doi.org/10.1002/9781444392296
9. Cottam, G., & Curtis, J. T. (1956). The Use of Distance Measurements in Phytosociological Sampling. Ecology, 37, 451-460. https://doi.org/ 10.2307/1930167
10. Duran, E., Meave, J. A., Lott, D. J., & Segura, G. (2006). Structure and Tree Diversity Patterns at Landscape Level in a Mexican Tropical Deciduous Forest. Boletin de Sociedad Botanica de Mexico, 79, 43-60.

11. Fickert, T., & Grüninger, F. (2010). Floristic Zonation, Vegetation Structure, and Plant Diversity Patterns within a Caribbean Mangrove and Swamp Forest on the Bay Island of Utila (Honduras). Ecotropica, 16, 73-92.
12. Field, R., O'Brien, E. M., & Whittaker, R. J. (2005). Global Models for Predicting Woody Plant Richness from Climate: Development and Evaluation. Ecology, 86, 2263-2277. https://doi.org/10.1890/04-1910
13. Francis, A. P., & Currie, D. J. (2003). A Globally Consistent Richness-Climate Relationship for Angiosperms. The American Naturalist, 161, 523-536. https://doi.org/10.1086/368223
14. Gentry, A. H. (1988). Changes in Plant Community Diversity and Floristic Composition on Environmental and Geographical Gradients. Annals of the Missouri Botanical Garden, 75, 1-34. https://doi.org/10.2307/2399464
15. Gentry, A. H. (1990). Floristic Similarities and Differences between Southern Central America and Upper and Central Amazonia. In A. H. Gentry (Ed.), Four Neotropical Rainforests (pp. 141-157). New Haven, CT: Yale University Press.
16. Gourlet-Fleury, G., Beina, D., Fayolle, A., Ouédraogo, D. Y., Mortier, F., Bénédet, F., Closset-Kopp, D., & Decocq, G. (2013). Silvicultural Disturbance Has Little Impact on Tree Species Diversity in a Central African Moist Forest. Forest Ecology and Management, 304, 322-332.
17. Huang, W., Pohjonen, V., Johansson, S., Nashanda, M., Katigula, M. I. L., & Luukkanen, O. (2003). Species Diversity, Forest Structure and Species Composition in Tanzanian Tropical Forests. Forest Ecology and Management, 173, 11-24.
18. Husch, B., Beers, T. W., & Kershaw, J. A. (2003). Forest Mensuration. New York: Wiley.
19. Igu, N. I., & Marchant, R. (2016). Aboveground Carbon Storage in a Freshwater Swamp Forest in the Niger Delta. Carbon Management, 7, 137-148. https://doi.org/10.1080/17583004.2016.1165355
20. Ihenyen, J., Okoegwale, E. E., & Mensah, J. K. (2009). Composition of Tree Species in Ehor Forest Reserve, Edo State, Nigeria. Nature and Science, 7, 8-18.
21. Ihuma, J. O., Chima, U. D., & Chapman, H. M. (2011). Tree Species Diversity in a Nigerian Montane Forest Ecosystem and Adjacent Fragmented Forests. ARPN Journal of Agricultural and Biological Science, 6, 17-22.
22. Keay, R. W. J. (1989). Trees of Nigeria. Oxford, UK: Clarendon Press.
23. Kelloff, C. L. (2008). Structure and Diversity of a Riparian Forest at Kaieteur National Park, Guyana. Journal of the Botanical Research Institute of Texas, 2, 521-545.
24. Kent, M., & Coker, P. (1992). Vegetation Description Analysis. Chichester: Wiley.
25. Knight, D. H. (1975). A Phytosociological Analysis of Species Rich Tropical Forest on Barro Colorado Islands, Panama. Ecological Monographs, 45, 259-284. https://doi.org/10.2307/1942424
26. Koponen, P., Nygren, P., Sabatier, D., Rousteau, A., & Saur, E. (2004). Tree Species Diversity and Forest Structure in Relation to Microtopography in a Tropical Freshwater Swamp Forest in French Guiana. Plant Ecology, 173, 17-32. https://doi.org/10.1023/B:VEGE.0000026328.98628.b8
27. Kurtz, B. C., Gomes, J. C., & Scarano, F. B. (2013). Structure and Phytogeographic Relationships of Swamp Forests of Southeast Brazil. Acta Botanica Brasilica, 27, 647-660. https://doi.org/10.1590/S0102-33062013000400002
28. Lopez, O. R., & Kursar, T. A. (2007). Interannual Variation in Rainfall, Drought Stress and Seedling Mortality May Mediate Monodominance in Tropical Flooded Forests. Oecologia, 154, 35-43. https://doi.org/10.1007/s00442-007-0821-0
29. Magurran, A. E. (1998). Ecological Diversity and Its Measurement. Princeton, NJ: Princeton University Press.
30. Magurran, A. E. (2004). Measuring Biological Diversity. Oxford: Blackwell.
31. Malhi, Y., Adu-Bredu, S., Asare, R. A., Lewis, S. L., & Mayaux, P. (2013). African Rainforests: Past, Present and Future. Philosophical Transactions of Royal Society B., 368.

32. Mori, S. A., Boom, B. M., De Carvalino, A. M., & Dos Santos, T. S. (1983). Southern Bahian Moist Forest. The Botanical Review, 49, 155-232. https://doi.org/10.1007/BF02861011
33. Ndah, N. R., Andrew, E. E., & Bechem, E. (2013). Species Composition, Diversity and Distribution in a Disturbed Takamanda Rainforest, South West, Cameroon. African Journal of Plant science, 7, 577-585. https://doi.org/10.5897/AJPS2013.1107
34. NDES (1997). The Niger Delta Environmental Survey. Environmental and Socio-Economic Characteristics. Lagos, Nigeria: Environmental Resources Managers Limited.
35. Nzewunwa, N. (1985). Man's Early Interest in, and Use of the Mangrove Ecosystem: The Archaeological Perspective. In B. H. R. Wilcox, & C. P. Powell (Eds.), The Mangrove Ecosystem of the Niger Delta: Proceedings of a Workshop at the University of Port Harcourt. Port Harcourt.
36. Ogbemudia, F. O., Etukudo, N. I., & Ubom, R. M. (2013). Plants Diversity along Watershed Environment: A Case Study at Ikot Uso Akpan Wildlife Sanctuary in Itu L.G.A. of Akwa Ibom State, Nigeria. Annual Review and Research in Biology, 3, 551-563.
37. Orians, G. H., Dirzo, R., & Cushman, J. H. (Eds.) (1996). Biodiversity and Ecosystem Processes in Tropical Forests. In Ecological Studies (Vol. 122). Berlin: Springer-Verlag.
38. Parmentier, I., Malhi, Y., Senterre, B. et al. (2007). The Odd Man Out? Might Climate Explain the Lower Tree Alpha-Diversity of African Rain Forests Relative to Amazonian Rain Forests? Journal of Ecology, 95, 1058-1071. https://doi.org/10.1111/j.1365-2745.2007.01273.x
39. Peh, K. S. H., Sonke, B., Séné, O., Marie-Noel, K., Djuikouo, K., Nguembou, C. K., Taedoumg, H., Begne, S. K., & Lewis, S. L. (2014). Mixed-Forest Species Establishment in a Monodominant Forest in Central Africa: Implications for Tropical Forest Invasibility. PLoS ONE, 9, e97585. https://doi.org/ 10.1371 /journal.pone.0097585
40. Philips, O. L., & Gentry, A. H. (1994). Increasing Turnover through Time in Tropical Forests. Science, 263, 954-958. https://doi.org/10.1126/science.263.5149.954
41. Richards, P. W. (1996). The Tropical Rain Forest: An Ecological Study. Cambridge: Cambridge University Press.
42. Scarano, F. R., Ribeiro, K. T., De Moraes, L. F. D., & De Lima, H. C. (1997). Plant Establishment on Flooded and Unflooded Patches of a Freshwater Swamp Forest in Southeastern Brazil. Journal of Tropical Ecology, 13, 793-803. https://doi.org/10.1017/S0266467400011007
43. Spalding, M., Kainamu, M., & Collins, L. (2010). World Atlas of Mangroves (319 p). Earthscan.
44. Udofia, S. I., Owoh, P. W., Attah, V. I., & Thomas, A. D. (2014). Assessment of Plant Species Composition in Ayan Sacred Forest of Akwa Ibom State, Nigeria. Nigerian Journal of Agriculture, Food and Environment, 10, 34-37.
45. Valencia, R., Balslev, H., & Paz Y Mino, G. C. (1994). High Tree Diversity Alpha-Diversity in Amazonian Ecuador. Biodiversity & Conservation, 3, 21-28. https://doi.org/10.1007/BF00115330
46. Van Andel, T. R. (2003). Floristic Composition and Diversity of Three Swamp Forests in Northwest Guyana. Plant Ecology, 167, 293-317. https://doi.org /10.1023/A:1023935326706
47. Whitmore, T. C. (1998). An Introduction to Tropical Rain Forests (2nd ed.). New York: Oxford University Press.

CHAPTER 10

Preserving Ecosystem Services on Indigenous Territory through Restoration and Management of a Cultural Keystone Species

Yadav Uprety[1,†], Hugo Asselin[1,*] and Yves Bergeron[2]

[1]*Chaire de recherche du Canada en foresterie autochtone, Université du Québec en Abitibi-Témiscamingue, 445 boulevard de l'Université, Rouyn-Noranda, QC J9X 5E4, Canada*

[2]*Chaire Industrielle CRSNG-UQAT-UQÀM en aménagement forestier durable, Institut de recherche sur les forêts, Université du Québec en Abitibi-Témiscamingue, 445 boulevard de l'Université, Rouyn-Noranda, QC J9X 5E4, Canada*

ABSTRACT

Eastern white pine (*Pinus strobus* L.) is a cultural keystone tree species in the forests of eastern North America, providing numerous ecosystem services to Indigenous people. White pine abundance in the landscape has considerably decreased over the last few centuries due to overharvesting, suppression of surface fires, extensive management, and plantation failure. The Kitcisakik Algonquin community of western Quebec is calling for restoration and sustainable management of white pine on its ancestral territory, to ensure provision of associated ecosystem services. We present five white pine restoration and management scenarios taking into account community needs and ecological types: (1) natural regeneration of scattered white pines to produce individuals of different sizes and ages used as medicinal plants; (2) protection of supercanopy white pines used as landmarks and for providing habitat for flagship wildlife species, and younger individuals left as regeneration and future canopy trees; (3) the uniform shelterwood system to create white pine-dominated stands that provide habitat for flagship wildlife species and support cultural activities; (4) under-canopy plantations to yield mature white pine stands for timber production; (5) mixed plantations to produce forests with aesthetic qualities that provide wildlife habitat and protect biodiversity.

Keywords: Aboriginal people; ecological restoration; ecosystem services; *Pinus strobus* L.; shelterwood; sustainable forest management; traditional ecological knowledge; white pine

1. INTRODUCTION

Eastern white pine (*Pinus strobus* L.) is a highly valuable tree species providing several ecosystem services, notably to Indigenous people [1,2,3]. Once an important component of northeastern North American forests, white pine has greatly decreased in abundance over the last few centuries due to overharvesting, fire suppression, and other factors [4,5,6,7]. Marked reduction in white pine abundance, coupled with the rising demand for high-quality timber and its many ecological, social, and cultural values [8], has increased interest in the restoration and sustainable management of white pine in a variety of ecosystems. However, white pine restoration and management are challenging for many reasons: specific site requirements for regeneration, slow initial growth rate, susceptibility to damage from white pine blister rust (*Cronartium ribicola* J.C. Fisch.) and white pine weevil (*Pissodes strobi* Peck), heavy browsing by herbivores, and suppression of surface fires that naturally created safe sites for seedling germination and establishment [9,10,11,12].

Indigenous people possess in-depth knowledge of ecosystem functioning that should be integrated into forest restoration and management strategies in order to fulfill cultural needs [13,14,15]. Moreover, industrial forest operations on public lands prioritize timber production, thus threatening Indigenous people's cultural values and access to ecosystem services on their ancestral territory. For example, white pine is a cultural keystone species for the Kitcisakik Algonquin community of western Quebec, Canada [3]. The community is concerned by the reduced abundance of white pine on its ancestral territory and has been calling for a restoration and sustainable management strategy to increase the provision of ecosystem services associated to this species. In this paper, we took into account both the cultural setting [3] and ecological constraints [16] to develop five culturally and ecologically-sound white pine restoration and management scenarios for the Kitcisakik ancestral territory. Implementing these scenarios would only have a marginal effect on the economic return of forestry operations compared to current operations, while favoring increased provision of several ecosystem services for Indigenous people, including medicinal plants, timber, cultural sites, landmarks, wildlife habitats, aesthetic value, and biodiversity. This integrated approach contributes to a growing body of literature on the inclusion of sociocultural factors in forest management and planning (e.g., [6,17,18,19,20,21,22,23]). It will thus be of interest to researchers and forest managers working in a variety of ecological and cultural contexts.

2. MATERIALS AND METHODS

2.1. Study Area

The study area is the Kitcisakik ancestral territory (ca. 5000 km^2) located within the boundaries of the La Vérendrye Wildlife Reserve in western Quebec, less than 300 km north of Ottawa (Figure 1). This area corresponds to white pine's northern limit of continuous distribution in the balsam fir (*Abies balsamea* (L.) Miller)–yellow birch (*Betula alleghaniensis*Britton.) bioclimatic domain [24]. The species reaches its absolute northern distribution limit in western Quebec ca. 150 km to the north-west, in the Lake Duparquet and Lake Abitibi regions where only small, scattered stands and individual trees are present [25,26].

The average annual temperature in the study area is 1.2–3.3°C, and average precipitation is 914–1014 mm/year, with 22–33% falling as snow (Val-d'Or and Mont-Laurier weather stations, Environment Canada: http: //www.climate.weatheroffice.gc.ca/climate_normals). Mixed forest types are dominant, with balsam fir and yellow birch mostly accompanied by sugar maple (*Acer saccharum* Marsh.), red maple (*Acer rubrum* L.), trembling aspen (*Populus tremuloides* Michx.), paper birch (*Betula papyrifera* Marsh.), black spruce (*Picea mariana* (Mill.) BSP.), white spruce (*Picea glauca* (Moench) Voss), red pine (*Pinus resinosa* Ait.), jack pine (*Pinus banksiana* Lamb.), and white pine.

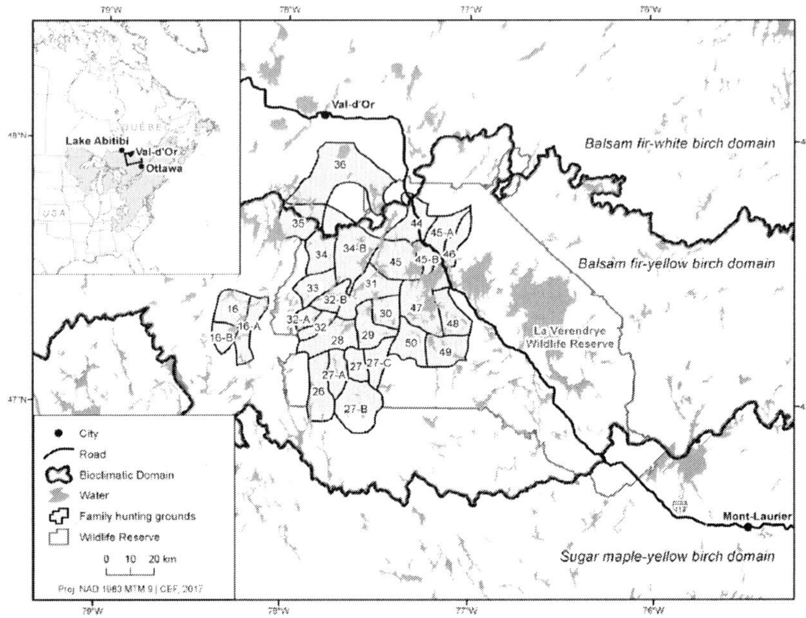

Figure 1. Location of the Kitcisakik ancestral territory in western Quebec. The inset shows the distribution of white pine in North America (after [27]).

Preindustrial forest dynamics were shaped by several disturbance factors including wildfire, spruce budworm outbreaks (*Choristoneura fumiferana* (Clem.)), windthrow [28], and selective logging [29]. Blister rust was introduced to North America in the beginning of the 20th century and extensive weevil damage was also first reported in the early 20th century, although the pest was first described in 1817. Crown fire was the main stand replacing disturbance controlling the landscape age structure in the preindustrial forests (fire cycle estimated at 257 years) [30]. The fire cycle of the industrial period has been estimated at more than 2000 years [30], and logging has replaced fire as the main disturbance in the study area.

2.2. The Kitcisakik Community

The Kitcisakik Algonquin community has ca. 480 members. Until the late 20th century, they maintained a lifestyle based on hunting, trapping, fishing, and gathering that was strongly dependent on the forest. Subsistence activities are still playing a key role in the community [31]. Young and old white pine trees and stands provide several ecosystem services to the Kitcisakik people and are a prominent part of cultural landscapes [3]. The Kitcisakik ancestral territory is on public land. Forest management is planned by the government (Ministry of Forests, Wildlife and Parks) and executed by licensed forestry companies. More than 60% of the productive forests on the Kitcisakik territory have been clear-cut over the last 40 years, following several decades of selective logging for large-diameter hardwoods and pines (white and red) [31]. Although the community is consulted during forest management planning, recurring requests for the implementation of restoration and management measures to increase white pine abundance have remained unaddressed.

2.3. Developing White Pine Restoration and Management Scenarios

This study was conducted in three phases. In the first phase, we compiled information obtained from two previous studies that investigated the cultural importance and traditional ecological knowledge related to white pine [3] and white pine natural regeneration dynamics [16] on this landscape. In the second phase, we reviewed the literature on white pine management within the northern portion of its natural range. In the third phase, we identified appropriate ecological types by their suitability for white pine restoration and management and developed culturally and ecologically-sound scenarios for the Kitcisakik territory. Forest inventory data have been used in Quebec since the 1980s to produce field guides to help foresters identify forest communities and ecological types [32]. The ecological type brings together all the cover types observed under similar environmental conditions (e.g., surficial deposit, drainage), linked among themselves by elements of forest dynamics [33] (Table A1). We briefly review the results of phases 1 and 2 (Section 3.1, Section 3.2 and Section 3.3) and then apply these findings as they relate to phase 3 (Section 3.4 and Section 3.5).

3. RESULTS AND DISCUSSION

3.1. Cultural Importance of White Pine to the Kitcisakik Algonquin

The ecosystem services that the Kitcisakik Algonquin derive from white pine [3] depend on the characteristics of trees or stands with which they are associated: (1) medicinal uses require scattered white pine individuals of different ages distributed evenly among the matrix of family hunting grounds; (2) geographic orientation on the land is facilitated by scattered supercanopy white pines that serve as landmarks, and which are also key habitat elements for some flagship wildlife species (e.g., bald eagle (*Haliaeetus leucocephalus* L.); black bear (*Ursus americanus* Pal.)); (3) cultural and spiritual activities (e.g., knowledge transmission, reconnection with oneself to overcome difficult times) are preferably held in pure white pine or white-pine dominated stands of natural origin, which are also key habitat for other flagship wildlife species (e.g., moose (*Alces americanus* Clin.); American marten (*Martes americana* Tur.); fisher (*Pekania penanti* Erxl.); wolverine (*Gulo gulo* L.)); (4) timber production, to be economically viable, requires a minimum volume per hectare which requires pure or white pine-dominated stands originating from natural regeneration or plantations; and (5) biodiversity protection requires mixed stands where white pine is a secondary or tertiary component, which are also valued aesthetically. Different management scenarios need to be developed to perpetuate these five sets of conditions to ensure sustained provision of ecosystem services by white pine.

3.2. Ecology of White Pine at Its Northern Limit of Continuous Distribution

In the study area, most white pine stands belong to mature age classes (70, 90, and >120 years) whereas younger stand age classes are scarce (regeneration, 30 and 50 years) [34]. White pine primarily occurs as scattered individuals in mixture with other tree species, typically black spruce, balsam fir, red maple, aspen, paper birch, and red pine. Natural white pine regeneration does occur in these stands, but in low amounts compared to the central part of the species' range [35]. Within this bioclimatic domain, recruitment success is strongly influenced by competitive or inhibitory effects of mature balsam fir, and moist substrates have a slightly positive effect on white pine regeneration [16].

Vegetative competition, blister rust, weevil, browsing (e.g., by snowshoe hare [*Lepus americanus* Erxl.], moose, or white-tailed deer [*Odocoileus virginianus* Zimm.]), seed predation by squirrels (*Tamiasciurus hudsonicus* Erxl.), and windthrow are potential damaging agents that can limit white pine establishment, growth, and survival. However, previous work indicated that these problems are less important in the study area than further south in the central part of white pine's range [16]. Blister rust and weevil damage only occurred in open areas (consistent with [36]) and were absent under partial to full canopy conditions producing moderate to deep shade. Blister rust could be problematic in hydric sites because of higher abundance of *Ribes* spp. (currant and gooseberries) that serve as alternative hosts for this pathogen [36,37]. Damage due to the white pine cone beetle (*Conophthorus coniperda* (Schwarz)) and Armillaria root disease were also reported in the Lake States [38,39,40,41], but not in the study area.

Competition with white pine regeneration by neighboring vegetation is more severe on nutrient rich, open sites [8,42,43], but the presence of an overstory canopy on such sites reduces understory light and the abundance of this competition [16]. Low browsing in the study area (reported in only 13% of the stands sampled by [16]) could mean that herbivore populations were low, or that preferred alternative food sources were available [44,45]. Seed predation was reported in 45% of the stands [16], but is probably not an important damaging agent, as mast seed years compensate for seed predation and minimize its influence on the potential for natural regeneration [35,46,47].

3.3. White Pine Restoration and Management Options

Natural regeneration and plantations can be used for white pine management and restoration. In either case, these efforts can benefit from: (1) proper site selection and planning; (2) knowledge of the local environment, including climate, soils, topography, vegetation, and animal populations; and (3) timely and application of appropriate silvicultural interventions [41].

3.3.1. Site Selection

Restoration efforts should focus on stands or sites known to have previously supported white pine [3,14]. In the absence of precise information regarding preindustrial distribution of white pine in the study area, an alternative strategy for site selection is to focus on ecological types where white pine is currently found. Ecological types are an ecological classification of the potential vegetation based on site conditions. Current presence of white pine in stands corresponding to a particular ecological type is an indication of this ecological type's suitability as a habitat for white pine. Although nutrient-rich, mesic sites have the greatest potential for white pine regeneration, these sites present a higher risk from competing vegetation and *Ribes* occurrence [41,48]. Therefore, dry to mesic sites and sites having poor to medium nutrient content have previously been suggested for white pine restoration [43,49].

More open canopies promote vegetation competition, and blister rust and weevil damage. Therefore, the best silvicultural practice is to manage white pine under an existing overstory [36,50,51]. Well-drained and well-aerated south-facing sites should be preferred. Such more open sites favor rapid evaporation of morning dew creating unsuitable conditions for blister rust.

Small canopy openings are more susceptible to blister rust, as they retain moisture [52,53,54]. Hence, topographic conditions and locations that favor persistent dew formation during cool, windless nights should be avoided [55]. Hollows or damp depressions should thus be avoided, as well as lower slopes, especially those with a northern exposure, small valleys or small openings surrounded by mature stands, and sites with dense vegetation where *Ribes* form large colonies [55]. A high abundance of broadleaf forbs, grasses, ferns, shrubs, and other woody vegetation typically develops rapidly after harvest disturbance where site preparation is not applied. Dense understory vegetation layers are known to limit air circulation and trap cool night air, thereby promoting high relative humidity and increased dew formation [56].

Intermediate understory light conditions (33–65%) had low amounts of blister rust and weevil damage in the study area, while allowing adequate stocking of white pine regeneration [16]. Near 50% light condition is an oft-suggested level for keeping balance between protection against pests and adequate growth [51,57,58,59,60]. Short-term volume losses of white pine grown under shaded conditions are compensated by long-term gain, as white pine grows taller and larger over time than other tree species present in the study area.

Although *Ribes* spp. can threaten restoration success, their abundance within a stand is not always correlated with local rust damage because the spores can be dispersed over long distances [61]. Past efforts to control *Ribes* mostly focused on eradication but were difficult, costly, and ineffective [37]. However, removal of *Ribes* from plantation sites and their close proximity could reduce the hazard [62]. New tactics emphasize biocides, biological agents, and use of silviculture to reduce *Ribes* regeneration [40]. Such methods are costly and are typically used only in high-value plantations aimed at timber production.

3.3.2. Site Preparation

Site preparation is used to enhance pine regeneration, creating a favorable seedbed by clearing competitive vegetation and exposing the mineral soil [41]. This was historically done by recurrent surface fires that have mostly been eliminated by fire suppression policies in several jurisdictions including Quebec. Site preparation can be done mechanically, with chemicals or prescribed burning (e.g., [51]), the latter two methods being prohibited in Quebec (but see [63] for a thorough discussion of the acceptability of such practices in Indigenous contexts). If natural regeneration is the selected strategy for white pine establishment, site preparation must coincide with the occurrence of mast seed years to maximize stocking.

3.3.3. Shelterwood System for White Pine Management

The uniform shelterwood system is an effective management tool for regenerating white pine in pine-dominated stands where adequate seed trees are present [12,50,64]. This silvicultural system emulates the surface fires that historically favored white pine regeneration by clearing the understory and some of the canopy trees, while retaining part of the canopy, including seed-producing canopy trees [65,66]. Residual canopy trees provide seeds for regeneration, as well as overstory protection to favor white pine establishment [59]. In this system, partial harvests (two to four passes) are done prior to final or near-complete overstory removal, depending on stand age [12,43,50,59] (Table 1).

Table 1. Step-by-step description of the uniform shelterwood system for white pine regeneration (after [59]).

Steps	Implications
Preparatory cut	Is used to improve the vigor of prospective seed-bearing trees. Low vigor trees are harvested while larger, healthy trees are retained.
Regeneration cut	Retains the largest, healthiest trees in the stand to be seed sources and to create conditions favorable for white pine establishment, survival, and growth, and limiting to blister rust and weevil damage. Additional trees are kept for wildlife habitat such as live cavity, mast, and supercanopy trees.
First removal cut	Applied to stands that have sufficient regeneration (at least 30 cm in height) in the understory to form a new white pine stand in the future. Some of the residual trees are harvested mimicking the eventual death of some trees after a natural disturbance such as a fire. Stands may be opened so that 50% crown closure remains after this cut. This creates conditions that reduce blister rust and weevil damage in white pine seedlings.
Final removal cut	Applied when white pine regeneration is about 3 m in height. Some parent trees (usually 10–20 per ha) are retained for ecological (e.g., veterans) and habitat (e.g., mast, supercanopy, and cavity trees) value.

Stands with a white pine basal area greater than 12 m^2/ha and a low component of red pine, spruce, or hardwood species can be managed under the shelterwood system [51]. The selection of seed trees should be based on the following criteria: trees in the dominant or co-dominant crown class; disease-free with clear, straight boles; well-formed crowns, with fine branching; and signs of good growth as evidenced by small, tight bark-flakes and good coverage of foliage on branches [50].

Assuming that a regeneration cut is properly timed to coincide with a mast year (for white pine, good seed crops occur every 3 to 10 years in the northern part of its range [27,35]), sites should be prepared using mechanical scarification to create suitable seedbed conditions. Where mast seed years do not coincide with site preparation, underplanting of white pine nursery stock will be needed to ensure the desired initial stocking is attained.

3.3.4. Underplanting White Pine

Underplanting white pine in hardwood, mixedwood, or conifer stands provides moderate shade needed to reduce blister rust and weevil damage with little growth loss. However, heavy shading can reduce growth, vigor, and survival of white pine [58]. The stands should thus be thinned initially if canopy is too dense to create adequate light for survival and competitive growth of underplanted pine. Some form of canopy needs to be retained until pines are >5 m tall for protection against weevil [12]. Underplanting in balsam fir stands should be avoided, as balsam fir shows inhibitory effects on white pine regeneration [16]. In aspen stands, the thinning level should be low, as the species can produce vigorous root suckers after logging [67]. Underplanting following strip cutting in deciduous stands (e.g., birch, aspen, maple, but also eastern larch [*Larix laricina* (Du Roi) K. Koch]) can be problematic, as such sites are warmer and drier in the spring, prior to leaf growth, and are thus more prone to weevil attack at the time when adult insects are more active [68].

3.3.5. Mixed Plantations

Mixed plantations (i.e., planting more than one species, or planting one species into an area with natural regeneration of other species) could be an option in open or partial canopy conditions and are the preferred restoration option of the Kitcisakik community [3] because they favor biodiversity [69]. The species to be planted and their proportion relative to white pine should be decided carefully. Mixedwood (conifer + hardwood) plantations are recommended, keeping the white pine proportion low (i.e., <33%). White pine can compete with thin-crowned species such as paper birch, but not with aspen or maple [70]. In a mixed Norway spruce (*Picea abies* (L.) H. Kars.)-white pine plantation trial (one-third white pine), weevil preferentially attacked Norway spruce [55]. Interestingly, weevil attacks did not severely affect spruce wood quality [55], hence maintaining good potential from a forestry perspective. Furthermore, in such plantations, commercially valuable spruce can be harvested during the first thinning operation [55].

3.3.6. Pure White Pine Plantations

Planting white pine in a clearcut or in an open field requires weighing the potential for greater tree growth against the threat of increased blister rust and weevil damage [41]. Pure white pine plantations in open conditions will require more intensive silviculture than in shelterwood, underplantating, or mixed plantations. An important practice in pure plantations is to maintain high densities (i.e., 2000–3000 stems/ha) until trees reach about 6 m in height [36]. This practice greatly improves white pine stem wood quality and growth [51]. Moreover, higher density creates competition and forces rapid height growth with minimal terminal diameter growth unfavorable to weevil. In addition, high density causes natural mortality of lower branches, which favors rust control [36]. Retaining more white pine trees at the time of pre-commercial thinning and also at harvest could compensate for later blister rust mortality and increase the likelihood of maintaining resistant trees on site [71].

Restoration in high quality sites that are usually prone to vegetation competition requires application of silvicultural techniques to promote regeneration establishment and growth [72]. Restoration efforts on clearcut sites should also focus on early management of understory vegetation and the gradual reduction of overtopping cover from woody vegetation [8,73]. As white pine grows relatively slow for the first five years, early control of competing vegetation is critical [43,59]. Competing vegetation control favors white pine seedling growth by increasing resource availability, while decreasing the risk of blister rust attack by altering the microclimate. Thinning can also be done to remove competition from overtopping aspen and birch [10]. Planting of large nursery stock (~50 cm) could be an option to increase the competitive potential of white pine.

Most fatal blister rust cankers occur in the lower portions of the trees, saplings, and seedlings, thus repeated pruning of lower branches can considerably reduce the likelihood of lethal cankers [36,53,74]. If the rate of infection is above 8% when the trees reach their sixth year, systematic pruning of lower branches is recommended [75]. According to [76], pruning should be performed yearly until trees reach 4.9 m height. Furthermore, pruning 2.7–5.2 m of the lower stem and maintaining 66% of the live crown of small white pine trees reduces the likelihood of lethal infection [41].

Although browsing impact is not important in our study area [16], damage severity could change with animal population cycles, availability of alternative food sources, or height of the surrounding vegetation [9,45,77]. Furthermore, while hunting controls herbivore populations in the study area, high population levels in other jurisdictions can inhibit pine regeneration. Browsing of terminal shoots and branches causes forked stems, growth loss, and seedling mortality, consequently delaying canopy recruitment, and decreasing wood quality [12,44,51].

3.4. White Pine Restoration and Management Scenarios

We propose five restoration and management scenarios to increase or maintain the provision of each of the five categories of ecosystem services provided by white pine, based on site conditions and white pine autecology (Table 2). There were a total of 179,489 productive forest stands (mapped polygons) in the study area according to the fourth decadal forest inventory of the Quebec Ministry of Forests, Wildlife and Parks, of which only 4507 (2.5%) had white pine as first, second, or third species in dominance importance. However, the corresponding ecological types represented 168,858 stands (94% of all stands), thus showing enormous potential for white pine restoration and management on Kitcisakik territory.

Table 2. White pine restoration and management framework for the Kitcisakik Algonquin territory.

Ecosystem Services Categories	White Pine Requirements	Restoration and Management Scenario	Scenario Number
Medicinal uses	Scattered young and mature trees	Protection of current young and mature trees and natural regeneration of new trees	I
Orientation on the land and wildlife habitat	Scattered supercanopy trees	Protection of current supercanopy trees and natural regeneration of new trees	II
Connection with the land and wildlife habitat	Natural white pine-dominated stands	Conservation of current natural stands and natural regeneration of new stands through uniform shelterwood	III
Timber production	Pure or white pine-dominated stands	Underplanting or pure/mixed plantations	IV
Biodiversity and aesthetics	Mixed stands	Natural regeneration or mixed plantations	V

To develop the restoration and management scenarios, we focused on relatively abundant ecological types (i.e., comprising >100 stands) for which white pine is currently found in ≥1% of the stands of the study area. Furthermore, we excluded ecological types with subhydric or hydric drainage, as they are susceptible to blister rust. The final list of 16 ecological types comprised 101,124 stands representing 56% of the productive forest stands (Table 3 and Table A1). Following ethical best practices [78], the scenarios were discussed with—and approved by—the Kitcisakik Land Department.

Table 3. Ecological types and suggested scenarios for white pine (WP) restoration and management.

Ecological Types *	No. of Stands	No. of Stands with WP as 1st, 2nd, or 3rd Component			Total Stands with WP	% Stands with WP	Scenarios **
		1st	2nd	3rd			
MJ10	296	20	2	19	41	13.85	I, V
MJ12	13,253	88	45	246	379	2.85	I, II, V
MJ20	1500	22	24	16	62	4.13	I, V
MJ21	4459	58	47	131	236	5.19	I, II, V
MJ22	39,958	547	265	718	1530	3.82	I, II, V
MS21	4856	83	35	121	239	4.92	I, II, V
MS22	5490	21	22	19	62	1.13	II, V
RP10	222	100	61	4	165	74.32	III
RP11	273	149	48	11	208	76.19	III
RP12	406	251	61	19	331	81.52	III
RS10	226	5	21	0	26	11.50	I, IV, V
RS11	302	1	22	0	23	7.61	I, II, IV, V
RS12	2901	22	124	0	146	5.03	II, IV
RS20	5520	9	85	0	94	1.70	I, IV, V
RS21	9086	107	102	5	214	2.35	I, II, IV, V
RS22	12,376	49	196	0	245	1.97	II, IV

* See Table A1 for definition of ecological types. ** See Table 2.

3.4.1. Scenario I: Scattered White Pine Individuals of All Ages

People from Kitcisakik use naturally-grown young, mature, and old white pine trees for medicinal purposes [3]. Ecological types that support white pine as a minor component could be targeted for this scenario (Table 3), as only scattered pines are needed. Hence, all that is needed is to protect current white pines and to make sure natural regeneration occurs and is let free to grow. Minimal silvicultural intervention is required. Because mature and old trees have special medicinal properties after having been struck by lightning [3], a special attention should be given to retaining such trees.

3.4.2. Scenario II: Scattered Supercanopy White Pines

Scattered supercanopy white pine trees provide landmarks and are a key habitat element for flagship species such as bald eagle and black bear. Therefore, in addition to protecting current supercanopy trees, naturally regenerated white pine seedlings or saplings should be protected and left to grow so that future landmark trees will be present on all family hunting grounds. As for scenario I, minimal silvicultural intervention is required. Ecological types that support white pine as a minor component could be targeted.

3.4.3. Scenario III: Mature, White Pine-Dominated Stands of Natural Origin

Mature white pine dominated stands are important for cultural activities and wildlife habitat and should be actively managed for natural regeneration, as plantations are less favorable for wildlife [79], and are negatively perceived by Indigenous people [80].

The uniform shelterwood system could be used to maintain the presence of white pine-dominated stands on the landscape. Three of the 16 selected ecological types are dominated by white and red pines, and white pine is currently present in 74–82% of these ecological types (Table 3). As these ecological types are spread evenly on the territory, they provide an opportunity to maintain white pine-dominated stands in all family hunting grounds, as well as near settlements, thus providing equal access to ecosystem services among Indigenous families [15,81].

3.4.4. Scenario IV: Pure or White Pine-Dominated Stands from Plantations

Pure or white pine-dominated stands are required for timber production. These can be obtained/maintained using either of two approaches depending on ecological type. One approach is to keep some of the canopy and underplant pine, while the other is to clearcut and plant pine. Undercanopy white pine plantations can be applied in ecological types with spruce (*Picea* spp.), or eastern white cedar (*Thuja occidentalis* L.) as major components (Table 3). Both approaches will achieve the same objective of timber production.

3.4.5. Scenario V: Mixed Stands where White Pine is a Secondary or Tertiary Component

Mixed stands are valued for aesthetic purposes, as habitat for wildlife, and for their potential to preserve biodiversity. Mixed plantations can be established in all ecological types that support white pine where timber production through plantation establishment is not an attractive option given site characteristics.

3.5. Spatial Distribution of the Scenarios

Ecological types supporting all scenarios are fairly evenly distributed among family hunting grounds, thus meeting a request by the Kitcisakik people that benefits and disadvantages of forest management should be equitably distributed [82]. Ecological types supporting scenarios I and II are relatively abundant and represent 44% and 52%, respectively, of productive forest stands on Kitcisakik territory. Preservation of scattered individuals (for medicinal uses or as supercanopy landmark trees) should thus not be problematic and would likely have minimal impact on the profitability of forest management operations, as they do not require silvicultural interventions. Scenario V (mixed plantations) would be possible on 47% of productive forest stands. While this shows interesting potential, it could nevertheless prove difficult to implement, as the Quebec forest industry favors pure plantations. Scenarios III and IV would require more intensive and more frequent silvicultural interventions and should thus be preferentially realized near the settlement, where opportunities currently exist, but only in a very limited number of stands. Scenario III is currently limited to less than 1% of productive stands, underlining the importance of protecting the remaining natural white pine stands and of facilitating the regeneration of new ones. About 17% of productive stands are suitable for scenario IV, providing opportunity for restoration and management of white pine stands for timber production. Options for white pine restoration and management are more limited in the northern and eastern parts of Kitcisakik territory, where balsam fir stands are abundant and where suitable ecological types are less frequent (Figure 2).

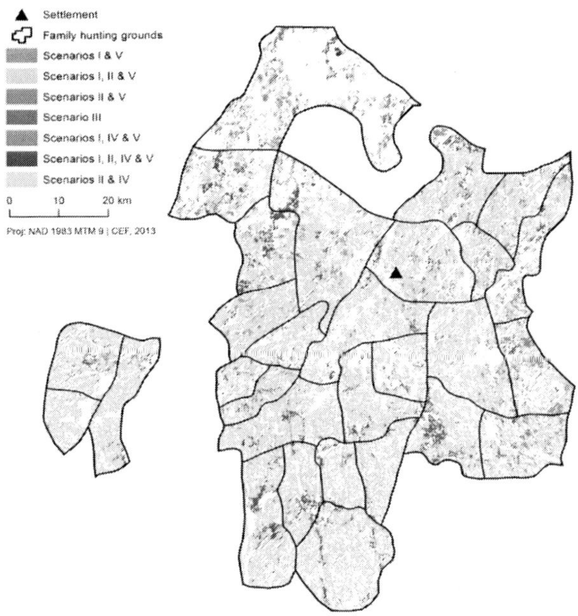

Figure 2. Spatial distribution of the ecological types supporting white pine restoration and management scenarios on Kitcisakik territory. See Figure 1 for details on the location of the territory, and Table 2 for details about the scenarios.

4. CONCLUSIONS

This paper addressed a request from the Kitcisakik Algonquin community to develop management recommendations for white pine on their ancestral territory. The aim is not to actively manage for white pine dominated stands everywhere on the territory for industrial purposes, but rather to maintain or increase the provision of white pine-associated ecosystem services to the Kitcisakik community. The five proposed scenarios would have low impact on profitability of the forest industry, as they require minimal silvicultural intervention (I and II), or would take place on a very small portion of the total productive forest stands (III, IV, and V). Moreover, the proposed scenarios correspond to the philosophy behind ecosystem-based forest management [83,84,85], as the shelterwood system emulates surface fires and scattered white pines in the landscape are maintained by small-scale canopy gaps. Ecosystem-based forest management has been favored on public lands of Quebec since 2010. Pure plantations, in this context, could not be considered to be emulating natural processes, but could be justified at a small scale and near settlements to favor wood production. Scenarios III and IV could be labor intensive and costly. However, if done at a small scale in a cultural context, people from Kitcisakik could contribute to site selection, preparation, management, and monitoring [14]. The responsibility of some of the restoration and management operations on family hunting grounds could be given to community members once guidelines and training are provided. Such community-based approaches have been shown to be efficient, to have increased legitimacy, and to be more sustainable [86].

Indigenous peoples' participation, and recognition and inclusion of their knowledge into restoration and management projects can contribute to building a strong partnership for successful implementation that significantly improves social acceptability, economic feasibility, and ecological viability of restoration projects [14,87,88]. Therefore, a shift from "just another stakeholder" to "shared decision makers" [89] is possible. The approach presented here, where restoration and management scenarios take into account cultural needs and ecological constraints, could find wide application in diverse forest settings, as it could help meet the objectives of certification standards (e.g., Forest Stewardship Council—FSC) with regards to the rights and needs of Indigenous people.

ACKNOWLEDGMENTS

Thanks to Mélanie Desrochers (Center for Forest Studies, Université du Québec à Montréal) for helping with Figure 1. The Chaire Desjardins en développement des petites collectivités, the Réseau de recherche et de connaissances relatives aux peuples autochtones (DIALOG), and the Centre de recherche sur la gouvernance des ressources naturelles et du territoire (CRGRNT) provided financial support for this project through scholarships to Y.U.

APPENDIX A

Table A1. Signification of inventory codes for ecological types, following the classification of the Quebec Ministry of Forests, Wildlife and Parks.

Code	Ecological Type
MJ10	Yellow birch, balsam fir, sugar maple on very thin deposits of varied texture with a xeric to hydric drainage
MJ12	Yellow birch, balsam fir, sugar maple on thin to thick medium deposits with a mesic drainage
MJ20	Yellow birch, balsam fir on very thin deposits of varied texture with a xeric to hydric drainage
MJ21	Yellow birch, balsam fir on thin to thick coarse deposits with a xeric to mesic drainage
MJ22	Yellow birch, balsam fir on thin to thick medium deposits with a mesic drainage
MS21	Balsam fir, paper birch on thin to thick coarse deposits with a xeric to mesic drainage
MS22	Balsam fir, paper birch on thin to thick medium deposits with a mesic drainage
RP10	White or red pine on very thin deposits of varied texture with a xeric to hydric drainage
RP11	White or red pine on thin to thick coarse deposits with a xeric to mesic drainage
RP12	White or red pine on thin to thick medium deposits with a mesic drainage
RS10	Balsam fir, eastern white cedar, on very thin deposits of varied texture with a xeric to hydric drainage
RS11	Balsam fir, eastern white cedar, on thin to thick coarse deposits with a xeric to mesic drainage
RS12	Balsam fir, eastern white cedar, on thin to thick medium deposits with a mesic drainage
RS20	Balsam fir, black spruce on very thin deposits of varied texture with a xeric to hydric drainage
RS21	Balsam fir, black spruce on thin to thick coarse deposits with a xeric to mesic drainage
RS22	Balsam fir, black spruce on thin to thick medium deposits with a mesic drainage

REFERENCES

1. Rogers, L.L.; Lindquist, E.L. Supercanopy white pine and wildlife. Proceedings of The White Pine Symposium: History, Ecology, Policy and Management, Duluth, MN, USA, 16–18 September 1992; Stine, R.A., Baughman, M.J., Eds.; Department of Forestry Resources, University of Minnesota: St. Paul, MN, USA, 1992; pp. 39–43
2. Schroeder, H.W. The tree of peace: symbolic and spiritual values of the white pine. Proceedings of The White Pine Symposium: History, Ecology, Policy and Management, Duluth, MN, USA, 16–18 September 1992; Stine, R.A., Baughman, M.J.,

Eds.; Department of Forestry Resources, University of Minnesota: St. Paul, MN, USA, 1992; pp. 73–83

3. Uprety, Y.; Asselin, H.; Bergeron, Y. Cultural importance of white pine (*Pinus strobus* L.) to the Kitcisakik Algonquin community of western Quebec, Canada. *Can. J. For. Res.* 2013, *43*, 544–551
4. Abrams, M.D. Eastern white pine versatility in the presettlement forest. *BioScience* 2001, *51*, 967–979
5. Thompson, I.D.; Simard, J.H.; Titman, R.D. Historical changes in white pine (*Pinus strobus* L.) density in Algonquin Park, Ontario, during the 19th century. *Nat. Areas J.* 2006, *26*, 61–71
6. Steen-Adams, M.M.; Mladenoff, D.J.; Langston, N.C.; Liu, F.; Zhu, J. Influence of biophysical factors and differences in Ojibwe reservation versus Euro-American social histories on forest landscape change in northern Wisconsin, USA. *Landsc. Ecol.* 2011, *26*, 1165–1178
7. Danneyrolles, V.; Arseneault, D.; Bergeron, Y. Pre-industrial landscape composition patterns and post-industrial changes at the temperate-boreal forest interface in western Quebec, Canada. *J. Veg. Sci.* 2016, *27*, 470–481
8. Pitt, D.G.; Morneault, A.; Parker, W.C.; Stinson, A.; Lanteigne, L. The effects of herbaceous and woody competition on planted white pine in a clearcut site. *For. Ecol. Manag.* 2009, *257*, 1281–1291
9. Tester, J.R.; Starfield, A.M.; Frelich, L.E. Modeling for ecosystem management in Minnesota pine forest. *Biol. Conserv.* 1997, *80*, 313–324
10. Burgess, D.; Robinson, C.; Wetzel, S. Eastern white pine response to release 30 years after partial harvesting in pine mixed wood forest. *For. Ecol. Manag.* 2005, *209*, 117–129
11. Steen-Adams, M.M.; Langston, N.; Mladenoff, D.J. White pine in the northern forests: An ecological and management history of white pine on the Bad River Reservation of Wisconsin. *Environ. Hist.* 2007, *12*, 614–648
12. Latremouille, C.; Parker, W.C.; McPherson, S.; Pinto, F.; Fox, B.; McKinnon, L. *Ecology and Management of White Pine in the Lake Abitibi (3E) and Lake Temagami (4E) Ecoregions of Ontario*; Science Development and Transfer Series No. 004; Ministry of Natural Resources: South Porcupine, ON, Canada, 2008
13. Parrotta, J.A.; Trosper, R.L. (Eds.) *Traditional Forest-Related Knowledge: Sustaining Communities, Ecosystems and Biocultural Diversity*; World Forest Series vol. 12; Springer: Dordrecht, The Netherlands, 2012
14. Uprety, Y.; Asselin, H.; Bergeron, Y.; Doyon, F.; Boucher, J.-F. Contribution of traditional knowledge to ecological restoration: Practices and applications. *Ecoscience* 2012, *19*, 225–237
15. Asselin, H. Indigenous forest knowledge. In *Routledge Handbook of Forest Ecology*; Peh, K., Corlett, R., Bergeron, Y., Eds.; Routledge: New York, NY, USA, 2015; pp. 586–596
16. Uprety, Y.; Asselin, H.; Bergeron, Y.; Mazerolle, M.J. White pine (*Pinus strobus* L.) regeneration dynamics at the species' northern limit of continuous distribution. *New For.* 2014, *45*, 131–147
17. Jiao, Y.M.; Li, X.Z.; Liang, L.H.; Takeuchi, K.; Okuro, T.; Zhang, D.D.; Sun, L.F. Indigenous ecological knowledge and natural resource management in the cultural landscape of China's Hani Terraces. *Ecol. Res.* 2012, *27*, 247–263
18. Mason, L.; White, G.; Morishima, G.; Alvarado, E.; Andrew, L.; Clark, F.; Durglo, M.; Durglo, J.; Eneas, J.; Erickson, J.; et al. Listening and learning from traditional knowledge and western science: A dialogue on contemporary challenges of forest health and wildfire. *J. For.* 2012, *110*, 187–193
19. Emery, M.R.; Wrobel, A.; Hansen, M.H.; Dockry, M.; Moser, W.K.; Stark, K.J.; Gilbert, J.H. Using traditional ecological knowledge as a basis for targeted forest inventories: Paper birch (*Betula papyrifera*) in the US Great Lakes region. *J. For.* 2014, *112*, 207–214

20. Armatas, C.; Venn, T.; McBride, B.; Watson, A.; Carver, S. Opportunities to utilize traditional phenological knowledge to support adaptive management of social-ecological systems vulnerable to changes in climate and fire regimes. *Ecol. Soc.* 2016, *21*
21. Tendeng, B.; Asselin, H.; Imbeau, L. Moose (*Alces americanus*) habitat suitability in temperate deciduous forests based on Algonquin traditional knowledge and on a habitat suitability index. *Ecoscience* 2016, *23*, 77–87
22. Lawler, J.H.; Bullock, R.C. A case for Indigenous community forestry. *J. For.* 2017, *115*, 117–125
23. Reo, N.J.; Whyte, K.P.; McGregor, D.; Smith, M.A.; Jenkins, J.F. Factors that support Indigenous involvement in multi-actor environmental stewardship. *AlterNative* 2017, *13*
24. Saucier, J.-P.; Bergeron, J.-P.; Grondin, P.; Robitaille, A. Les régions écologiques du Québec méridional (troisième version). *L'Aubelle* 1998, *124*, S1–S12
25. Bergeron, Y.; Leduc, A.; Li, T. Explaining the distribution of *Pinus* spp. in a Canadian boreal insular landscape. *J. Veg. Sci.* 1997, *8*, 37–44
26. Engelmark, O.; Bergeron, Y.; Flannigan, M.D. Age structure of eastern white pine, *Pinus strobus* L., at its northern distribution limit in Quebec. *Can. Field Nat.* 2000, *114*, 601–604
27. Wendel, G.W.; Smith, H.C. *Pinus strobus* L., Eastern white pine. In *Silvics of North America*; Agriculture Handbook 654; Conifers; Burns, R.M., Honkala, B.H., Eds.; U.S.D.A. Forest Service: Washington, DC, USA, 1990; Volume 1
28. Boucher, Y.; Bouchard, M.; Grondin, P.; Tardif, P. *Le registre des états de référence : intégration des connaissances sur la structure, la composition et la dynamique des paysages forestiers naturels du Québec méridional. Mémoire de recherche forestière n° 161*; Direction de la recherche forestière, Ministère des Ressources naturelles et de la Faune, Gouvernement du Québec: Quebec City, QC, Canada, 2011
29. Asselin, M. L'Abitibi-Témiscamingue: Trois sous-région. In *Histoire de l'Abitibi-Témiscamingue*; Vincent, O., Ed.; Institut québécois de recherche sur la culture: Quebec City, QC, Canada, 1995; pp. 21–65
30. Bergeron, Y.; Cyr, D.; Drever, C.D.; Flannigan, M.; Gauthier, S.; Kneeshaw, D.; Lauzon, È.; Leduc, A.; Le Goff, H.; Lesieur, D.; et al. Past, current, and future fire frequencies in Quebec's commercial forests: implications for the cumulative effects of harvesting and fire on age-class structure and natural disturbance-based management. *Can. J. For. Res.* 2006, *36*, 2737–2744
31. Saint-Arnaud, M.; Asselin, H.; Dubé, C.; Croteau, Y.; Papatie, C. Developing criteria and indicators for aboriginal forestry: Mutual learning through collaborative research. In *Changing the Culture of Forestry in Canada: Building Effective Institutions for Aboriginal Engagement in Sustainable Forest Management*; Stevenson, M.G., Natcher, D.C., Eds.; Canadian Circumpolar Institute Press: Edmonton, AB, Canada, 2009; pp. 85–105
32. Blouin, J.; Berger, J.-P. *Guide de reconnaissance des types écologiques de la région écologique 5b coteaux du réservoir Gouin*; Ministère des Ressources Naturelles, Forêt Québec, Direction des inventaires forestiers, Division de la classification écologique et productivité des stations: Quebec City, QC, Canada, 2001
33. Grondin, P.; Saucier, J.-P.; Blouin, J.; Gosselin, J.; Robitaille, A. *Information écologique et planification forestières au Québec, Canada*; Note de recherche forestière 118; Ministère des Ressources naturelles, de la Faune et des Parcs, Direction de la recherche forestière: Quebec City, QC, Canada, 2003
34. Delisle, S. Modèle prédictif d'occurrence du pin blanc sur le territoire des Anicinapek de Kitcisakik. Ph.D. Thèse, Faculté de foresterie, de géographie et de géomatique, Université Laval, Quebec City, QC, Canada, 2011
35. Parker, W.C.; Noland, T.L. ; Morneault. A.E. Comparative mast seed production in unmanaged and shelterwood white pine (*Pinus strobus* L.) stands in central Ontario. *New For.* 2013, *44*, 613–628

36. Katovich, S.; Mielke, M. *How to Manage Eastern White Pine to Minimize Damage from Blister Rust and White Pine Weevil*; NAFR-01-93; USDA Forest Service: Newtown Square, PA, USA, 1993
37. Zambino, P.J. Biology and pathology of *Ribes* and their implications for management of white pine blister rust. *For. Pathol.* 2010, *40*, 264–291
38. Katovich, S.A.; O'Brien, J.G.; Mielke, M.E.; Ostry, M.E. Restoration and management of eastern white pine within high blister rust hazard zones in the Lake States. In *Silviculture in Special Places*; Sheppard, W.D., Askew, L.G., Eds.; RMRS-P-34; USDA Forest Service: Fort Collins, CO, USA, 2004; pp. 135–145
39. Zenner, E.K.; Puettmann, K.J.; Krueger, J.A. Early growth responses of naturally regenerated eastern white pine (*Pinus strobus* L.) to partial release from juvenile aspen and pathological pruning. *North J. Appl. For.* 2005, *22*, 27–34
40. Hunt, R.S.; Geils, B.W.; Hummer, K.E. White pines, *Ribes*, and blister rust: Integration and action. *For. Pathol.* 2010, *40*, 402–417
41. Ostry, M.E.; Laflamme, G.; Katovich, S.A. Silvicultural approaches for management of eastern white pine to minimize impacts of damaging agents. *For. Pathol.* 2010, *40*, 332–340
42. Gillespie, A.R.; Hocker, H.W. The influence of competition on individual white pine thinning response. *Can. J. For. Res.* 1986, *16*, 1355–1359. [Google Scholar] [CrossRef]
43. Burgess, D.; Wetzel, S. Nutrient availability and regeneration response after partial cutting and site preparation in eastern white pine. *For. Ecol. Manag.* 2000, *138*, 249–262
44. Pastor, J. Browsing damage to white pine. Proceedings of The White Pine Symposium: History, Ecology, Policy and Management, Duluth, MN, USA, 16–18 September 1992; Stine, R.A., Baughman, M.J., Eds.; Department of Forestry Resources, University of Minnesota: St. Paul, MN, USA, 1992; pp. 137–144
45. Saunders, M.R.; Puettmann, K.J. Use of vegetational characteristics and browsing patterns to predict deer damage in eastern white pine (*Pinus strobus*) plantations. *North. J. Appl. For.* 1999, *16*, 96–102
46. Smith, C.C. The coevolution of pine squirrels (*Tamiasciurus*) and conifers. *Ecol. Monogr.* 1970, *40*, 349–371
47. Gurnell, J. Squirrel numbers and the abundance of tree seeds. *Mammal Rev.* 1983, *13*, 133–148
48. Stiell, W.M. Characteristics of eastern white pine and red pine. In Proceedings of the White and Red Pine Symposium, Chalk River, ON, Canada, 20–22 September 1977; Cameron, D.A., Ed.; Department of Environment, Canadian Forest Service: Sault Ste. Marie, ON, Canada, 1978; pp. 7–51
49. Kotar, J. Managing white pine: Finding the proper niche. Proceedings of The White Pine Symposium: History, Ecology, Policy and Management, Duluth, MN, USA, 16–18 September 1992; Stine, R.A., Baughman, M.J., Eds.; Department of Forestry Resources, University of Minnesota: St. Paul, MN, USA, 1992; pp. 150–156
50. Pinto, F. Silvicultural practices in Ontario's white pine forests. Proceedings of The White Pine Symposium: History, Ecology, Policy and Management, Duluth, MN, USA, 16–18 September 1992; Stine, R.A., Baughman, M.J., Eds.; Department of Forestry Resources, University of Minnesota: St. Paul, MN, USA, 1992; pp. 170–178
51. Ontario Ministry of Natural Resources (OMNR). *A Silvicultural Guide for the Great Lakes-St Lawrence Conifer Forest in Ontario*; Queens's Printer: Ottawa, ON, Canada, 1998
52. Van Arsdel, E.P. Environment in relation to white pine blister rust infection. In *Biology of Rust Resistance in Forest Trees: Proceedings of a NATO-IUFRO Advanced Study Institute, 17–24 August, 1969*; Hoff, R.J., McDonald, G.I., Eds.; Misc. Pub. No. 1221; USDA Forest Service: Washington, DC, USA, 1969; pp. 479–493
53. French, D.W. White pine blister rust can be controlled. Proceedings of The White Pine Symposium: History, Ecology, Policy and Management, Duluth, MN, USA, 16–18 September 1992; Stine, R.A., Baughman, M.J., Eds.; Department of Forestry Resources, University of Minnesota: St. Paul, MN, USA, 1992; pp. 123–125

54. White, M.A.; Brown, T.N.; Host, G.E. Landscape analysis of risk for white pine blister rust in the Mixed Forest Province of Minnesota, U.S.A. *Can. J. For. Res.* 2002, *32*, 1639–1650
55. Coulombe, C.; Bélanger, G.; Lavallée, R.; Laflamme, G. *A Simple Effective Tool for Controlling White Pine Weevil and Blister Rust on Norway Spruce and Eastern White Pine*; Natural Resources Canada, Canadian Forest Service, Laurentian Forestry Centre: Sainte-Foy, QC, Canada, 2004
56. Hodge, J.C.; Juzwik, J.; Gross, H.L.; Retnarkan, A. *White Pine Blister Rust and White Pine Weevil Management: Guidelines for White Pine in Ontario*; Ontario Ministry of Natural Resources: Ottawa, ON, Canada, 1989
57. Logan, K.T. *Growth of Tree Seedlings as Affected by Light Intensity. II. Red Pine, White Pine, Jack Pine and Eastern Larch*; Tech. Note 121; Department of Forestry: Ottawa, ON, Canada, 1966
58. Messier, C.; Parent, S.; Chengaou, M.; Beaulieu, J. Juvenile growth and crown morphological plasticity of eastern white pine (*Pinus strobus* L.) planted along a natural light gradient: Results after six years. *For. Chron.* 1999, *75*, 275–279
59. Burgess, D.; Pinto, F.; Wetzel, S. *Some Management Implications from an Eastern White Pine Regeneration Experiment*; Technology Transfer Note 28; Forestry Research Applications Pacific Forestry Centre: Saanich, BC, Canada, 2002
60. Boucher, J.-F.; Bernier, P.Y.; Margolis, H.A.; Munson, A.D. Growth and physiological response of eastern white pine seedlings to partial cutting and site preparation. *For. Ecol. Manag.* 2007, *240*, 151–164
61. Van Arsdel, E.P.; Riker, A.J.; Kouba, T.F.; Suomi, V.E.; Bryson, R.A. *The Climatic Distribution of Blister Rust on White Pine in Wisconsin*; Stn. Pap. 87; U.S. Department of Agriculture, Forest Service, Lake States Forest Experiment Station: St. Paul, MN, USA, 1961
62. Robbins, K.; Jackson, W.A.; McRoberts, R.E. White pine blister rust in the eastern Upper Peninsula of Michigan. North. *J. Appl. For.* 1988, *5*, 263–264
63. Kayahara, G.J.; Armstrong, C.L. Understanding First Nations rights and perspectives on the use of herbicides in forestry: A case study from northeastern Ontario. *For. Chron.* 2015, *91*, 126–140
64. Lancaster, K.F.; Leak, W.B. *A Silvicultural Guide for White Pine in the Northeast*; Gen. Tech. Rep. NE-41.; U.S. Department of Agriculture, Forest Service, Northeastern Forest Experiment Station: Broomall, PA, USA, 1978
65. Heinselman, M.L. Fire in the virgin forests of the Boundary Waters Canoe Area, Minnesota. *Quat. Res.* 1973, *3*, 329–382
66. Quinby, P.A. Self-replacement in old-growth white pine forests of Temagami, Ontario. *For. Ecol. Manag.* 1991, *41*, 95–109
67. Stiell, W.M. *Seeding and Planting Red Pine and White Pine*; Tech. Note No. 80; Canada Department of Northern Affairs and National Resources, Forestry Branch Forest Research Division: Ottawa, ON, Canada, 1959
68. Stiell, W.M.; Berry, A.B. Limiting white pine weevil attacks by side shade. *For. Chron.* 1985, *61*, 5–9
69. Paquette, A.; Messier, C. The role of plantations in managing the world's forests in the Anthropocene. *Front. Ecol. Environ.* 2010, *8*, 27–34
70. Engle, L.G. *Releasing White Pine from Oak and Aspen*; Tech. Note No. 346; USDA Forest Service, Lake States Forest Experiment Station: St. Paul, MN, USA, 1951
71. Conklin, D.A.; Fairweather, M.L.; Ryerson, D.E.; Geils, B.W.; Vogler, D.R. *White Pines, Blister Rust and Management in the Southwest*; R3-FH-09-01; U.S. Department of Agriculture, Forest Service, Southwest Region: Albuquerque, NM, USA, 2009
72. Stiell, W.M.; Robinson, C.F.; Burgess, D. 20-year growth of white pine following commercial improvement cut in pine mixedwoods. *For. Chron.* 1994, *70*, 385–394
73. Pitt, D.G.; Meyer, T.; MacDonald, L.; Buscarini, T.; Thompson, D.G. Application of slow-release tablets to enhance white pine regeneration: Growth response and efficacy against white pine blister rust. *Can. J. For. Res.* 2006, *36*, 684–698

74. Lavallée, A. *The Spread of White Pine Blister Rust in Young White Pine Plantations*; Info. Rep. LAU-X-101; Canadian Forestry Service, Laurentian Forestry Centre: Quebec City, QC, Canada, 1992
75. Canadian Forest Service (CFS). *White Pine Blister Rust: Knowledge, Control and Innovation*; Canadian Forest Service, Laurentian Forestry Centre: Sainte-Foy, QC, Canada, 2012
76. Laflamme, G.; Blais, R.; Desgagné, D. Control of *Cronartium ribicola* in *Pinus strobus* plantations. In Proceedings of the 1st IUFRO Rusts of Forest Trees Working Party Conference, Saariselkä, Finland, 2–7 August 1998; Jalkanen, R., Crane, P.E., Walla, J.A., Aalto, T., Eds.; Finnish Forest Research Institute: Helsinki, Finland, 1998; pp. 299–303
77. Krebs, C.; Boutin, S.; Boonstra, R.; Sinclair, A.; Smith, J.; Dale, M.; Turkington, R. Impact of food and predation on the snowshoe hare cycle. *Science* 1995, *269*, 1112–1115
78. Asselin, H.; Basile, S. Éthique de la recherche avec les peuples autochtones: Qu'en pensent les principaux intéressés? *Éthique publique* 2012, *14*, 333–345
79. Roy, C.; Imbeau, L.; Mazerolle, M.J. Transformation of abandoned farm fields into coniferous plantations: Is there enough vegetation structure left to maintain winter habitat of snowshoe hares? *Can. J. Zool.* 2010, *88*, 579–588
80. Germain, R. Évaluation de l'acceptabilité sociale par la communauté algonquine de Pikogan d'une stratégie d'aménagement écosystémique. M.Sc. Thesis, Département des sciences appliquées, Université du Québec en Abitibi-Témiscamingue, Rouyn-Noranda, QC, Canada, 2012
81. Deutsch, N.; Davidson-Hunt, I. Pikangikum family hunting areas and traplines: Customary lands and aboriginal landuse planning in Ontario's Far North. In *Planning Co-Existence: Aboriginal Considerations and Approaches in Land Use Planning*; Stevenson, M.G., Natcher, D.C., Eds.; Canadian Circumpolar Institute Press: Edmonton, AB, Canada, 2010; pp. 149–170
82. Asselin, H.; Larouche, M.; Kneeshaw, D. Assessing forest management scenarios on an Aboriginal territory through simulation modeling. *For. Chron.* 2015, *91*, 426–435
83. Attiwill, P.M. The disturbance of forest ecosystems: The ecological basis for conservative management. *For. Ecol. Manag.* 1994, *63*, 247–300
84. Bergeron, Y.; Harvey, B.; Leduc, A.; Gauthier, S. Forest management guidelines based on natural disturbance dynamics: Stand-and forest-level considerations. *For. Chron.* 1999, *75*, 49–54
85. Drever, C.R.; Peterson, G.; Messier, C.; Bergeron, Y.; Flannigan, M. Can forest management based on natural disturbances maintain ecological resilience? *Can. J. For. Res.* 2006, *36*, 2285–2299
86. Ribot, J.C.; Agrawal, A.; Larson, A.M. Recentralizing while decentralizing: How national governments reappropriate forest resources. *World Dev.* 2006, *34*, 1864–1886
87. Garibaldi, A.; Turner, N. Cultural keystone species: Implications for ecological conservation and restoration. *Ecol. Soc.* 2004, *9*
88. Higgs, E.S. What is good ecological restoration? *Conserv. Biol.* 1997, *11*, 338–348
89. Stevenson, M.G.; Webb, J. Just another stakeholder? First Nations and sustainable forest. In *Towards Sustainable Management of the Boreal Forest*; Burton, P.J., Messier, C., Smith, D.W., Adamowicz, W.L., Eds.; NRC Research Press: Ottawa, ON, Canada, 2003; pp. 65–112

CHAPTER 11

Finding Alternatives to Swidden Agriculture: Does Agroforestry Improve Livelihood Options and Reduce Pressure on Existing Forest?

Syed Ajijur Rahman, Jette Bredahl Jacobsen, John Robert Healey, James M. Roshetko, Terry Sunderland

Department of Food and Resource Economics, Section of Environment and Natural Resources, University of Copenhagen, 1958 Frederiksberg, Denmark

School of Environment, Natural Resources and Geography, Bangor University, Bangor LL57 2UW, UK

Center for International Forestry Research (CIFOR), Bogor Barat 16115, Indonesia

Center for Macroecology, Evolution and Climate, University of Copenhagen, Universitetsparken 15, Building 3, 2100 Copenhagen, Denmark

World Agroforestry Centre (ICRAF), Bogor Barat 16115,Indonesia

ABSTRACT

Swidden cultivation can contribute to deforestation and land degradation, which can subsequently result in a number of serious environmental problems. This paper examines the economic and social potential of agroforestry systems and the barriers to their widespread adoption, as a land use alternative to swidden cultivation, which may potentially help protect local forest. The Gunung Salak valley in West Java, Indonesia is presented as a case study. Based on farmers' and experts' assessment, costs and benefits have been estimated, which show that the two investigated agroforestry systems have higher net present value and benefit-cost ratio (B/C) than the two swidden cultivation systems. Tree ownership also creates more permanent rights to farmland and is prestigious in the community. Agroforestry products (fruit, vegetables etc.) have high monetary value and help strengthen social cohesion when shared with neighbors. However, farmers are reluctant to implement agroforestry. Stated reasons are related to both culture and capacity. Farmers practicing agroforestry are less involved in forest clearing and forest products collection than swidden farmers indicating that it may contribute positively to conservation of local forests. Increasing the adoption of agroforestry farming in the study area will require support to overcome capacity constraints.

Keywords: Agroforestry adoption Income Social potential Forest protection Policy support

INTRODUCTION

Swidden agriculture, also known as slash-and-burn farming is a widespread subsistence practice in the tropics (Peng et al. 2014; Schuck et al. 2002). Swidden is mainly practiced in the mountainous and hilly parts of Latin America, Central Africa and Southeast Asia by smallholder farmers (Munthali 2013; Van et al. 2012), and often drives deforestation as well as forest degradation (Rahman et al. 2012; Styger et al. 2006).

Multiple terms are used to refer to swidden cultivation and related systems in the scientific literature: swidden agriculture, shifting cultivation, slash-and-burn farming, as well as regional terms *jhum* in South Asia or *ladang* in Indonesia (Van et al. 2012; Mertz et al. 2009; Imang et al. 2008). 'Swidden' was first proposed as a term by the Swedish anthropologist K.G. Izikovitz in 1951 in the sense of burning woody vegetation to clear land for agriculture (Peng et al. 2014; Russell 1988). 'Shifting cultivation' is often used more broadly to refer to agricultural activities where fields are cultivated for crop production for a number of years and then left fallow for a number of years (Vongvisouk et al. 2014; Therik 1999). However, others define it more narrowly to refer to systems in which the entire livelihoods of farmers are shifted with the cultivation within the forest landscape (Aweto 2013; Inoue 2000; Adimihardja 1992). Our focus is on swidden that does not necessarily refer to shifting fields but only to land cleared by burning (Peng et al. 2014; Marten 1986), as is the case in our research site in Gunung Salak.

In Gunung Salak valley, West Java, Indonesia swidden cultivation practices are deeply rooted in communities' culture and provide various subsistence products mostly to local poor farmers (Galudra et al. 2008). However, this system can have serious negative environmental consequences by contributing to deforestation and land degradation (Peng et al. 2014; Rahman et al. 2012; Barraclough and Ghimire 1995; Gupta 1993). The most severe environmental impacts occur in two ways, firstly, when the swidden cultivators clear forests to prepare land for cultivation and, secondly, from the forest clearing process fire can escape and burn uncontrolled in adjacent forest areas (Rahman et al. 2012; Mai 1999). Loss of forest cover and degradation of remaining forest can greatly increase the incidence of soil erosion in areas on steep slopes (Shoaib et al. 1998; Sfeir-Younis and Dragun 1993). Soil erosion and landslides have negative effects on a range of ecosystem services including food provisioning from agriculture in both uplands and lowlands, and can negatively affect farm families' standards of living (Rahman et al. 2012).

In order to overcome the negative consequences of swidden, farmers would need to adopt new practices that serve multiple purposes including conserving forest resources as well as producing food and supporting sustainable development (Leakey 2010; Roshetko et al. 2008; Sunderland et al. 1999). Agroforestry, and specifically the practice of growing trees on farmland alongside crops, has well-established research evidence of its potential to reduce deforestation and forest degradation at a landscape scale (Rahman et al. 2014; Idol et al. 2011; Garrity 2004). One definition of agroforestry is 'a dynamic, ecologically-based, natural resources management system that, through the integration of trees on farms and in the agricultural landscape, diversifies and sustains production for increased social, economic and environmental benefits for land users at all levels' (Mead 2004). In response to both environmental concerns and the need to ensure the sustenance of livelihoods, there are many examples where agroforestry is advocated as a potential solution to the need to develop a more sustainable form of land use that improves farm productivity while, at same time, improving the welfare of the community (Roshetko et al. 2013; Leakey et al. 2012; Ahmed and Rahman 2000). Agroforestry can be more financially profitable to local farmers than traditional monoculture systems, and support the transition to permanent cultivation (Rahman et al. 2014; Franzel and Scherr 2002; Predo 2002; Mai 1999). Agroforestry is not only financially, but also environmentally, promising compared with simpler systems, by ameliorating the agroclimate and increasing biodiversity (Jessica et al. 2014; Swallow et al. 2006; Huxley 1993), protecting soil organic matter and increasing nutrient cycling (Elevitch and Wilkinson 1998; Wu 1996; Sae-Lee et al. 1992).

If agroforestry is really as beneficial as scientific studies suggest, the logical prediction would then be that this system would be adopted by a high proportion of farmers. However, this is not the case in large areas of the tropics (Meijer et al. 2015; Jepma 2013; Dahlquist et al. 2007; Kiptot et al. 2007; Craswell et al. 1998). The research reported in this paper addresses this issue by analyzing the value of existing agroforestry systems, investigating their economic and social potential relative to swidden farming. We also seek to identify what factors are barriers to widespread agroforestry adoption. This information would be valuable for the development of appropriate strategies to encourage more farmers to adopt agroforestry and to improve management of existing agroforestry systems (Fisher and Bunch 1996; Saxena and Ballabh 1995; Nair and Dagar 1991). We also assess the evidence that agroforestry may better conserve forest, by comparing forest products' extraction and land clearing between agroforestry and swidden farmers.

MATERIALS AND METHODS

Study site

The study area lies between 6º32′11.31″S and 6º40′08.94″S latitudes and between 106º46′12.04″E and 106º47′27.42″E longitudes, and is located in the Gunung Salak valley, Bogor District, west Java, Indonesia. The reason for selecting this site is that both agroforestry and swidden cultivation are practiced by farmers in the same communities and environments. Thus their economic and social potential can be compared and the barriers to agroforestry adoption can be investigated with precision. The sustainability of livelihoods in the study area, like much of Indonesia, is threatened by overall poverty with low income and poor infrastructure development (Badan Pusat Statistik 2013), and the expansion of subsistence agriculture (especially swidden) due to rapid population growth is a major contributing factor to forest loss and environmental degradation (EST 2015; Galudra et al. 2008). Moreover, restrictions on the harvest of some products (e.g. timber) from natural forest provide an economic incentive for smallholders to integrate trees into their farming systems. All of these characteristics of the study area are representative of a large proportion of Indonesian and tropical Asian agricultural landscapes.

The climate in this region is equatorial with two distinct seasons,[1] i.e. dry (April–October) and rainy (November–March). The region is more humid and rainy than most parts of west Java, the average relative humidity and annual precipitation are 70 % and 1700 mm respectively. The average temperature is 25.9 °C, and the diurnal range is 9–10 °C, rather high for Indonesia (Badan Pusat Statistik 2013; Wiharto et al. 2008). The soils are highly fertile and dominated by volcanic sedimentary rocks. Given the proximity of large active volcanoes, the area is considered highly seismic.

Field data were collected during January–August, 2013 from two purposively selected[2] sample villages, i.e. Sukaluyu and Tamansari located in the northern valley of Gunung Salak, where the total population is approximately 8200, living in 1200 households. The study site has poor infrastructural facilities, and the local economy is mainly based on agricultural and forest products (Badan Pusat Statistik 2013). Our survey showed that in the two villages most community members have small land holdings (<1 ha) and carry out subsistence agriculture. Upland rice, irrigated rice, maize, and varieties of vegetables and fruits are the main agricultural crops. Land is used in various ways, such as rice fields (sawah), gardens (kebun), mixed gardens (kebun talun), mixed forests (talun) and swidden cultivation fields (huma/ladang) (Kleden et al. 2009). Private land use rights are granted by the government but farmers have no formal rights to state forest land. In the agroforestry farms, people cultivate various fruits, e.g. durian (*Durio zibethinus*), mango (*Mangifera indica*), rambutan (*Nephelium lappaceum*) and menteng (*Baccaurea racemosa*), and timber trees, e.g. teak (*Tectona grandis*), sengon (*Albizia falcataria*) and Jabon (*Anthocephalus chinensis*), with various understory crops, e.g. cassava (*Manihot esculenta*), maize (*Zea mays*), pineapple (*Ananas comosus*) and cincau (*Cylea barbata*). In the swidden fields, people commonly cultivate upland rice (*Oryza javanica)*, maize (*Zea mays*), yam (*Dioscorea* spp.), beans (*Dolichos lablab*) and cassava (*Manihot utilissima*). Fruits, vegetables, bamboo, rattan and firewood are also collected from nearby forests. Agricultural and forest products are sold in the local and district markets, and are an important source of household income, besides wage labor, and retailing.

Data collection

Primary data were collected by rapid rural appraisal (RRA) for the basic socio-economic and geographical information of the research site using village mapping and key informant interviews (FAO 2015; Angelsen et al. 2011). These sessions were conducted by involving village heads in the purposive selection of farmers based on their knowledge about the village and surrounding areas.

Two focus group discussion (FGD) sessions (one in each village[3]) and field observation methods were used to identify the types of local cultivation systems and their contribution to local livelihoods. The village heads and local farmer representative groups (consisting of eight to twelve farmers[4]) were present in the FGD sessions. Field observations were carried out in 25 locations which were decided based on the information gathered from RRA and FGD. During the observation period, several pictures of local cultivation systems were taken, and relevant information was noted with the help of an expert local informant.[5]

In-depth interviews of farmers were conducted to obtain the data needed for cost-benefit analysis of agroforestry and swidden. Two agroforestry farms of contrasting types (i) durian and cassava (agroforestry 1) and (ii) teak, yam and maize (agroforestry 2); and two swidden farms of contrasting types (i) upland rice (swidden 1) and (ii) maize (swidden 2), were selected. Based on the output of FGDs and field observations, these four farm types were purposefully selected by the first author as being popular (commonly adopted at a wider range) and providing the highest incomes among the farm populations in the agroforestry and swidden farming categories. During the interviews, the farmers were asked several questions about the actual and envisaged costs and benefits of each cultivation system, i.e. establishment cost, total yields, total labor requirement, cost of irrigation, pesticides, and fertilizer. The data collected from the four cultivation systems were checked with a local government agriculture officer to verify that the absolute values were in the expected range based on his experience of farming systems in the study area.

Twenty agroforestry and 20 swidden farmers were selected for semi-structured questionnaire interviews to collect information about their land holding area, income, farming benefits to their livelihood, forest products (FPs) collection, the area of forest that they cleared,[6] and the barriers to agroforestry adoption that they faced. Due to the range of land use practices and the unequal distribution of farms in the study area, purposive sampling was used to select farms that adequately represented the full development of the system type into which they were classified within the range of local land use practices.[7] We estimate that they represent about 30 and 40 % of the farmer populations who are practicing agroforestry and swidden respectively. A number of questions were refined with the help of the expert local informant and during FGD sessions to make sure that they elicited the information required. The product value of crops was calculated with the key informant farmers during the interview based on the amount harvested in one production year (the most recent year).

Other data were gathered from the local government forestry office, the Southeast Asian regional office of ICRAF and CIFOR headquarters located in Bogor, west Java, to corroborate the primary data that were collected from the research site, and for background information and qualitative inputs for the study.

Analysis

Qualitative analysis was carried out using the narrative analysis technique, particularly to investigate the social potential of existing agroforestry systems. For cost benefit analysis, the *net present value* (NPV), *benefit-cost ratio* (B/C) and *payback period* were calculated and compared following Stocking et al. (1990). The NPV determines the present value of net benefits by discounting the streams of benefits and costs back to the beginning of the base year (Disney et al. 2013; Stocking et al. 1990). The NPV is calculated by the following formula:

$$NPV = \sum_{t=0}^{T} \frac{(B_t - C_t)}{(1+r)^t} \tag{1.1}$$

where B_t is the benefits of production by a cultivation practice, C_t is the costs of production by a cultivation practice, t is the time, running until the end of the investment at T, r is the discount rate.

The B/C compares the discounted benefits with discounted costs. A B/C of greater than 1 means the cultivation is profitable, whilst a B/C of less than 1 means that it generates losses. The B/C is calculated as follows:

$$\frac{B}{C} = \frac{\sum_{t=0}^{n} \frac{B}{(1+r)^t}}{\sum_{t=o}^{n} \frac{C_t}{(1+r)^t}} \tag{1.2}$$

The *payback period* measures the number of years it will take for the undiscounted net benefits to repay the investment (Stocking et al. 1990).

Assumptions

Land and establishment cost

The market for agricultural land is underdeveloped in the study area, therefore the price of land is difficult to identify. However, as mentioned by Macdicken and Vergara (1990), there is no need to value the land separately if farmers want to change the use of their existing land to agroforestry. Thus, in our analysis the land value is omitted from the calculation. Establishment costs include: i) labor cost for land preparation, and ii) the price of seeds, seedlings and fertilizer which are required to start a project.

Yields

Crop components included in calculations for the selected cultivation systems are summarized in Table 1. The values of yields were calculated on an annual basis. Yields of durian (from grafted seedlings) are calculated under three categories, i) low yields during the fourth to sixth year, ii) medium yields during the seventh to eighth year, iii) high yields from the ninth year onwards. The market value of timber for the teak, yam and maize agroforestry system is calculated in ten-year rotation periods, after which it is assumed that teak is replanted.

Table 1 Brief description of selected cultivation systems for analysis

Cultivation system	Component	Cultivation type
Agroforestry 1	Durian, cassava	Permanent
Agroforestry 2	Teak, yam, maize	Rotational @ 10 year
Swidden 1	Upland rice	Semi-permanent
Swidden 2	Maize	Semi-permanent

Labor

Farmers often use family labor for farm work, but hired labor is also important in the study area. Family labor is not a cash expenditure from the farmer's perspective, and it is complicated to identify the amount of family labor contributed to each cultivation system, as farmers have different household size and labor availability. Therefore, all calculations were conducted based on the total amount of labor^{-day} required for each cultivation system.

Pesticides, fertilizer, irrigation

Even though pesticides and fertilizers are minimally used in swidden and for understory crops in agroforestry, the costs are calculated based on the amount used in one production year as reported during the interviews.

The cost of irrigation is ignored as high intensity rainfall occurs throughout the year, thus irrigation is not a cash expenditure for farmers.

Time horizon for analysis

Once forest trees are included in the agroforestry system the lifespan of this project can be considered indefinite. However, for simplicity, in our analysis the project life is considered to be 30 years as this may be a realistic lifetime for one rotation of durian trees in agroforestry system 1, which has the longest cycle. The consequence is that trees planted for timber in agroforestry system 2 can have three rotations (harvest cycles) and other crops have 30 annual cultivation cycles during the project lifespan. A similar time horizon is used in other comparable studies (e.g. Rahman et al. 2008, 2014).

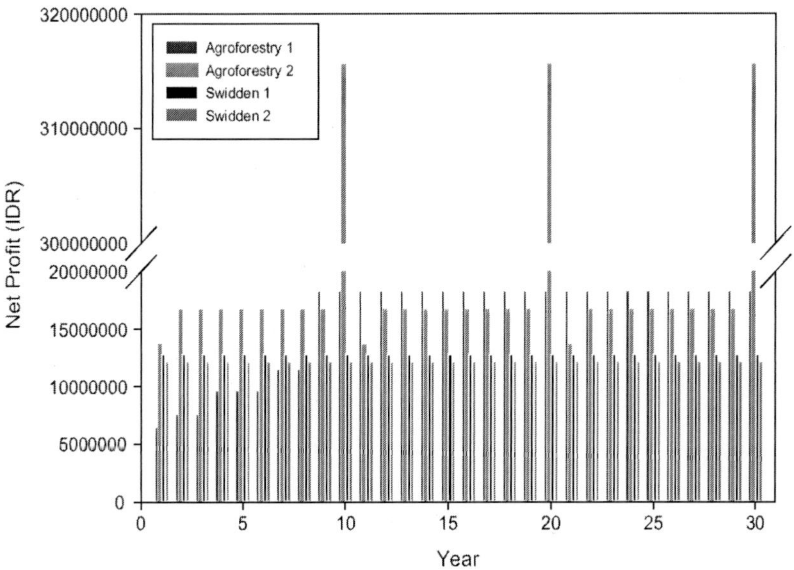

Figure. 1 Annual cash flow (net profit) of different cultivation systems (IDR/ha)

Table 2 Annual cost and revenue of selected cultivation systems in Indonesian Rupiah (IDR) per hectare

Type of operation	Year	Agroforestry 1	Agroforestry 2	Swidden 1	Swidden 2
Site preparation	0	375.000	500.000	180.000	180.000
Operational cost, i.e. labor, seeds, seedlings, fertilizer, pesticide	1	2161.667	5130.150		
	2–3	1461.667			
	4–6	1581,667*			
	7–8	1641,667*			
	9–30	1701,667*			
	………				
	2–9, 12–19, 22–29		2630.150		
	10, 20, 30		3750.000*		
	11, 21		5730.150*©		
	………				
	1–30			2171,000	2861.000
Annual crop yields	1–3	9025.000			
	4–6	11.225,000**			
	7, 8	13,125.000**			
	9–30	20,025.000**			
	………				
	1–30		19,348,333	15,000,000	15,000,000
Revenue from selling timber	10,20,30	n/a	300,000,000***		
NPV (r = 10 %)		122,077,993	330,154,427	120,937,885	114,433,314
B/C		10.36	16.19	6.91	5.24
Payback period (year)		1	1	1	1

Agroforestry 1 (durian and cassava) = no cost for cassava seeds/seedlings from years 2–30 as farmers produce it from the previous year: additional labor cost* for the durian harvesting** in the years 4–6 (trees first bearing fruit/low production), 7–8 (medium production) and 9–30 (full production)

Agroforestry 2 (teak, yam, maize) = additional labor cost * for timber harvesting*** in years 10, 20, 30, and land preparation (e.g. stump clearing) in years 11, 21: extra cost for seedlings© in years 11, 21

Cost and revenue are estimated to be the same for years 1–30 for swidden 1 (upland rice) and swidden 2 (maize)

RESULTS

The cash flow of the four different cultivation systems (Fig. 1), and the calculations of NPV and especially B/C, show that both agroforestry systems are more profitable than the two swidden systems (Table 2). Whereas profitability measured by NPV is similar in three of the systems, for agroforestry system 2 (with teak) it is almost three times higher. This is driven by the high output prices of the teak timber production from this system. Even though teak-based agroforestry requires some additional costs during rotations, these are offset by the return from selling timber. Furthermore, the value of intensively managed diversified understory crop yields in the teak-based system is higher than for both swidden systems, thus agroforestry system 2 is the superior land use option in the study site.

Risk factors should be accounted for in the financial analysis, as agricultural projects may face a wide variety of risks.[8] Furthermore, it is important to consider the assumptions in the calculations. Therefore, sensitivity analysis was conducted on changes in discount rate[9](Table 3), and variation in yields (Table 4). Regardless of the discount rate used, agroforestry 2 remains the most profitable system, whereas agroforestry 1 provides a lower NPV than the two swidden systems for discount rates of 20 % and above. In the case of decrease in yields, the NPV of both agroforestry systems are always positive and higher than swidden cultivation (Table 4).

Table 3 Sensitivity of profitability (NPV) to change in discount rate of agroforestry and swidden cultivation systems

Discount rate (r) (%)	NPV			
	Agroforestry 1	Agroforestry 2	Swidden 1	Swidden 2
5	221,438,725	616,502,476	197,213,174	186,606,183
10	122,077,993	330,154,427	120,937,885	114,433,314
20	53,588,283	137,580,080	63,874,775	60,439,308
30	31,582,414	76,579,723	42,747,011	40,447,889
40	21,803,255	50,274,014	32,071,174	30,346,246
50	16,499,245	36,674,210	25,657,866	24,277,873
60	13,228,225	28,714,082	21,381,650	20,231,651
70	11,026,191	23,599,254	18,327,140	17,341,426
80	9447,968	20,065,665	16,036,249	15,173,749
90	8263,065	17,482,427	14,254,444	13,487,777

No difference in payback period was found between the four systems (Table 2). A 1 year payback period for the agroforestry systems indicates that within a year the undiscounted net benefit is high enough to repay the comparatively higher investment in establishing this system.

Table 4 Sensitivity of profitability (NPV) to varying the yields of agroforestry and swidden cultivation systems

Decrease of production (%)	NPV (r = 10 %)			
	Agroforestry 1	Agroforestry 2	Swidden 1	Swidden 2
0	122.077.993	330.154.427	120.937.885	114.433.314
5	115.974.093	313.646.706	114.890.991	108.711.648
10	109.870.193	297.138.984	108.844.097	102.989.983
20	97.662.394	264.123.542	96.750.308	91.546.651
30	85.454.595	231.108.099	84.656.519	80.103.320
40	73.246.795	198.092.656	72.562.731	68.659.988
50	61.038.996	165.077.213	60.468.942	57.216.657
60	48.831.197	132.061.771	48.375.154	45.773.325
70	36.623.397	99.046.328	36.281.365	34.329.994
80	24.415.598	66.030.885	24.187.577	22.886.662
90	12.207.799	33.015.442	12.093.788	11.443.331

Through the semi-structured questionnaire interviews and FGDs, it was identified that agroforestry not only creates production capacity, but also tree planting establishes more permanent land rights for farmers, with those rights transferring to future generations. In contrast, fallow or swidden systems may weaken tenure security. One of the respondents established his durian-based agroforestry farm in 2001, and he remembered that before practicing agroforestry 'I left my land abandoned and one of my neighbors used it to stack his logs to sell that he had harvested'. During FGDs, it was reported that cultivation categories defined as 'agroforestry' are prestigious in the community, owing to the high value of tree products (e.g. teak, durian) which have higher monetary value than do products from swidden agriculture. Additionally, agroforestry farmers share their fruit and vegetable products with neighbors, providing direct benefits to others and strengthening social cohesion. Agroforestry also creates various jobs, such as traders and regular or seasonal wage-laborers for harvesting, transporting, sorting etc. of fruit and timber, thus supporting the emergence of farm-related rural employment and specialization.

Despite agroforestry systems being more profitable, more prestigious and better for securing land use rights, some farmers in the study area still persist with the less profitable swidden cultivation. The semi-structured questionnaire interviews with 20 key informant swidden farmers revealed some of the factors underlying non-adoption of agroforestry (Table 5). Adoption is hampered by capacity (2, 3, 4) and motivational (1) factors. Capacity constraints were mentioned 27 times by the farmers, while motivational factors were mentioned 18 times. 'No interest' in agroforestry practice is deeply rooted in their tradition, whereas swidden practice has been practiced by generations. 'Lack of capital' is also a serious constraint on initial investment in agroforestry. This is particularly true for swidden farmers as their cultivation practices are largely subsistence-oriented and yield insufficient capital to invest in agroforestry, i.e. it requires about half of their annual household income to invest in agroforestry (Tables 2and 6). Lack of technical assistance is another major constraint as government programs to promote agroforestry do not exist in the study site. There is no agroforestry extension, no technical or market information, no price guarantees and no supply of high quality seedlings.

Table 5 Constraints on the adoption of agroforestry, as mentioned by 20 swidden farmers. The motivational factor is marked with M and factors related to capacity are marked with C

Reasons	Number of farmers	Per cent
(1) No interest (M)	18	90
(2) Lack of sufficient knowledge (C)	7	35
(3) Lack of capital (C)	16	80
(4) Lack of technical assistance (C)	4	20

The interviews with the 40 key informant farmers revealed that most of the swidden lands in the study site are semi-permanent with cultivation interspersed with either short or long fallow periods, whereas other agricultural land is cultivated continuously without fallow periods. Swidden farmers occupy less land than agroforestry farmers because (i) low household income restricts them from investing in new land and (ii) limited labor is available for agriculture as a high proportion of household labor is required for off-farm work which accounts for a high proportion of their income[10] (Table 6). Eight-five per cent of swidden farmers were involved in forest clearing whereas only 30 % of agroforestry farmers were involved in this activity. As a result, on average a swidden farming household cleared a larger area (0.29 ha) of forest than an agroforestry farming household (0.09 ha). Among swidden farmers, 45 % of them cleared forest for the establishment (by slash-and-burn) of swidden farming, whereas a relatively low number of agroforestry farmers (15 %) cleared forest for agroforestry purposes.

Table 6 Farm size, income, forest clearing activity and collecting of forest products by swidden farmers and agroforestry farmers

Description	Swidden farmers (n = 20)	Agroforestry farmers (n = 20)
Total swidden land (ha)	0.46	–
Total agroforestry land (ha)	–	0.85
Total other cropland (ha)	0.29	0.11
Total homestead land (ha)	0.02	0.02
Total land area (ha)	0.77	0.98
Total annual income from all sources (million IDR)	12.07	20.15
Total annual income from swidden/agroforestry (million IDR)	1.04	3.25
Total annual income from other cropland (million IDR)	2.52	1.66
Forest area cleared per household (last 5 years) (ha)	0.29	0.09
Reason for clearing	Swidden: 45 % Permanent monoculture: 35 % Plantation: 0.5 % Not cleared: 15 %	Agroforestry: 15 % Permanent monoculture: 15 % Not cleared: 70 %
Distance to the edge of nearest forest (minutes of walking)	24.0	10.6
Firewood collected from forest per household (kg month^{-1})	33	5.60
Fodder collected from forest per household (kg month^{-1})	1.65	3.15
Forest food[a] collected per household (kg month^{-1})	4.85	1.70

[a] Forest food mainly constitutes bamboo shoots, mushrooms, tubers and other leafy vegetables, nuts and fruit including rambutan, menteng and wild bananas

Swidden farmers collect, on average, more firewood from forests than do agroforestry farmers (Table 6). In interviews, the respondents said that this difference is because there is a big stock of firewood available in the agroforestry farms, especially from tree pruning and thinning. Also, their relatively higher farm income enables agroforestry farmers to buy gas cylinders, thereby reducing their need for firewood. Cattle rearing is not common in the research site, thus the rate of fodder collection from forest is low. Swidden farmers collect more forest food than agroforestry farmers. This was due to the diversity of crop species in agroforestry systems providing various types of food, and at the same time the higher farm income of agroforestry farmers enabled their households to buy food from local markets. There are a total of 4 timber, 15 fruit and nut, and 23 other understory crop species cultivated in the agroforestry systems.

DISCUSSION

As an alternative to swidden farming, in the Gunung Salak study site agroforestry systems were found to be financially profitable and have good potential to secure sustainable livelihoods through diversified food sources and strengthened land tenure. Durian- and teak-based agroforestry systems are the most popular in the study site. The B/C indicated that total monetary gain is much higher in both of these systems than the total costs required to undertake the project, and much higher than for swidden systems. In addition, the payback period showed that there was no notable problem of delayed cash returns for those farmers adopting either agroforestry system; it was equal to the 1 year period of the swidden systems. However, NPV showed only one agroforestry system (the teak-based one) to be notably more profitable than both the swidden systems. Both sensitivity analyses confirmed that it is the teak-based agroforestry system that is more profitable over a range of conditions than are the durian-based agroforestry or swidden cultivation systems.

Smallholder teak production in Java is an important source of cash income for rural families (Roshetko et al. 2013) and has become part of many farmers' culture (Perdana et al. 2012), whereas swidden has retained this cultural status in the study area. There are 1.5 million smallholder farmers in Java managing 444,000 ha of tree-based agroforestry systems, where teak is the dominant tree crop. In other parts of Indonesia, there is an additional 800,000 ha of smallholder agroforestry, where teak is one component of multispecies, tree-based systems, favored because of its high market price (Departemen Kehutanan 2005). In Central and East Java, smallholder farmers see tree farming systems as a 'living savings account' that diversifies production, reduces risk, and builds assets to enhance family incomes and security (van Noordwijk et al. 2008). De Foresta et al. (2004) found that the average annual income from mature fruit and timber agroforestry systems in Krui, Lampung were IDR 2,410,000 $ha^{-1} yr^{-1}$. Tree farming systems in the Philippines provided a range of annual incomes equivalent to IDR 2,374,802–163,553,043 $ha^{-1} yr^{-1}$, which greatly exceed incomes provided by annual crop systems, and the imperata grassland shifting cultivation system (Predo 2002). Tree-based production systems are also promoted in government policies because of their perceived biological, economic and social resilience in the context of anthropogenic climate change and other production challenges (Alfaro et al. 2014; Steffan-Dewenter et al. 2007; Thorlakson and Neufeldt 2012).

In our research site, through active tree planting, agroforestry creates permanent rights to farm land that transfer to future generations. Practicing this permanent form of cultivation is also prestigious in the community, because the tree products have high monetary and social values. From a social and institutional point of view, agroforestry is an important element in smallholder farmers' land security strategies in Indonesia (Michon and de Foresta, 1999), giving farmers the opportunity to secure tenure, as the recognized tree planter, with the property being legally transferred to descendants as patrimony (Michon 2005).

Furthermore, swidden farmers have capacity constraints on agricultural cultivation of a large land area, thus they are only able to use less land than agroforestry farmers. Low household income limits the capacity to invest in cultivating new land, due to the importance of off-farm income in the livelihoods of the studied households. Available labor to cultivate agricultural crops is the most limiting resource for them. On the other hand, more permanent sustainable agroforestry practices require less labor input. Thus, smallholder tree cultivation is recognized as a viable livelihood strategy in various agroforestry and community forestry programs (e.g. FAO 2006; Sales et al. 2005; ICRAF 2003).

The debates on the underlying causes of tropical deforestation and the drivers of agents' behaviour are complex, and the relationships between forest clearing and household and contextual variables vary depending on the setting (VanWey et al. 2005). Even with a limited land holding capacity, swidden farming households at our study site cleared a larger area of forest than did agroforestry farming households. Even though the average distance of swidden farm household to the nearest forest is relatively far, they collected more firewood from forests than did agroforestry households. This is because agroforestry farms have a good supply of firewood, and relatively higher farm income allowing a larger proportion of agroforestry farmers to buy gas cylinders. Recent studies in different locations around the tropics indicate that one important reason for deforestation is crop growing (Babigumira et al. 2014), and swidden farming is often held to be the principle driving force for that (Fox et al. 2000; Angelsen 1995; Myers 1992). However, Heltberg et al. (2000) reported that one of the main drivers of forest degradation in rural India is unsustainable firewood collection. A study in the buffer zone of the Kerinci Seblat National Park, Indonesia highlighted the relationship between farm diversification and reliance on adjacent national park resources (Murniati and Gintings 2001). Factors associated with a higher tendency to extract forest products from protected areas were low farm income and low supply of on-farm tree-based products. A study by Garrity et al. (2002) around the Mount Kitanglad Range National Park in Mindañao, the Philippines provides support for a link between adoption of agroforestry and reduction in pressure on forest.

Even though agroforestry systems have major economic benefits for farmers, several factors constrain agroforestry adoption. The major one in the study area is lack of investment capital and the higher traditional cultural value of swidden farming, which has been practiced by many generations, within the local communities. There is an absence of government assistance which could help to overcome these barriers to adoption of agroforestry. Several other studies have also found that tradition and customs are still a decisive factor influencing farmers' choice to practice swidden cultivation (Padoch et al. 2014; Peng et al. 2014; Predo 2002) and that lack of capital and government backing[11] are crucial constraints on agroforestry adoption (Rahman et al. 2012; Van et al. 2012; Mai 1999). Institutional innovation theory pioneered by economists (Hayami and Ruttan 1971; Schultz 1964) argues that physical constraints can be compensated by knowledge and institutional influence. Empirical evidence from Sumatra, Indonesia illustrated that with a supportive local institutional influence, tree culture has extended greatly into the landscape of swidden cultivation fields where young trees are cultivated with crops (Michon 2005). Swidden cultivation eventually disappeared when the agroforestry silvicultural system had sufficiently matured and started to function as a productive and profitable tree-based system. When agroforestry systems fit local biophysical and socioeconomic conditions, they can rapidly become part of local culture (Perdana et al. 2012). There is potential for this intensification to be achieved in our study area through a smooth adaptation of tree-based farming practices with necessary government backup, thus the association of 'agro' and 'forest' components will occur at the level of the farming system itself, and if adopted at a sufficient scale it will significantly contribute to increasing tree cover in agricultural landscapes (see also Michon 2005).

CONCLUSIONS

Communities in Gunung Salak have created a cultivated landscape which their livelihoods depend on. Their traditional swidden cultivation practices provide various subsistence products, but they can have serious negative environmental consequences by contributing to deforestation and land degradation. Agroforestry is an alternative cultivation strategy that has been adopted by some farmers within the communities. It does increase average farm income, making it more resilient to changes in market and economic conditions, and reduce pressure on adjacent forest for conversion to agriculture and as a source of firewood, fruits, vegetables and other products. These agroforestry systems also enable farmers to secure permanent land tenure and can improve social cohesive in communities. Adoption of agroforestry by farmers in the Salak valley can be increased by the implementation of supportive policies and measures (including capital support and technical assistance) by government and non-government organizations. These measures are most likely to be effective if they are sensitive to the strong local tradition of swidden cultivation and underlying systems of local knowledge. Effective policies should be propagated not by temporary projects but by permanent, government-backed institutions that are focused on agroforestry practices and the needs for their adaptation to meet new opportunities and constraints (see also Rahman et al. 2008). The successful adoption of durian-and teak-based agroforestry by many farmers in the study area indicates the high potential for success of such a programme.

FOOTNOTES

1. In the study site rainfall occurs throughout the year, but based on its intensity, two seasons are recognised, where heavy rainfall occurs in the "rainy" season.

2 Villages were selected to represent two contrasting watershed locations, i.e. mid-stream (Sukaluyu) and up-stream (Tamansari); and having the largest sample size of farm households, i.e. agroforestry and swidden.

3 One semi-structured questionnaire interview (village survey, consisting of a set of questions concerning basic information about the village, e.g. demographic, infrastructure, land use) was also conducted during the FGD.

4 Farmers in each group were purposively selected based on their knowledge of local cultivation systems.

5 One resident of the study site, who had considerable knowledge of local land use systems, products, markets and institutions, was employed as an expert local informant. This informant was present during the whole period of fieldwork, and helped check the validity of information obtained.

6 Households were asked whether or not in the last five years they had cleared any forest, and if yes, we also asked how much, and for what purpose it was cleared. We have used FAO's forest definition (FAO 2000), which defined forest as lands of more than 0.5 hectares, with a tree canopy cover of more than 10 %, where the trees should be able to reach a minimum height of 5 m in situ, and which are not primarily under agricultural land use.

7 For example, some farmers started agroforestry farming but after a few years gave up planting the understory, for various reasons (e.g. lack of management interest or capital). Thus many agroforestry farms were converted to simple tree orchards, and we have excluded them from our sample. In fact very few farmers had developed the system type in full, and this was the only basis for the selection of farms who met that criterion.

8 Many natural risk factors are site specific (e.g. landslides, lava flows) whereas others are more widespread (e.g. storms). Some threats are induced by humans, such as fire, pest introductions and price fluctuation (e.g. if supply is increased due to increases in output due to expansion of farm production).

9 One method to include risk into analysis is to use an increased discount rate, which reflects the added yearly risk of a project (see Elevitch and Wilkinson 2000).

10 Annual household off-farm income is calculated to be 8.5 and 15.2 million IDR, i.e. 70 and 75 % of total household income for swidden and agroforestry farmers respectively, and much greater than the total farm income. During FGDs it was reported that households allocate a high proportion of their labor to this off-farm work.

11 Other studies conducted in West Java, Sumatra, and Sulawesi also indicate that technical assistance is an important factor for agroforestry farm intensification and farmer motivation (Martini et al. 2012; Manurung et al. 2008; Roshetko et al. 2007).

ACKNOWLEDGMENTS

This work is funded by Forest and Nature for Society (FONASO, initiated by the Erasmus Mundus programme of the European Commission to enhance and promote European higher education throughout the world) and Center for International Forestry Research (CIFOR). The authors are grateful to the scientists and staff of Bangor University, University of Copenhagen, CIFOR and ICRAF who provided support and guidance. Many thanks are also extended to the people at the study sites where the field investigation was undertaken, who shared their precious time, knowledge and concerns.

REFERENCES

1. Adimihardja K (1992) Kasepuhan yang Tumbuh di Atas yang Luruh: Pengelolaan Lingkungan secara Tradisional di Kawasan Gunung Halimun Jawa Barat. Tarsito, Bandung
2. Ahmed MFU, Rahman SML (2000) Agroforestry farming system in the homesteads of Gazipur, Central Bangladesh. Texas A & M University, Dallas
3. Alfaro R, Fady B, Vendramin GG, Dawson IK, Fleming RA, Saenz-Romero C, Lindig-Cisneros RA, Murdock T, Vinceti B, Navarro CM (2014) The role of forest genetic resources in responding to biotic and abiotic factors in the context of anthropogenic climate change. For Ecol Manag 333:76–87
4. Angelsen A (1995) Shifting cultivation and deforestation: a study from Indonesia. World Dev 23(10):1713–1729
5. Angelsen A, Larsen HO, Lund JF, Smith-Hall C, Wunder S (eds) (2011) Measuring livelihoods and environmental dependence: methods for research and fieldwork. CIFOR, Bogor
6. Aweto AO (2013) Shifting cultivation and secondary succession in the tropics. CABI, Oxfordshire
7. Babigumira R, Angelsen A, Buis M, Bauch S, Sunderland T, Wunder S (2014) Forest clearing in rural livelihoods: household-level global-comparative evidence. World Dev 64:67–79
8. Badan Pusat Statistik (2013) Statistik Daerah Kabupaten Bogor Tahun 2013. Komplek Perkantoran Pemkab Bogor, Bogor
9. Barraclough SL, Ghimire KB (1995) Forests and Livelihoods: the Social dynamics of deforestation in developing countries. International Political Series. UNRISD, London
10. Coulter JK (1972) Soil management systems. National Academy of Sciences, Washington D.C.
11. Craswell ET, Sajjapongse A, Howlett DJB, Dowling AJ (1998) Agroforestry in the management of sloping lands in Asia and the Pacific. Agrofor Syst 38:121–137
12. Dahlquist RM, Whelan MP, Winowiecki L, Polidoro B, Candela S, Harvey CA, Wulfhorst JD, McDabiel PA, Bosque-Perez NA (2007) Incorporating livelihoods in biodiversity conservation: a case study of cacao agroforestry systems in Talamanca, Costa Rica. Biodivers Conserv 16(8):2311–2333
13. DANIDA (Danish International Development Agency) (2000). Identification report. Watershed development project, Chittagong Hill Tracts, Bangladesh
14. de Foresta H, Michon G, Kusworo A, Levang P (2004) Damar agroforest in Sumatra, Indonesia: domestication of forest ecosystem through domestication of *Dipterocarps* for resin production. In: Kusters K, Belchers B (eds) Forest products, livelihoods and conservation, case studies of non-timber forest product systems, vol 1. CIFOR, Bogor, pp 202–221
15. Departemen Kehutanan (2005) Data Potensi Hutan Rakyat. Direktorat Jenderal Rehabilitasi Lahan dan Perhutanan Sosial dan Biro Pusat Statistik. Ministry of Forestry, Jakarta
16. Disney SM, Warburton RDH, Zhong CQ (2013) Net present value analysis of the economic production quantity. J Manag Math 24:423–435
17. Elevitch C, Wilkinson KM (1998) A guide to orchard alley cropping for fertility, mulch and soil conservation. AgroForester, Hawaii

18. Elevitch CR, Wilkinson KM (2000) Economics of farm forestry: financial evaluation for landowners. agroforestry guides for Pacific Islands # 7. Permanent Agricultural Resources (PAR), Hawaii
19. EST (Endangered Species Team) (2015) Ecological study Halimun-Salak corridor Mount Halimun-Salak national park, GHSNPMP-JICA. http://www.sci.kagoshima-u.ac.jp/~itp/student/siryo/Corridor%20Study-Eng2005.pdf. Accessed 10 Apr 2015
20. FAO (2000). Forest Resource Assessment 2000: on definitions of forest and forest change. Working Paper 33. FAO, Rome
21. FAO (2006) Better forestry, less poverty: a practitioner's guide. FAO, Rome
22. FAO (2015) What is rapid rural appraisal?. http://www.fao.org/ docrep /006/W2352E/W2352E03.htm. Accessed 9 Jan 2015
23. Fisher L, Bunch R (1996) Challenges in promoting forest patches in rural development efforts. In: Schelhas J, Greenberg R (eds) Forest patches in tropical landscapes. Island Pressure, Washington D.C., pp 381–400
24. Fox J, Truong DM, Rambo AT, Tuyen NP, Cuc LT, Leisz AS (2000) Shifting cultivation: a new old paradigm for managing tropical forests. Bioscience 50(6):521–528
25. Franzel S, Scherr SJ (2002) Introduction. In: Franzel S, Scherr SJ (eds) Trees on the farm: assessing the adoption potential of agroforestry practices in Africa. CABI, Wallingford, pp 1–11
26. Galudra G, Nurhawan R, Aprianto A., Sunarya Y, Engkus, (2008) The last remnants of mega biodiversity in West Java and Banten: an in-depth exploration of RaTA (Rapid Land Tenure Assessment) in Mount Halimun-Salak National Park, Indonesia, Working Paper no. 69. World Agroforestry Centre, Bogor
27. Garrity DP (2004) Agroforestry and the achievement of the millennium development goals. Agrofor Syst 61:5–17
28. Garrity D, Amoroso VB, Koffa S, Catacu-tan D, Buenavista G, Fay P, Dar W (2002) Landcare on the poverty-protection interface in an Asian watershed. Conserv Ecol 6(1):12
29. Gupta A (1993) Ecology and Development in the Third World. Routledge, London
30. Hayami Y, Ruttan VW (1971) Agricultural development: an international perspective. The Johns Hopkins University Press, USA
31. Heltberg R, Arndt TC, Sekhar NU (2000) Fuelwood consumption and forest degradation: a household model for domestic energy substitution in rural India. Land Econ 76(2):213–232
32. Huxley PA (1993) Tree crop interface orientation experiment with *Grevillea robusta* hedgerows and maize. Agrofor Syst 26:23–45
33. ICRAF (2003) Agroforestry for poverty reduction: realizing the potential. Strategic plan for ICRAF East and Central Africa Program 2000–2010. ICRAF, Nairobi
34. Idol T, Haggar J, Cox L (2011) Ecosystem services from smallholder forestry and agroforestry in the tropics. In: Campbell WB, Ortiz SL (eds) Integrating agriculture, conservation and ecotourism: examples from the field, issues in agroecology—present status and future prospectus, vol 1. Springer, New York, pp 209–270
35. Imang N, Inoue M, Sardjono MA (2008) Tradition and the influence of monetary economy in swidden agriculture among the Kenyah people of East Kalimantan, Indonesia. Int J Soc For (IJSF) 1(1):61–82
36. Inoue M (2000) Mechanism of changes in the Kenyah's swidden system: explanation in terms of zensification theory. In: Guhardja E, Fatawi M, Sutisna M, Mori T, Ohta S (eds) Rainforest Ecosystems of East Kalimantan—El Niño, drought, fire and human impacts. Springer-Verlag, Tokyo, pp 167–184
37. Jepma CJ (2013) Tropical Deforestation: a socio-economic approach. Routledge, New York
38. Jessica LB, Stewart AWD, James PG, Stephen VS, Jorge MV (2014) Implications of Mayan agroforestry for biodiversity conservation in the Calakmul biosphere reserve, Mexico. Agrofor Syst 88:269–285

39. Kiptot E, Hebinck P, Franzel S, Richards P (2007) Adopters, testers or pseudo-adopters? dynamics of the use of improved tree fallows by farmers in Western Kenya. Agric Syst 94(2):509–519
40. Kleden OE, Chidley L, Indradi Y (eds.) (2009) Forests for the future: indigenous forest management in a changing world. Indigenous peoples alliance of the Archipelago, and down to earth, Jakarta
41. Leakey RRB (2010) Agroforestry: a delivery mechanism for multi-functional agriculture. In: Kellimore LR (ed) Handbook on agroforestry: management practices and environmental impact, environmental science, engineering and technology series. Nova Science Publishers, Hauppauge, pp 461–471
42. Leakey RRB, Weber JC, Page T, Cornelius JP, Akinnifesi FK, Roshetko JM, Tchoundjeu Z, Jamnadass R (2012) Tree domestication in agroforestry: progress in the second decade (2003–2012). In: Nair PKR, Garrity D (eds) Agroforestry: the future of global land use, advances in agroforestry series, vol 9. Springer, Dordrecht, pp 145–173
43. Macdicken KG, Vergara NT (1990) Agroforestry: classification and management. Wiley, New York
44. Mai PT (1999) Socio-economic analysis of shifting cultivation versus agroforestry system in the upper stream of lower Mekong watershed in Dak Lak province. MA Thesis in economics of development. College of Economics, National University-HCMC, Viet Nam, and Institute of Social Study (ISS), The Netherlands
45. Manurung GES, Roshetko JM, Budidarsono S, Kurniawan I (2008) Dudukuhan tree farming systems in West Java: how to mobilize self-strengthening of community-based forest management? In: Snelder DJ, Lasco RD (eds) Smallholder tree growing for rural development and environmental services: lessons from Asia. Advances in agroforestry series, vol 5. Springer, Dordrecht, pp 99–116
46. Marten GG (1986) Traditional agriculture in Southeast Asia: a human ecology perspective. Westview Press, Boulder
47. Martini E, Tarigan J, Purnomosidhi P, Prahmono A, Surgana M, Setiawan A, Megawati, Mulyoutami E, Meldy BD, Syamsidar, Talui R, Janudianto, Suyanto S, Roshetko JM (2012) Agroforestry and forestry in Sulawesi series: agroforestry extension needs at the community level in AgFor project sites in South and Southeast Sulawesi, Indonesia. Working paper 159. ICRAF, Bogor
48. Mead D.J (2004) Agroforestry. Encyclopedia of life support systems. Vol. 1, pp 324–355
49. Meijer SS, Catacutan D, Ajayi OC, Sileshi GW, Nieuwenhuis M (2015) The role of knowledge, attitudes and perceptions in the uptake of agricultural and agroforestry innovations among smallholder farmers in sub-Saharan Africa. Int J Agric Sustain 13(1):40–54
50. Mertz O, Padoch C, Fox J, Cramb R, Leisz S, Lam N, Vien T (2009) Swidden change in Southeast Asia: understanding causes and consequences. Hum Ecol 37:259–264
51. Michon G (ed) (2005) Domesticating forests: how farmers manage forest resources. IRD, CIFOR and ICRAF, Nairobi
52. Michon G, de Foresta H (1999) Agro-forests: incorporating a forest vision in agroforestry. In: Buck LE, Lassoie JP, Fernandes ECM (eds) Agroforestry in sustainable agricultural systems. Lewis Publishers, New York, pp 318–406
53. Munthali KG (2013) Modelling deforestation in dzalanyama forest reserve, Lilongwe, Malawi: using multi-agent simulation approach. Ph.D. Dissertation, University of Tsukuba, Tsukuba
54. Murniati Garrity DP, Gintings AN (2001) The contribution of agroforestry systems to reducing farmers' dependence on the resources of adjacent national parks: a case study from Sumatra. Agrofor Syst 52:171–184
55. Myers N (1992) Tropical forests: the policy challenge. Environmentalist 12(1):15–27
56. Nair PKR, Dagar JC (1991) An approach to developing methodologies for evaluating agroforestry systems in India. Agrofor Syst 16:55–81

57. Padoch C, Coffey K, Mertz O, Leisz SJ, Fox J, Wadley RL (2014) The demise of swidden in southeast Asia? local realities and regional ambiguities. Dan J Geogr 107(1):29–41
58. Peng L, Zhiming F, Luguang J, Chenhua L, Jinghua Z (2014) A review of swidden agriculture in Southeast Asia. Remote Sens 6:1654–1683
59. Perdana A, Roshetko JM, Kurniawan I (2012) Forces of competition: smallholding teak producers in Indonesia. Int For Rev 14(2):238–248
60. Predo CD (2002) Bioeconomic analysis of land use options for grassland areas in Claveria, Misamis Oriental. Leyte State University, Leyte
61. Rahman SA, Groot W, Snelder DJ (2008) Exploring the agroforestry adoption gap: financial and socioeconomics of litchi-based agroforestry by smallholders in Rajshahi (Bangladesh). In: Snelder DJ, Lasco RD (eds) Smallholder tree growing for rural development and environmental services: lessons from Asia. Advances in agroforestry series, vol 5. Springer, Dordrecht, pp 227–24
62. Rahman SA, Rahman MF, Sunderland T (2012) Causes and consequences of shifting cultivation and its alternative in the hill tracts of eastern Bangladesh. Agrofor Syst 84(2):141–155
63. Rahman SA, Rahman MF, Sunderland T (2014) Increasing tree cover in degrading landscapes: 'integration' and 'intensification' of smallholder forest culture in the alutilla valley, Matiranga, Bangladesh. Small-scale For 13:237–249
64. Roshetko JM, Martini E, Tarigan J, Manurung G, Budidarsono S, Wijaya K, Tukan JC, Kurniawan I, Galudra G, Nugroho DK, Ekadinata A, SDewi S, Lusiana DHB, van Noordwijk M, Purba J (2007) Agroforestry on the interface of orangutan conservation and sustainable livelihoods in Batang Toru (North Sumatra). Working Paper no. 56. ICRAF, Bogor
65. Roshetko JM, Snelder SD, Lasco RD, van Noordwijk M (2008) Future challenge: a paradigm shift in the forestry sector. In: Snelder DJ, Lasco RD (eds) Smallholder tree growing for rural development and environmental services: lessons from Asia. Advances in agroforestry series, vol 5. Springer, Dordrecht, pp 453–485
66. Roshetko JM, Rohadi D, Perdana A, Sabastian G, Nuryartono N, Pramono AA, Widyani N, Manalu P, Fauzi MA, Sumardamto P, Kusumowardhani N (2013) Teak agroforestry systems for livelihood enhancement, industrial timber production, and environmental rehabilitation. For Trees Livelihoods 22(4):251–256
67. Russell W (1988) Population, swidden farming and the tropical environment. Popul Environ 10:77–94
68. Sae-Lee S, Vityakon P, Prachaiyo B (1992) Effects of trees on paddy bund on soil fertility and rice growth in northeast Thailand. Agrofor Syst 18:213–223
69. Sales RF, Lasco RD, Bonaticla R (2005) Carbon storage and sequestration potential of smallholder tree farms on Leyte Island, The Philippins. In: Harrison S, Herbohn J, Mangaoan E, Vanclay J (eds.) ACIAR smallholder forestry project ASEM 200/008, redevelopment of a timber industry following extensive land clearing. Proceedings from the end of Project Workshop, 19–21 August 2004, Ormoc City
70. Saxena NC, Ballabh V (1995) Farm forestry and the context of farming systems in South Asia. In: Saxena NC, Ballabh V (eds) Farm forestry in South Asia. Sage Publications, New Delhi, pp 23–50
71. Schuck EC, Nganje W, Yantio D (2002) The role of land tenure and extension education in the adoption of slash and burn agriculture. Ecol Econ 43:61–70
72. Schultz TW (1964) Transforming traditional agriculture. Yale University Press, USA
73. Sfeir-Younis A, Dragun AK (1993) Land and soil management: technology, economics and institutions. West View Press, Boulder
74. Shoaib JU, Mostafa G, Rahman M (1998) Soil erosion hazard in Chittagong Hill tracts: a case study. Annual report. Soil Resources Development Institute, Dhaka
75. Steffan-Dewenter I, Kessler M, Barkmann J, Bos MM, Buchori D, Erasmi S, Faust H, Gerold G, Glenk K, Gradstein SR, Guhardja E, Harteveld M, Hertel D, Höhn P, Kappas M, Köhler S, Leuschner C, Maertens M, Marggraf R, Migge-Kleian S, Mogea J, Pitopang R, Schaefer M, Schwarze S, Sporn SG, Steingrebe A, Tjitrosoedirdjo SS, Tjitrosoemito

S, Twele A, Weber R, Woltmann L, Zeller M, Tscharntke T (2007) Tradeoffs between income, biodiversity, and ecosystem functioning during tropical rainforest conversion and agroforestry intensification. Proc Natl Acad Sci USA 104(12):4973–4978

76. Stocking M, Bojo J, Abel N (1990) Financial and economic analysis of agroforestry: key issues. In: Prinsley RT (ed) Agroforestry for sustainable production: economic implications. The Commonwealth Secretariat, London, pp 13–119

77. Styger E, Rakotondramasy HM, Pfeffer MJ, Fernandes ECM, Bates DM (2007) Influence of slash-and-burn farming practices on fallow succession and land degradation in the rainforest region of Madagascar. Agric Ecosyst Environ 119:257–269

78. Sunderland TCH, Clark LE, Vantomme P (eds) (1999) The non-wood forest products of Central Africa: current research issues and prospects for conservation and development. FAO, Rome

79. Swallow B, Boffa JM, Scherr SJ (2006) The potential for agroforestry to contribute to the conservation and enhancement of landscape biodiversity. In: Garrity DP, Okono A, Grayson M, Parrott S (eds) World agroforestry into the future. World Agroforestry Centre (ICRAF), Nairobi, pp 95–101

80. Therik T (1999) The role of fire in swidden cultivation: a timor case study. In: Russell SJ, Hill GJE, Djoeroemana S, Myers BA (eds) Fire and sustainable agricultural and forestry development in Indonesia and northern Australia. Australian Centre for International Agricultural Research (ACIAR), Canberra, pp 77–79

81. Thorlakson T, Neufeldt H (2012) Reducing subsistence farmers' vulnerability to climate change: evaluating the potential contributions of agroforestry in Western Kenya. Agric Food Secur 1:1–15

82. van Noordwijk M, Roshetko JM, Angeles MD, Suyanto Fay C, Tomich TP (2008) Farmer tree planting barriers to sustainable forest management. In: Snelder DJ, Lasco RD (eds) Smallholder tree growing for rural development and environmental services: lessons from Asia. Advances in agroforestry series, vol 5. Springer, Dordrecht, pp 429–451

83. Van VN, Mertz O, Heinimann A, Langanke T, Pascual U, Schmook B, Adams C, Schmidt-Vogt D, Messerli P, Leisz S (2012) Trends, drivers and impacts of changes in swidden cultivation in tropical forest-agriculture frontiers: a global assessment. Glob Environ Change 22:418–429

84. VanWey LK, Ostrom E, Meretsky V (2005) Theories underlying the study of human environment interactions. In: Moran E, Ostrom E (eds) Seeing the forest and the trees: human-environment interactions in forest ecosystems. MIT Press, Cambridge

85. Vongvisouk T, Mertz O, Thongmanivong S, Heinimann A, Phanvilay K (2014) Shifting cultivation stability and change: contrasting pathways of land use and livelihood change in Laos. Appl Geogr 46:1–10

86. Wiharto M, Kusmana C, Prasetyo LB, Partomihardjo T (2008) Tree diameter class distribution in various vegetation types on mount Salak, Bogor, west Java. J Ilmu Pertan Indones 13(2):95–102

87. Wu Y (1996) Paulownia agroforestry in China: a contribution to adaptive research. PhD Thesis. Cranfield University, Silsoe

CHAPTER 12

Assessing Soil-Related Black Spruce and White Spruce Plantation Productivity

Shane Furze[1], Mark Castonguay[1], Jae Ogilvie[1], Mina Nasr[1], Pierre Cormier[1], Rolland Gagnon[2], Greg Adams[2], Paul A. Arp[1]

[1]*Forest Watershed Research Center, UNB Faculty of Forestry and Environmental Management, Fredericton, Canada*

[2]*J.D. Irving Ltd., Sussex, Canada*

ABSTRACT

This article focuses on modelling and mapping the productivity of black (Picea mariana) and white spruce (Picea glauca) plantations across the Black Brook forest management area in northwestern New Brunswick, Canada, encompassing about 200,000 ha. This effort involved establishing 3500 50 m^2 survey plots, each informing about: plantation age (15 to 43 years), planted species type, stem count, tree height, basal area, and wood volume. All of this was supplemented with location-specific productivity predictors, i.e., xy location and specifications pertaining to soil type, soil drainage (established through digital elevation modelling by way of the depth-to-water index DTW), and years since thinning (pre-commercial and commercial), and. The DTW index, as it emulates the elevation rise away from open water features such as streams, rivers and lakes, allowed the re-mapping of existing soil borders by topographic position and drainage association. Non-linear regression analysis revealed that plantation height, basal area and volume all increased with plantation age, as to be expected. Pre-commercial thinning in plantations <30 years old had a positive while the more recent commercial thinning still had the negative effect on standing wood volume and mean annual volume increment (MAI). White spruce MAI generally exceeded black spruce (MAI) by a factor of 1.25. Poor and excessive soil drainage reduced MAI. Best growth performances occurred on plantations established on well-drained calcareous soils. The best-fitted results so obtained allowed for generating black and white spruce MAI maps across the forest management area by ridge-to-valley soil and DTW location at 10 m resolution. These maps were subsequently used for site-by-site silvicultural evaluation and ranking purposes.

Keywords: Forest Plantation Productivity Mapping, Black Spruce, White Spruce, Soil Type & Drainage, Digital Terrain Modelling

1. INTRODUCTION

In general, of forest plantation productivities are not only affected by forest management actions including species selection and stocking level to promote plantation growth and yields but also by climate and soil conditions. This article focuses on aspects pertaining to how differences in soil quality and drainage affect the growth of black and white spruce plantations, with soil type referring to differences in landforms and related surface expressions and geological substrates. Likely soil-related growth and yield affecting differences refer to rooting depth, organic matter and coarse fragment content, and nutrient availabilities as modified by rate of soil weathering, litter decomposition, soil moisture regime and climate. Climate related factors refer to length of growing season, the extent of atmospheric deposition in the form of precipitation (rain, snow, and nutrients such as N, S, Ca, Mg and K), and actual evapotranspiration as dictated by monthly precipitation and air temperature patterns (Keys et al., 2016).

Highest soil productivities can be expected to occur on deep, friable and organically enriched soils with well-functioning root systems. In other soils, root growth is often restricted by high coarse fragment content, soil compaction, lack of aeration, low temperature, and/or the presence of toxic substances (Table 1, Arp & Krause, 2002). Typically, plantation growth is better maintained by organically enriched soils on well-drained locations due to prolonged retention of plant available water. On poorly to imperfect drainage locations, insufficient soil aeration due to persisting water saturation restricts root development (Kozlowski, 1985 ; Gale and Grigal, 1987 ; Drever & Lertzman, 2001 ; Crow, 2005 ; Angstmann et al., 2013). For example, Douglas fir and white spruce do not tolerate low soil aeration, but black spruce, red alder, western cedar, Sitka spruce and western hemlock are adapted to shallow water tables. Rooting is also affected

Table 1. General relationships between soil properties and site quality, including optimal conditions.

Soil property	Relation to site quality	Optimal conditions*
Lithology	Affects K, Mg and Ca supply	Coarse mafic; calcareous inclusions
pH	Affects nutrient availability	pH 5 to 7
Soil organic matter content	Source of N and P, promotes aggregation	Deep accumlations, >10%
Texture	Affects drainage, water and nutrient storage	Sandy loam, loam, silt loam
Structure	Promotes aeration, counteracts compaction and high soil strength	Granular
Consistency	Affects soil strength	Friable
Drainage	Determines soil moisture regime and aeration status	Moderate to well (fresh)
Depth of Ah horizon	Location of high biological activity and nutrient availability	>10 cm
Stone content	Improves thermal regime, but limits soil volume for rooting	<15%
Depth to least permeable layer	Limits rooting space, improves moisture regime in coasre soils	>80 cm
Mottling	Reveals poor soil aeration	>75 cm
Thickness of B horizon	Enhances growth through increased clay, organic matter and nutrient content within rooting space	>30 cm

*Classes as defined by the Agriculture Canada Expert Committee on Soil Survey (1987).

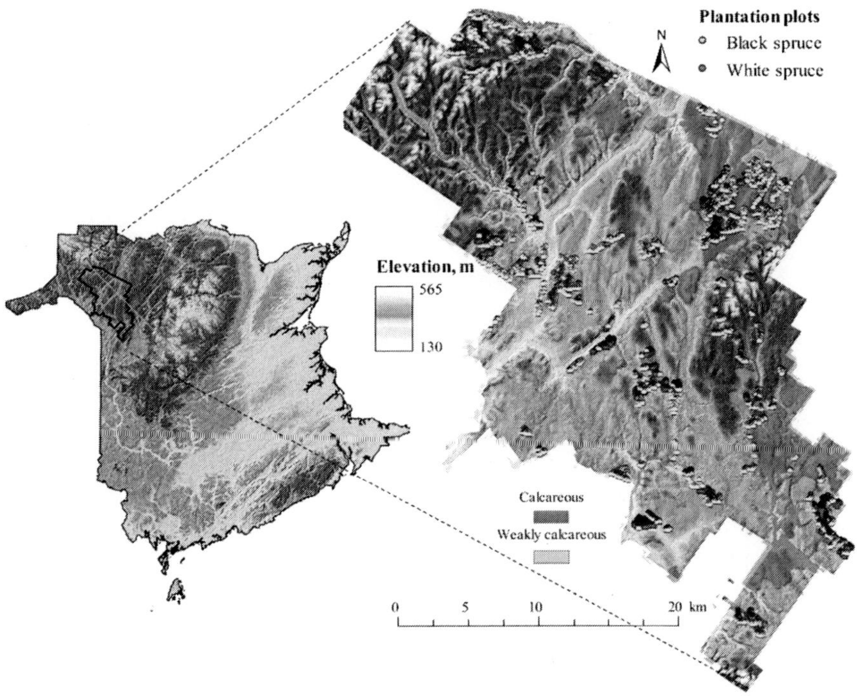

Figure 1. Black Brook forest management area within northwestern New Brunswick, Canada: location, extent, terrain elevation, black and white spruce plantation survey plots, and distribution of soils with calcareous content.

by soil temperatures (Alvarez-Uria and Körner, 2007), being slow and delayed in soils that remain cool as seasons transit from winter to spring and summer. In addition, fine roots are sensitive to acidification, mainly due to acid-induced Al mobilization (Cronan and Grigal, 1994

This article focuses on stem count, tree height, basal area and wood volume growth in select black and white spruce plantations across a 250,000 ha forest management area in northwestern New Brunswick (Figure 1). The research involved discerning how these growth metrics are affected by species type (white versus black spruce), soil type and soil drainage. Soil type was defined by type-specific landform and geological substrate combinations, also referred to as soil associations or catenas. Soil drainage type was adjusted to conform to the DTM-derived depth-to-water pattern (DTW) across the land as generated from a 10-m resampled LiDAR-generated digital terrain model (DTM) for the area.

2. METHODS

2.1. Study Area

The forest management area in northwestern New Brunswick refers to the Black Brook management zone (longitude −67.66; latitude 47.231) spans elevations from 128 to 569 above sea level (Figure 1). Mean annual precipitation amounts to 1104 mm; mean annual air temperature is 3.5°C; mean degree growing days above 5°C amount to 1532.6 (Environment Canada, 2017). The dominant upland species are Sugar Maple (Acer saccharum) (30%), Yellow Birch (Betula alleghaniensis) (11%), and Balsam Fir (Abies balsamea) (10%). Planted stands cover 33% of the management zone and consist off White Spruce (Picea glauca, 42%), Black Spruce (Picea mariana, 40%), and Norway Spruce (Picea abies, 14%). Surface deposits vary from alluvium within floodplains, glacio-fluvial along the floodplains, residuals on ridges and steep slopes, and combinations of ablation and/or basal till along generally undulating to rolling terrain (Langmaid et al. 1976; Colpitts et al., 1995; Fahmy et al., 2002). An overview of forest soil units and associated bedrock formations is provided in Table 2.

2.2. DTM-Based DTW Delineations

The DTM raster at 1-m resolution was generated from the 2014 LiDAR point- cloud data for the area, through systematic point-of-last-return classification, followed by natural neighbour interpolation. This DTM was then re-sampled at 10 m resolution and used to generate the cartographic depth-to-water delineation as developed from the DTM-generated datalayers (Murphy et al., 2011). This delineation involved the following raster-based algorithms: fill, flow direction, flow accumulation, stream network classification (4 ha for stream flow initiation), and determining the cartographic depth-to-water (DTW) pattern as dictated by the least elevation rise away from the stream network pattern for which DTW = 0. Floodplains were delineated in the same way, by setting the upslope flow-initiation area for floodplain initiation requiring about 400 ha of upstream flow accumulation areas. Each floodplain so located was subsequently delineated by limiting the floodplain elevation rise to DTW = 4 m away from the DTW = 0 defining floodplain channels.

Table 2. Overview of forest soils and associated bedrock geology, mode of deposition, and lithology of parent material for the Black Brook management zone, northwestern New Brunswick, Canada. Bedrock Geology retrieved from (New Brunswick Department of Natural Resources N.B. DNR, 2008; Fahmy et al., 2010).

Mode of Deposition	Soil Association	Topography	Texture	Rooting zone depth, cm	Lithology	Number of plots
Residual	Undine	Ridges and hillsides	SiL	55	Calcareous shales, soft	68
	Harquail	Below ridges	L, ClL	30	Partially calcareous, soft	6
Residual to Ablation till	Glassville	Strongly rolling to hilly	L, SiL; slaty	55	Metaquartzites, slates	472
	Quisibis	Ridges and slopes	SiL, L	40	Phyllite, slightly calcareous	14
	Boston Brook	Rolling	L; gravelly	40	Argillites, slates, fine-grained sandstones	22
Ablation till	Monquart	Gently undulating to rolling	SL; gravelly	100	Calcareous shales, quartizte, argillites	26
	Caribou	Gently undulating to rolling	L, SiL, ClL	75	Calcareous shales	204
	Jardine	Gently undulating to gently rolling	L, ClL; gravelly	80	Calcareous shales	370
	Thibault	Well-drained ridges & slopes	SL, L, Sl	140	Weakly calcareous quartzites	16
	Holmesville	Undulating to strongly rolling	L, SiL	45	Quartzites, sandstones (fine-grained)	1291
Basal till	Siegas	Gently rolling	ClL, Cl	40	Quartzite, sandstones, slates, argillites, weakly calcareous in subsoil	291
	Kedgwick	Gently rolling topography	ClL, Cl	50	Calcareous shales	91
Colluvial, water reworked	Victoria	Undulating to steep slopes	SL, SiL; gravelly	40	Shales, sandstones and quartzites	324
	McGee	Slopes towards valleys	SL, SiL; gravelly	80	Shales, sandstones and quartzites	147
Glaciofluvial outwash	Muniac	River terraces	SL; gravelly	50	Calcareous shales	50
	Grand Falls	River terraces	SL; gravelly	100	Gaciofluvial gravel	108

SL: sandy loam, L: loam; SiL: silt loam; ClL: clay loam; Cl: clay.

2.3. Soil Association and Mode of Surface Deposition: Shapefile Adjustments

The DTM-derived slopes and DTW patterns away from the stream and floodplain flow channels were used to hydro-topographically correct the existing soil catena borders to conform to the lake, river, and wetland features of the area. In so doing, open-water features were excluded from the soil polygons, and organic soils would be relocated and/or limited to the extent of the corresponding image-recognized wetland features. Residual soil formations were constrained to occur along steep ridges and slopes only. The lower-lying borders of the till and glacio-fluvial formations were set to coincide with their adjacent DTM-recog- nized floodplains borders. The existing soil drainage specifications within each soil catena were replaced by the stream-channel based DTW-pattern using the following specifications: <10 cm, very poor; 10 to 25 cm, poor; 25 to 50 cm, imperfect; 50 to 100 cm, moderate; 1 to 20 m well, >20 m excessive. In addition, the geological bedrock formation outlines were adjusted to conform to the lithology descriptions per soil association polygon. The soil types so adjusted are listed in Table 2, by mode of surface deposition, texture, and by lithology with emphasis on calcareous versus siliceous content.

2.4. Plantation Survey

In 2007, 3500 50-m^2 circular forest development survey plots were used to generate data for stem count, tree height, basal area and stem volume for select black and white spruce plantations across the Black Brook forest management area. These plots were systematically located within select plantation along a 100 × 100 m sampling grid. All planted and non-planted trees were tallied for each plot by species, diameter at breast height. Plot height was set to be the height of tree closest to each grid centre. These determinations were used to estimate stem density (stems/ha), basal area (m^2/ha), total volume (m^3/ha) and mean annual increment (m^3/ha/year) for each plot. Also noted was plantation age and years since pre-commercial (n = 791) and commercial (n = 203) thinning. Each plot was further characterized by planted species, landform by mode of deposition, soil type, elevation, slope, aspect, and DTW. The number of plots per soil type are listed in Table 2. Due to the nature of the plot survey and the area limitations per soil type, observations per soil type varied widely, from 6 to 1291. In addition, white spruce plantations (n = 1050) were generally established on well drained sites.

2.5. Data Management

All plot-based numerical and categorical data were compiled into a single spreadsheet, with column headings identifying all the above variables, and each row providing the data entries, one plot at a time. Planted species and soil type per plot were binary coded (present 1; not present 0). The plot-specific numbers for DTW numbers were transformed into $\log_{10}$DTW to capture the changing drainage conditions from very wet to dry in a linear fashion. Also added where the DEM-extracted values for slope and elevation for each plot at centre.

2.6. Regression Model and Analysis

The compiled data were subject to multivariate nonlinear regression analysis (stepwise backward). Plot-estimated per hectare stem counts, height, basal area, total volume and mean annual total volume increment (MAI = total wood volume / planation age) were used as dependent variables. All the other numerical and binary variables were entered as independent variables. The regression model was formulated as follows:

for stems:

$$y = a\text{age}^{-b}(1 + c\text{WS})[\text{F}(\text{DTW})](1 + f\text{CL} + g\text{CT})\text{SF};\qquad (1a)$$

for tree height, basal area and total wood volume:

$$y = a[1 - \exp(-b\text{age})]^3 (1 + c\text{WS})[\text{F}(\text{DTW})]$$
$$(\text{Stem\_Count}/2500)^e (1 + f\text{CL} + g\text{CT})\text{SF}\qquad (1b)$$

where WS refers to white spruce (coded 1 when present, zero otherwise for black spruce),

$$\text{F}(\text{DTW}) = x^d (1 - x)^{1/d} \text{ with } x = [\log_{10}(\text{DTW}) + h]/k,\ \text{DTW in m}\qquad (2)$$

a, b, c e, d, f, g, h, k are adjustable coefficients, CL and CT referring to plantations with pre-commercial and commercial thinning operations, and SF as the soil factor, given by

$$\text{SF} = s_1^{\text{soil1}} s_2^{\text{soil2}} s_3^{\text{soil3}} \cdots\ ,\qquad (3)$$

with soil types numbered soil1,soil2,soil3···, coded 1 or 0 for presence and absence, respectively, and with s_1,s_2,s_3··· as adjustable parameters.

Equations ((1a) and (1b)) differ because stem counts tend to decrease with increasing age due to self-thinning and other processes, while tree height, basal area and total stem volume increase asymptotically with age towards upper species-, climate- and soil-moderated limits. The step-wise regression analysis systematically eliminated variables with non-significant predictor coefficients due to low probability ($P < 0.001$) and t-values ($t < 3$). The best-fitted regression models served to generate the predictor equations for stem count, tree height, basal area and total tree volume.

2.7. Plantation Productivity Mapping, by Plantation Type

The best-fitted model for MAI was used to map MAI across the entire forest management area, using ArcMap procedures guided as follows:

1) Two maps were generated, one for black spruce and one for white spruce.
2) Plantation age was fixed at 20 years.
3) Stem count was fixed at 2500stems.ha^{-1}, to reflect typical planting stock density.
4) The best-fitted soil coefficients were assigned to each soil type (catena) where present.
5) The effects of pre-commercial and commercial thinning were not considered.
6) The DTW factor was calculated as specified above (Equation (2)).

2.8. Analytical Limitations

Even with the availability of the 3500 plot determinations, there is the issue of non-equal representation by plantation type (n = 1050 for black spruce; n = 2450 for white spruce) and soil type (Table 2). In addition, there is an uneven plot distribution across the DTW classes, with the majority located within the 1 < DTW < 10 m. As result, it was not possible to determine how plantation growth would be affected by species-specific sensitivities to soil type and topographic position outside this DTW range. For that reason, the h and k parameters in Equation (2) were set equal to h = 2.6, and k = 4.8 for both species, to enable growth predictions across the entire DTW range from essentially open water conditions to steep ridge tops. In addition, all soil parameters were kept common not only by species but also for stem count, tree height, basal area and total wood volume.

3. RESULTS

In total, there were 2450 black spruce and 1050 white spruce plots (Figure 1). The plot-generated variables are presented in Table 3 by listing the basic statistics for plantation age, height, basal area, total volume, MAI, and for the DEM-derived variables $\log_{10}$DTW, slope, elevation. The histograms of these variables (Figure 2) show wide variations for each these variables, and especially so for the stem-counts: from 20 to 20,400 per hectare. The higher stem count numbers were due to dense non-planted species regenerations. Low stem counts occurred on wet spots, on shallow soils with rock exposures, and densely machine-trafficked areas. A similar stem-count range was reported by Pitt and Lanteigne (2008) in their work on un-thinned and thinned balsam fir and red spruce thinning trials in northwestern New Brunswick. These trials were conducted 13 to 16 years

Table 3. Variables informing about site-specific attributes based on plot observations (n = 3500).

Plot-generated variables	Units	Mean	Std. Dev.	Minimum	Maximum
Age	year (a)	23.2	4.7	15	43
Since PC-T	year (a)	2.1	4.2	0	18
Since C-T	year (a)	0.37	1.91	0	14
Stems	ha^{-1}	3805	21,805	200	20,400
Height	m	12.3	2.9	4.7	23.34
Basal area	$m^2\ ha^{-1}$	28.5	11.7	0.3	73.4
Total wood volume	$m^3\ ha^{-1}$	122	69	1	428
MAI	$m^3\ ha^{-1}\ a^{-1}$	5.1	2.4	0	15.2
Elevation	m	273	49	184	5010
Slope	%	8.7	6.8	0	53.4
$\log_{10}$DTW	$\log_{10}$m	0.58	0.46	−1.87	1.9

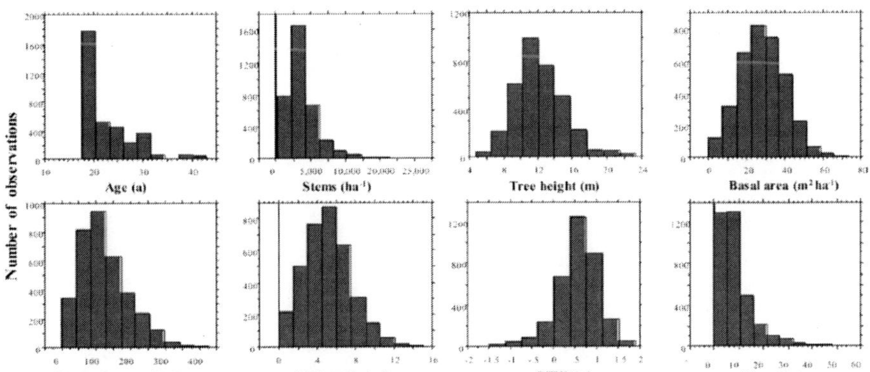

Figure 2.Histograms for the plot-generated plantation variable attributes (n = 3500).

within harvest blocks that were clear-cut between 1946 and 1955, with post-thinning stem counts, basal areas, total volumes and other forest metrics recorded at 5-year intervals up to 43 years after thinning. In comparison with these trials, the above plot-estimated values for basal areas and total volumes from plantation age 15 to 43 correspond to the Green River results 5 to 33 years after thinning. Hence, the plantations already reached similar basal area and total volume growth at 15 years after establishment than what was noted for the naturally re-generating un-thinned and thinned Green River blocks 20 years after harvesting.

3.1. Best-Fitted Regression Coefficients

The best-fitted regression coefficients for stem count tree height, basal area and total volume by way of Equation (1) are listed in Table 4 and Table 5. Also shown are their least-squares coefficient estimates, and associated errors of esti mation and t-values.

Table 4. Best-fitted regression coefficients for stem count tree height, total wood volume, and basal area by plantation age, white versus black spruce, soil moisture regime (quantified by way of DTW), and thinning intervention (all *p* values <0.001 except larger where absolute *t*-values <4).

	Parameters	Stem Count			Tree Height			Basal Area			Total Wood Volume		
		ha^{-1}			m			$m^2\ ha^{-1}$			$m^3\ ha^{-1}$		
		Coeff.	St. Error	t-value	Coeff.	St. Error	t-value	Coeff.	St. Error	t-value	Coeff.	St. Error	t-value
	a	**3.00**	0.40	7.5	**85.5**	1.4	61.5	**472**	42	11.3	**3782**	597	6.3
b	Age adjustor	**0.46**	0.04	10.7	**0.090**	0.001	64.0	**0.044**	0.002	20.2	**0.029**	0.002	13.7
c	White spruce adjustor	**−0.24**	0.01	−17.0	**0.035**	0.006	5.8	**0.107**	0.016	6.8	**0.25**	0.03	9.6
d	f(DTW) exponent	**1.24**	0.02	55.1	**1.43**	0.01	106.7	**1.60**	0.04	39.6	**1.57**	0.03	53.0
e	Stem count exponent	**not applicable**			**−0.42**	0.015	-27.7	**-0.15**	0.04	-3.9	**0.09**	0.05	1.8
f	Pre-commercial thinning adjustor	**−0.068**	0.019	−3.5	**−0.042**	0.006	−6.6	**0.017**	0.020	0.9	**0.074**	0.025	3.0
g	Commercial thinning adjustor	**−0.58**	0.03	−19.2	**−0.25**	−0.02	16.5	**−0.50**	0.03	−16.4	**−0.25**	0.05	−5.0
	R^2	0.239			0.583			0.414			0.535		

Table 5. Best-fitted regression coefficients for stem count tree height, total wood volume, and basal area by soil type (all p-values < 0.001).

Soil type	Stem Count			Tree Height			Basal Area			Total Wood Volume		
	ha^{-1}			m			$m^2\ ha^{-1}$			$m^3\ ha^{-1}$		
	Coeff.	St. Error	t-value	Coeff.	St. Error	t-value	Coeff.	St. Error	t-value	Coeff.	St. Error	t-value
Muniac	0.51	0.06	8.5	0.79	0.03	26.6	0.87	0.08	10.8	0.28	0.05	5.2
Siegas	0.53	0.04	14.9	0.92	0.02	36.9	0.75	0.06	13.1	0.41	0.03	12.3
Victoria	0.61	0.04	15.0	0.92	0.03	36.4	0.75	0.06	13.2	0.50	0.04	12.2
Monquart	0.65	0.08	8.3	0.93	0.04	24.0	0.83	0.10	8.7	0.51	0.08	6.7
Glassville	0.72	0.04	16.3	0.99	0.03	37.8	0.77	0.06	13.4	0.58	0.04	13.1
McGee	0.67	0.05	14.6	0.99	0.03	35.5	0.84	0.07	12.6	0.60	0.05	12.7
Boston Brook	0.84	0.08	10.9	0.95	0.04	24.8	0.62	0.09	7.1	0.66	0.07	8.9
Holmesville	0.74	0.04	16.8	1.00	0.03	38.4	0.84	0.06	14.0	0.67	0.05	14.3
Grand Falls	0.73	0.05	13.9	0.96	0.03	34.3	0.86	0.07	12.2	0.67	0.06	11.9
Quisibis	0.85	0.11	7.5	0.94	0.05	18.9	0.93	0.12	7.8	0.72	0.13	5.5
Harquail	0.87	0.16	5.3	0.91	0.07	13.1	0.99	0.16	6.3	0.77	0.20	3.8
Thibault	0.88	0.10	8.6	0.97	0.05	20.0	1.09	0.12	9.3	0.80	0.12	6.7
Jardine	0.83	0.05	16.4	1.10	0.03	37.5	1.02	0.07	13.6	0.87	0.06	14.5
Caribou	1.04	0.06	16.7	1.10	0.03	37.7	1.11	0.08	13.8	1.09	0.08	14.3

Stem counts within the black and white spruce decreased with age, as quantified by the age exponent, i.e.,b = 0.45. In comparison, white spruce generally had lower stem counts than black spruce, i.e., the stem count adjustment for white spruce was negative, i.e.c = 0.240. This would be due to the difference in canopy extent, being wider and therefore more vegetation suppressant and less light transmitting than black spruce (Canham et al., 1994, Power et al., 2012). Not surprisingly, thinning had a negative effect on stem count, but this effect was quite small as per the pre-commercial thinning adjustment factor given by (1 – fCL)=0.932. In contrast, the commercial thinning operations reduced the stem counts by a factor of (1–gCT)= 0.421, respectively. This difference may be due to continuing in-growth of natural regenerating species, and this ingrowth would have occurred over a longer time period since the pre- commercial than the commercial thinning operations.

Tree height also increased with plantation age in a non-linear fashion, i.e., starting slow at first before accelerating for about forty years. Thereafter, height growth is expected to slow down, as formulated by the $\left[1-\exp(-b\text{Age})\right]^3$ expression in Equation (1b), for which the best-fitted age adjustor for tree height equals 0.090 year^{-1} (Table 4). High stem counts had a depressing effect on tree height as indicated by the negative stem-count exponent for tree height, i.e., $(Stem\_count/2500)^{-0.42}$. This would likely be due to conspecific neighbourhood competition as reported by Smith et al. (2015). In addition, height was slightly taller for white than for black spruce: white spruce height = 1.035 black spruce height. Hence, for same soil conditions, a 20 m tall black spruce tree would be equivalent to a 20.7 m tall white spruce, all other conditions (age, time since thinning, stem counts) being the same. Generally, tree height was slightly depressed by the thinning operations, with the commercial thinning operations having a somewhat larger effect than the pre-commercial thinning operations. This could be due to a removal of some of the taller non-planted species. Generally, thinning operations have no effect on tree height, but there are positive and negative exceptions as well (Lanner, 1985; Masaka et al., 2013; McKinnon et al., 2006).

Both basal area and total wood volume accelerated with age within the plantation age range up to 40 years, with white spruce outperforming black spruce in total basal area and wood volume accumulations by a factor of 1.07 and 1.25, respectively (Table 4). The latter number is higher than what was found by Mallik et al. (2008) by way of white versus black spruce experimentation, but similar to what was reported by Homagain et al. (2011) for same aged black and white spruce plantations in northwestern Ontario. Hence, the wood volume productivity is 25% higher for white spruce than for black spruce plantations, all other conditions (age, soil type, time since thinning) being the same. These differences, however, may increase or decrease as the plantations age further (Homagain et al., 2011). In terms of the post-thinning effects, there were small but significant basal area and wood volume benefits stemming from the pre-commercial but not yet from the more recent commercial thinning operations. This would be due to the intended basal area and wood volume reductions at the time of commercial thinning, and could in part also be due to post-com- mercial thinning mortality (Thorpe et al., 2008).

With respect to soil drainage, stem count, tree height, basal area and total wood volume were all similarly affected, being optimal within the $1 < DTW < 10$ m range, with tree height, basal area and wood volume productivity somewhat more sensitive to extreme soil moisture regimes than stem count. This is in agreement with Wang and Klinka (1996) who provided the following metric tree height index assessment of Ontario in terms of the soil moisture regime variations: moderately dry, 15.8 m; slightly dry, 20 m; fresh, 20.5 m; moist, 20 m; very moist 16.3 m; wet, 13 m; very wet, 5.8 m.

With respect to soil type, there were also significant differences for stem count, tree height, basal area and total wood volume, as listed in Table 5. In particular, this list revealed that the effects of soil type on wood volume productivity were most strongly correlated to the soil type contributions pertaining to basal area and tree height, but were only weakly correlated to the soil contributions pertaining to stem count (Table 6). Sorting the soil contributions to wood volume (or MAI) from low to high indicated that the greatest gains in wood volume occurred on well-drained and medium texture calcareous soils (Figure 3). The lowest gains occurred on coarse-textured and gravelly outwash plain soils (Muniac) and fine-textured basal till soils (Siegas). Among the soils on siliceous substrate, plantation productivity decreased with increasing stoniness: least for Holmesville, and most pronounced for Victoria. The ranking of the best-fitted regression coefficients between total wood volume and soil type generally agrees with the soil property descriptions in Table 2, whereby the Caribou, Undine and Kedgwick soils encompass most if not all of the optimal soil conditions as per Table 1. The opposite applies to the Muniac and Siegas soils.

The multivariate regression results varied in the amount of plot-to-plot variance explained, i.e., $R^2 = 0.228$ for stem counts, 0.588 for tree height, 0.399 for basal area, and 0.530 for tree volume (Table 4). Hence, much of the plot-to-plot variations remain unresolved by way of Equations (1)-(3), and especially so for stem count. Choosing MAI as dependent variable explained 33% of the overall MAI variations. The spread of the regression-generated best-fitted residuals for tree height, basal area, total wood volume and MAI are plotted in Figure 4 and Figure 5 by way of scatterplots and histograms. These results are similar to what is normally found across plot-based modelling efforts, as documented by, e.g., Pinjuv et al. (2007) in Dykstra (2007). Using actual instead of modelled stem counts improved the analytical results slightly by increasing regression-captured variations towards R^2 values amounting to 0.613 for basal area, 0.676 for total wood volume, and 0.434 for MAI. This small increase is not surprising since the plot-based basal area and volume estimates already include the plot-

Table 6. Best-fitted multiple regression results for the best-fitted soil type contributions to total wood volume (Table 5), using the best-fitted soil type contributions for basal area, tree height and stem count (Table 5) as predictor variables.

Variable	Coeff.	Std. Error	t-Value	P-Value
Intercept	−1.00	0.09	−10.6	<0.0001
Basal Area	0.89	0.07	12.5	<0.0001
Tree height	0.81	0.12	6.8	<0.0001
Stem Count	0.23	0.06	3.9	0.0024

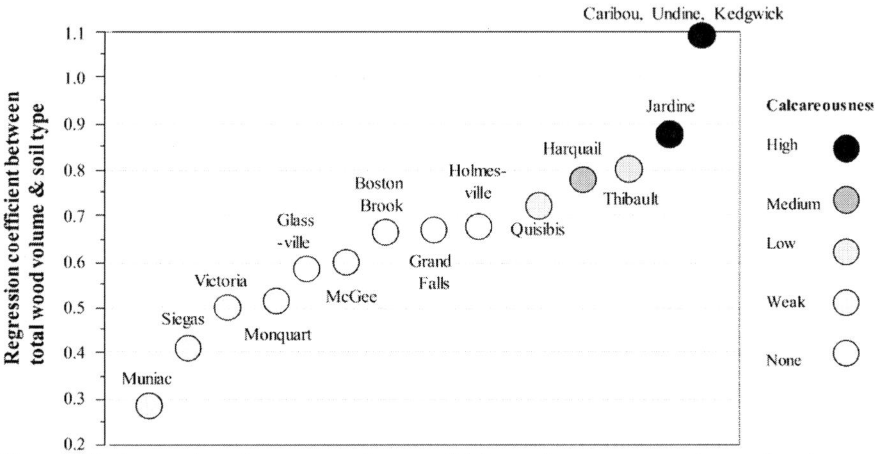

Figure 3. Arranging the regression coefficients between total wood volume and soil type (Table 5) from low to high.

generated stem-count estimates. By plot-to-model comparisons, the following 80% conformance results were obtained: ±1300 stems ha^{-1}, ±3.8 m for tree height, ±11 m · ha^{-1} for basal area, ±55 m^3 · ha^{-1} for total wood volume, and 2.2 m^3 · ha^{-1} · $year^{-1}$ for MAI (Figure 6).

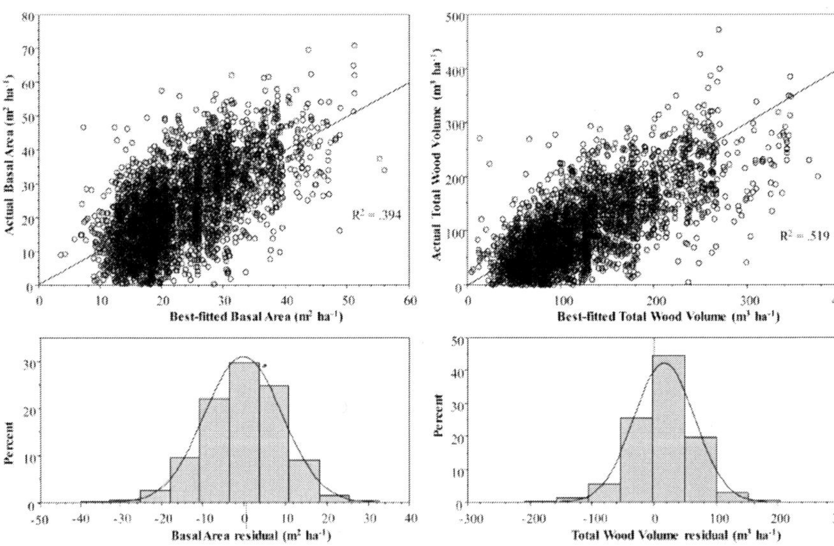

Figure 4. Best-fitted scatterplots and histograms for the best-fitted residuals for basal area and total wood volume (n = 3500).

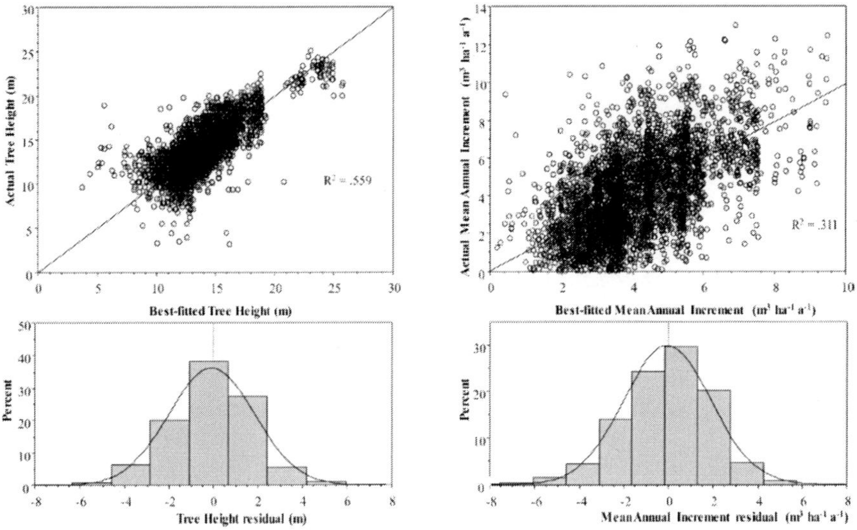

Figure 5. Best-fitted scatterplots and histograms of the best-fitted residuals for plantation height, and MAI (n = 3500).

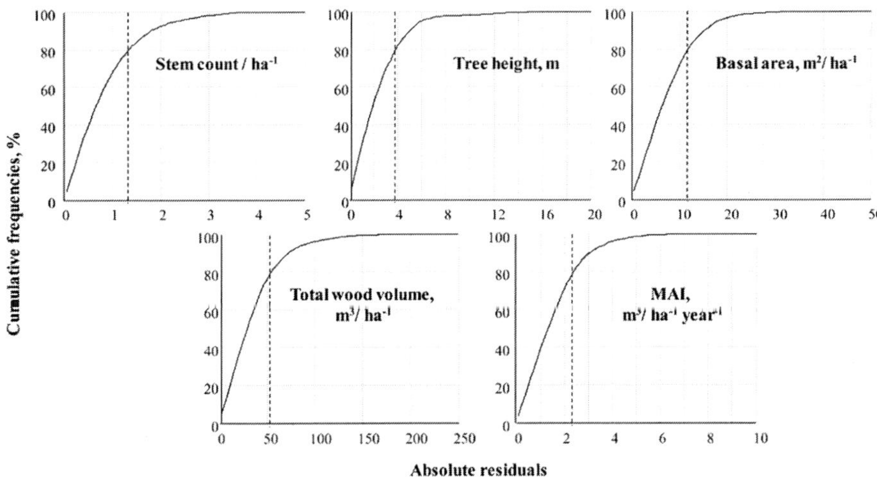

Figure 6. Conformance plots showing the cumulative frequencies of the absolute residual between the individual plot determinations and corresponding best-fitted model results for stem count, tree height, basal area, total wood volume, and mean annual increment (MAI).

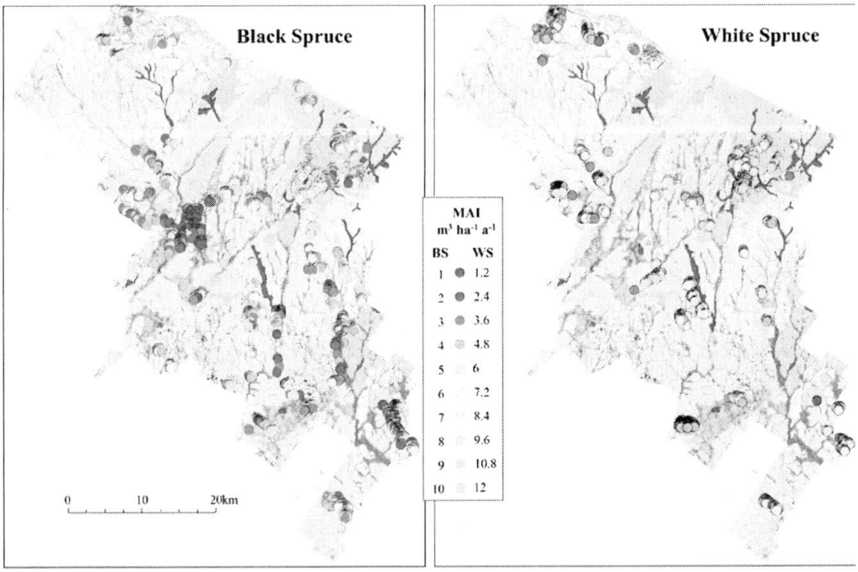

Figure 7. Mean Annual Productivity (MAI) projections [Equations (1b)-(3)] for 20-year old black (left) and white (right) spruce plantations within the Black Brook Forest Management area, with plot-generated MAI estimates (dots) overlaid.

3.2. Limitations

The above results do not match the precision that is generally needed for forest related growth and yield predictions. This is in part due to:

i. Procedural issues, whereby the plots were selected for a one-time grid evaluation of the overall plantation growth response variations. For best growth and yield results, plots need to be re-measured at pre-determined intervals.

ii. The age-limited distribution of the plantations. This limitation does not allow for calibrating Equations (1)-(3) beyond 40 years. The exponent of the age-dependent part, i.e. $\left[\left(1-\exp\left(1-b\ \text{age}\right)\right)\right]^3$, was therefore set at 3 to allow for the curvilinear fit for height, basal area and total wood volume within the 0 to 40-year range. The extent to which this determination remains valid beyond age 40 requires further research.

iii. The unequal distribution of the plots by age, species and soil type and drainage conditions. This limits the extent to which the above results fully represent the entire variations across these conditions.

For example, the results obtained for the less frequent soil types would be more influenced by chance occurrences and outliers than the results obtained for the more frequent soil types.

iv. The need to map growth-affecting soil properties, from ridge tops to valleys mapping, with soil properties referring to soil rooting depth, texture, organic matter, coarse fragment content and mineralogy. Doing this requires further in-depth studies and model calibrations.

v. For generalization purposes, the evaluations need to be placed into the context of regionally varying climate and atmospheric deposition conditions (Keys et al., 2016).

In addition to the above plot-based evaluations, one could use LiDAR-genera- ted forest metrics for stem counts, tree height, basal and total wood volume at the tree-by-tree level, with, e.g., 20 × 20 or 25 × 25 m^2 summations (Strunk et al. 2008; Woods et al., 2011; Treitz et al., 2012; Wulder et al., 2012; Latifi et al., 2015). In doing this, Tompalski et al., 2016 reported the following relative root mean square differences between air-borne laser scanned and stand-matched yield curve projections regarding select forest stands on Vancouver Island: for maximum tree height, 31.1%; for basal area, 19.8%; for total stem volume, 21.8%; all at a stand-level age of 80 years. This was done through single-tree summations for 9559. 25 × 25 m^2 grid plots representing four western species groups (hemlock, cedar, fir/spruce, alder). Overall tree height and stem wood volumes amounted to 33 m, and 576 $m^3 \cdot ha^{-1}$.

Since productivity may range from low to high on imperfectly to poorly drained soil based on micro-topographic mound to pit variations, it would be important to examine this aspect as well. However, whether this detail can be generated from LiDAR-DTMs remains to be seen, because point densities for LiDAR-generated ground reflections can be quite low underneath densely spaced forest plantations. It may, however be possible to infer the extent of mound-and pit surface expressions based on soil type characteristics that lead to wind-in- duced tree uprooting due to shallow rooting.

3.3. MAI Productivity Mapping

The best-fitted Equations (1)-(3) were used to produce white and black spruce MAI productivity maps at age 20 across the entire Black Brook forest management area, as shown in Figure 7, with excerpts in Figure 8(a) and Figure 8(b). Overlaid on these maps are the plot-based black and white spruce plantation MAI determinations. In general, these maps illustrate that the overall landscape dependence of MAI is—apart from plantation age and

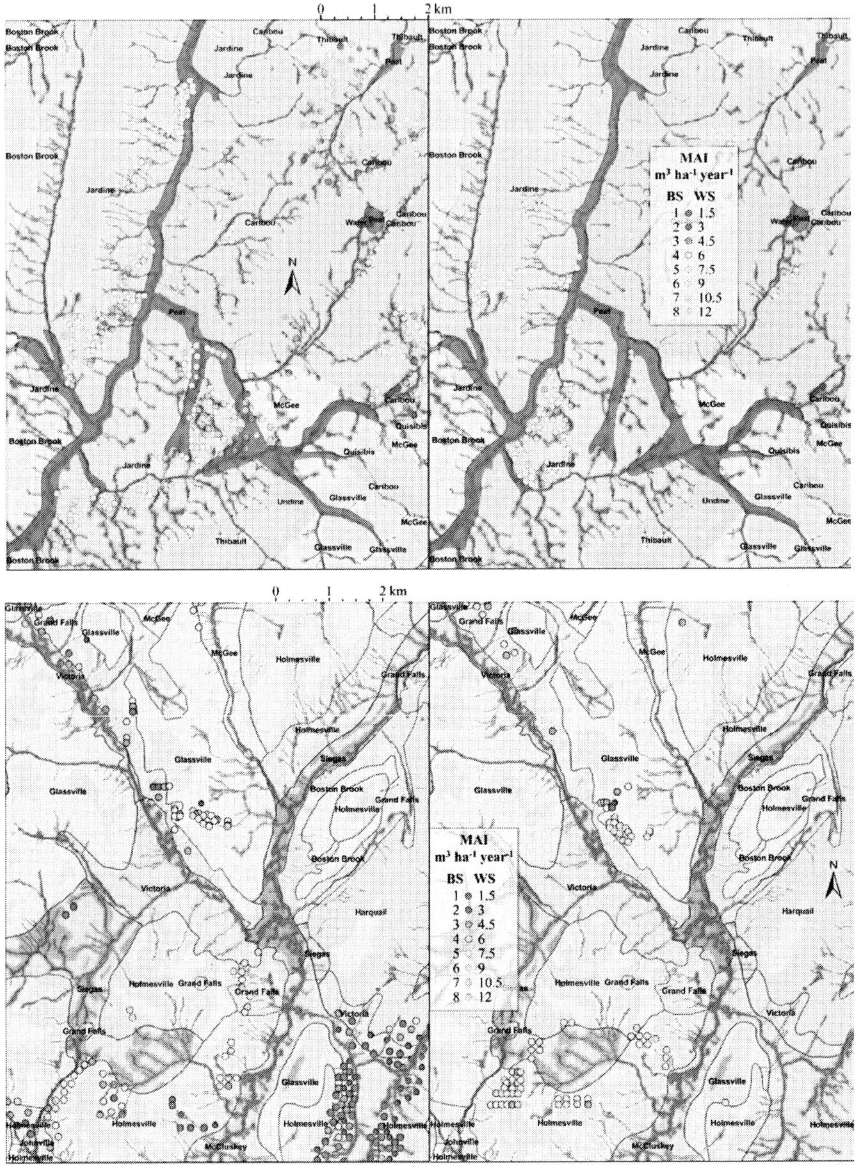

Figure 8. Actual versus mapped MAI conformance details from Figure 7 for the black and and white spruce plantations (left and right sides of panels, respectively) within the Black Brook Forest Management area. Top: part of the top eastern management area, centered on mostly calcareous soils (Caribou, Jardine). Bottom: part of the mid-western management area, centered on siliceous soils (Glassville, Holmesville, Victoria, Grand Falls).

species—mainly affected by soil type and drainage. In detail, MAI is lowest in wet spots, is best along the lower portions of rolling to hummocky slopes, and gradually decreases from there towards steeper ridge tops. The positive effect of soil type on MAI is most pronounced on deep and well-drained soils with calcareous content, such as the Caribou, Jardine, Kedgwick and Undine soils.

4. CONCLUDING REMARKS

Examining the results of the 3500 forest development survey plots in relation to plantation age, planting stock, thinning operations, soil type and drainage allowed for the development of a generalizable model (Equations (1)-(3)). The resulting best-fitted model was used to estimate and predict the productivities of black and white spruce plantations by soil type, from ridge tops to valleys based on theoretical considerations: e.g., no growth on wet soils and fully exposed bedrock, and varying growth expectations by soil type as affected by substrate lithology, texture and accessible rooting depth. In view of currently forthcoming forest metrics that detail tree-by-tree performances in terms of stem count, height, basal area and wood volume within, e.g., each 20 × 20 m^2 square across entire forest management areas, more of this could be done at a much finer resolution. Doing so is promising because—even in view of the above limitations—about one third to one half of the plot-based productivity variations for tree height, basal area and total wood volume can be quantified using Equations (1)-(3). While the results so generated are still insufficient for precise growth and yield predictions across the plantations and beyond the existing age range, they are providing a starting point for further soil factor quantifications in forest management. At this stage, the mapped white and black spruce projections in Figure 6 and Figure 7 provide a semi-quantitative bases for stand- and site- specific ranking purposes, and—as such—are being used for directing silvicultural operations (harvesting, site preparations, seedling stock selection planting, thinning) away from low productivity sites that are either too wet or too dry.

ACKNOWLEDGEMENTS

This work was supported by NSERC by way of a Discovery grant, contributions from the NSERC-supported AWARE project, and from the Canadian Fibre Centre (Dynamic Forest Inventory Project). Special thanks go to J.D. Irving, Ltd. for data access (forest inventory, plot data, and LiDAR DEM).

REFERENCES

1. Alvarez-Uria, P., & Korner, C. (2007). Low Temperature Limits of Root Growth in Deciduous and Evergreen Temperate Tree Species. Functional Ecology, 21, 211-218. https://doi.org/10.1111/j.1365-2435.2007.01231.x
2. Angstmann, J. L., Ewers, B. E., Barber, J., & Kwon, H. (2013). Testing Transpiration Controls by Quantifying Spatial Variability along a Boreal Black Spruce Forest Drainage Gradient. Ecohydrology, 6, 783-793.
3. Canham, C. D., Finzi, A. C., Pacala, S. W., & Burbank, D. H. (1994). Causes and Consequences of Resource Heterogeneity in Forests: Interspecific Variation in Light Transmission by Canopy Trees. Canadian Journal of Forest Research, 24, 337-349. https://doi.org/10.1139/x94-046
4. Colpitts, M. C., Fahmy, S. H., MacDougall, J. E., Ng, T. T., McInnis, B. G., & Zelazny, V. F. (1995). "Forest Soils of New Brunswick" and Maps, CLBRR Contribution (pp. 95-38). Fredericton: New Brunswick Dept. of Nat. Resources and Energy, Agriculture and Agri-Food Canada and Canadian Forestry Service, Natural Resources Canada.
5. Cronan, C. S., & Grigal, D. F. (1994). Use of Calcium/Aluminum Ratios as Indicators of Stress in Forest Ecosystems. Journal of Environmental Quality, 24, 209-226. https://doi.org/10.2134/jeq1995.00472425002400020002x
6. Crow, P. (2005). The Influence of Soils and Species on Tree Root Depth (7 p). Edinburgh: Forestry Commission.
7. Drever, C. R., & Lertzman, K. P. (2001). Light-Growth Responses of Coastal Douglas-Fir and Western Red Cedar Saplings under Different Regimes of Soil Moisture and Nutrients. Canadian Journal of Forest Research, 31, 2124-2133. https://doi.org/10.1139/x01-149
8. Environment Canada (2017). Canadian Climate Normals 1981-2010 Station Data: Edmunston, New Brunswick. http://climate.weather.gc.ca/climate_ normals/index_e.html#1981
9. Fahmy, S. H., Hann, S. W. R., & Jiao, Y. (2010). Soils of New Brunswick: The Second Approximation. Technical Publication Number: NBSWCC-PRC 2010-01, Fredericton: Agriculture and Agri-Food Canada.
10. Gale, M. R., & Grigal, D. F. (1987). Vertical Root Distributions of Northern Trees in Relation to Successional Status. Canadian Journal of Forest Research, 17, 829-834. https://doi.org/10.1139/x87-131
11. Homagain, K., Shahi, C. K., Carmean, W. H., Mathew Leitch, M., & Bowling, C. (2011). Growth and Yield Comparisons for Red Pine, White Spruce and Black Spruce Plantations in Northwestern Ontario. The Forestry Chronicle, 87, 494-503. https://doi.org/10.5558/tfc2011-048
12. Keys, K., Noseworthy, J. D., Ogilvie, J., Burton, D. L., & Arp, P. A. (2016). A Simple Geospatial Nutrient Budget Model for Assessing Forest Harvest Sustainability across Nova Scotia. Open Journal of Forestry, 6, 420-444. https://doi. org/10.4236/ojf.2016.65033
13. Kozlowski, T. T. (1985). Soil Aeration, Flooding, and Tree Growth. Journal of Arboriculture, 11, 85-96.
14. Langmaid, K. K, MacMillan, J. K., & Losier, J. G. (1976). Soils of Northern Victoria County New Brunswick (152 p.). Seventh Report of the New Brunswick Soil Survey, Canada Department of Agriculture.
15. Lanner, R. M. (1985). On the Insensitivity of Height Growth to Spacing. Forest Ecology and Management, 13, 143-148.

16. Latifi, H., Fassnacht, F. E., Müller, J., Tharani, A., Dech, S., & Heurich, M. (2015). Forest Inventories by LiDAR Data: A Comparison of Single Tree Segmentation and Metric-Based Methods for Inventories of a Heterogeneous Temperate Forest. International Journal of Applied Earth Observation and Geoinformation, 42, 162-174.
17. Mallik, A. U., Kamal Hossain, M. K., & Lamb, E. G. (2008). Species and Spacing Effects of Northern Conifers on Forest Productivity and Soil Chemistry in a 50-Year-Old Common Garden Experiment. Journal of Forestry, 106, 83-90.
18. Masaka, K., Sato, H., Torita, H., Kon, H., & Fukuchi, M. (2013). Thinning Effect on Height and Radial Growth of Pinus thunbergii Parlat. Trees with Special Reference to Trunk Slenderness in a Matured Coastal Forest in Hokkaido, Japan. Journal of Forest Research, 18, 475-481. https://doi.org/10.1007/s10310-012-0373-y
19. McKinnon, L. M., Kayahara, G. J., & White, R. G. (2006). Biological Framework for Commercial Thinning Even-Aged Single-Species Stands of Jack Pine, White Spruce, and Black Spruce in Ontario (Report TR-046). Northeast Science and Information Section: Ontario Ministry of Natural Resources.
20. Murphy, P. N. C., Ogilvie, J., Meng, F.-R., White, B., Bhatti, J. S., & Arp, P. A. (2011). Modelling and Mapping Topographic Variations in Forest Soils at High Resolution: A Case Study. Ecological Modelling, 222, 2314-2332.
21. New Brunswick Department of Natural Resources (2008). Bedrock Geology of New Brunswick. Minerals Policy and Planning Division. Map NR-1. Scale 1: 500,000 (2008 ed.).
22. Pinjuv, G., Mason, E. G., & Watt, M. (2007). Incorporating an Index of Root Zone Water Balance into and Existing Growth and Yield Model Can SPBL (1.2). In D. D. Dykstra (Ed.), Forest Growth and Timber Quality: Crown Models and Simulation Methods for Sustainable Forest Management (pp. 81-97). Collingdale, PA: DIANE Publishing.
23. Pitt, D., & Lateigne, L. (2008). Long-Term Outcome of Pre-Commercial Thinning in Northwestern New Brunswick: Growth and Yield of Balsam for and Red Spruce. Canadian Journal of Forest Research, 38, 592-610. https://doi.org/10.1139/X07-132
24. Power, H., LeMay V., Berninger, F., Sattler, D., & Kneeshaw, D. (2012). Differences in Crown Characteristics between Black (Picea mariana) and White Spruce (Picea glauca). Canadian Journal of Forest Research, 42, 1733-1743. https://doi.org/10.1139/x2012-106
25. Smith, J., Harvey B. D., Koubaa, A., Brais, S., & Mazerolle, M. K. (2016). Sprucing up the Mixed Woods: Growth Response of White Spruce (Picea glauca) to Partial Cutting in the Eastern Canadian Boreal Forest. Canadian Journal of Forest Research, 46, 1205-1215. https://doi.org/10.1139/cjfr-2015-0489
26. Strunk, J. L., Reutebuch, S. E., & Foster, J. R. (2008). LiDAR Inventory and Monitoring of a Complex Forest. In ASPRS 2008 Annual Conference Portland (8 p.).
27. Thorpe, H. C., Thomas S. C., & Caspersen, J. P. (2008). Tree Mortality Following Partial Harvests Is Determined by Skidding Proximity. Ecological Applications, 18, 1652-1663. https://doi.org/10.1890/07-1697.1
28. Tompalski, P., Coops, N. C., White, J. C., & Wulder, M. A. (2016). Enhancing Forest Growth and Yield Predictions with Airborne Laser Scanning Data: Increasing Spatial Detail and Optimizing Yield Curve Selection through Template Matching. Forests, 7, 255. https://doi.org/10.3390/f7110255
29. Treitz, P., Lim, K., Woods, M., Pitt, D., Nesbitt, D., & Etheridge, D. (2012). LiDAR Sampling Density for Forest Resource Inventories in Ontario, Canada. Remote Sensing, 4, 830-848. https://doi.org/10.3390/rs4040830
30. Wang, G. G., & Klinka, K. (1996). Use of Synoptic Variables in Predicting White Spruce Site Index. Forest Ecology and Management, 80, 95-105.
31. Woods, M., Pitt, D., Penner, M., Lim, K., Nesbitt, D., & Treitz, P. (2011). Operational Implementation of a LiDAR Inventory in Boreal Ontario. The Forestry Chronicle, 87, 512-528. https://doi.org/10.5558/tfc2011-050

32. Wulder, M. A., White, J. C., Nelson, R. F., Næsset, E., Ørka, H. O., Coops, N. C., Hilke, T., Bater, C. W., & Gobakken, T. (2012). Lidar Sampling for Large-Area Forest Characterization: A Review. Remote Sensing of Environment, 121, 196-209.

CHAPTER 13

Post-Fire Restoration Plan for Sustainable Forest Management in South Korea

Soung-Ryoul Ryu[1], Hyung-Tae Choi[2], Joo-Hoon Lim[2], Im-Kyun Lee[3] and Young-Sang Ahn[4,*]

[1]*Department of Renewable Resources, University of Alberta, Edmonton, AB T6G 5H5, Canada*

[2]*Division of Forest Restoration, National Institute of Forest Science, Seoul 02455, Korea*

[3]*Division of Forest Genetic Resources, National Institute of Forest Science, Suwon 16631, Korea*

[4]*Division of Forest Resources, College of Agriculture and Life Sciences, Chonnam National University, 77, Yongbong-ro, Buk-gu, Gwangju 61186, Korea*

ABSTRACT

This review was to determine a standard post-fire restoration strategy for use in South Korea according to the magnitude of the damage and the condition of the affected site. The government has strongly enforced reforestation in deforested areas as well as fire prevention and suppression since the 1960s. These efforts have successfully recovered dense even-aged forests over the last five decades. However, high fuel loading and the homogeneous structure have made forests vulnerable to large fires. In recent years, large forest fires have occurred in the eastern coastal region of Korea. Forest fires can significantly influence the economic and social activities of the residents of such affected forest regions. Burned areas may require urgent and long-term restoration strategies, depending on the condition of the affected site. Erosion control is the most important component of an urgent restoration and should be completed before a rainy season to prevent secondary damage such as landslides and sediment runoff in burned areas. Long-term restoration is necessary to renew forest functions such as timber production, water conservation, ecosystem conservation, and recreation for residents. Sound restoration for burned areas is critical for restoring healthy ecological functions of forests and providing economic incentives to local residents.

Keywords: ecosystem function; large burned area; long-term restoration; urgent restoration

1. INTRODUCTION

Climate change is expected to increase dryness globally because of the warmer temperatures and reduced precipitation [1]. Overall, this tendency will lead to increasing climatic fire risk. Widespread increases in fire activity, including the areas that are burned [2,3], number of large fires [4,5], and fire season length [6,7], have been apparent worldwide over the past half century [8,9]. However, this trend is also influenced by fuel availability, which is determined by both the quantity of fuel and climate-driven fuel moisture [1,4]. Although lower air humidity and fuel water content are positively related to fire ignition and propagation, a drier climate will eventually lead to a decrease in fuel load due to lower productivity [10,11].

Fire affects soil properties, because any organic matter (OM) that is located on or near the soil surface, is combusted [4]. These results depend on fire severity and intensity. The changes in OM, in turn, affect several chemical, physical, and microbiological properties of the underlying soil. Although some nutrients are volatilized and lost, most nutrients become more available. Fire acts as a rapid mineralizing agent [12] that releases nutrients instantaneously, in contrast to natural decomposition processes, which may require years or even decades. Fire is integral in maintaining the stability and sustainability of forest health and thereby facilitating the architecture of overstory trees and the composition of vegetation communities [4,13]. The interactions of historic land use and climate change have resulted in high levels of fuel loading, which have produced historically uncommon frequent, large, and severe stand-replacing crown fires [4,14,15]. The effects of these fires include increased post-fire tree mortality [4], initial decreases in the understory plant cover [16] and the subsequent susceptibility of the landscape to invasion of non-native species, soil erosion, and flooding [16,17]. Highly severe fires generally cause greater and longer-lasting effects [4,18] and significantly reduce the timber supply and forest by-products [19]. Forest fires can also decrease the income of farmers [4,18,20]. Forest fires can change the habitat quality for wildlife by altering the abundance, distribution, productivity, and diversity of the animals occupying these habitats [21,22,23,24]. Many animals often require more than one type of habitat due to their different requirements for each stage of their life cycles. A mosaic of habitats is thus ideal for a rich biodiversity of wildlife [25,26].

The growing stock (m^3 of wood ha^{-1}) of Korean forests increased from 9.6 (1960) to 142.2 (2014) owing to advanced silvicultural practices and fire prevention efforts [27]. However, higher forest growing stock means high fuel loading, which may lead to more intense and larger forest fires [28]. Despite both the central and local governments' fire suppression efforts in recent years [4], the size and frequency of forest fires have been increasing in South Korea and are expected to continue to increase under the changing global climate. This, consequently, requires developing a standard protocol for prompt and sound restorations. A sound restoration protocol should be built on thorough understandings of the effects of forest fires on forest ecosystems and the post-fire recovery processes. The Korea Forest Research Institute (KFRI) has designated areas burned by the Goseong forest fire in 1996 (100 ha) and the Donghaean forest fire in 2000 (400 ha) as long-term ecological research (LTER) sites [29]. Multidisciplinary joint studies have been conducted at the LTER sites to better understand the impact of forest fires on the ecosystems and the recovery process [4,20]. As a result, standard restoration principles have been established by integrating ecological, social, and economic factors to support the restoration of burned forest [4,29]. The objective of this study was to summarize and describe a standard restoration plan based on the identified principles to maintain sound and sustainable forest ecosystems in Korea. In addition, we provide an update on the current state of restoration projects and suggest improved restoration methods. In particular, we discuss approaches with apparent adaptive value in fire-prone environments and methods for implementing ecologically sound post-fire forest restoration practices.

2. FORMATION OF MODERN KOREAN FORESTS

Korean forests were degraded during the Japanese colonial rule period (1910–1945). Korea only had 13.2 m^3 ha^{-1}growing stock (main trunk only) over 16.1 m ha of forests in 1943 [30]. After the establishment of the Korean Government in 1948, the Ministry of Agriculture and Forestry legislated and enforced three major forest policies: (1) planting five saplings for each harvested mature tree; (2) substituting firewood with charcoal firewood; and (3) banning illegal logging [27]. However, these efforts did not lead to any tangible results due to the Korean War between 1950 and 1953. After the Korean War, the South Korean government initiated the establishment of forest protection areas in 1955 under the supervision of the Ministry of Home Affairs and the Ministry of Defense [27,30]. However, this was not successful due to social instability and a lack of responsibility among government bodies.

Forests were protected properly only after the 1960s, when social and political conditions were stabilized [27]. However, South Korea only had 4.1 m ha of forested land in 1952. Several decades of illegal harvesting and the Korean War left 2.3 m ha deforested [27]. These deforested areas were susceptible to soil erosion and landslides due to the steep topography and concentrated rainfalls in the summer. A strong forest management policy and artificial planting were desperately needed for successful forest rehabilitation. The 'Forest Law' and 'Law of Enforcement of Illegal Forest Products' were implemented in 1961. The South Korean government strictly prohibited illegal forest harvesting and illegal shifting cultivation [27]. The government also implemented the 'First National Forest Development Plan' between 1973 and 1978 and applied a stricter forest protection policy. There were two noteworthy enforcements: (1) allowing fine fuel (e.g., leaves and branches) collection within a limited time and only under the supervision of forest officers and village leaders and (2) putting the full responsibility on the local governments to minimize the anthropogenic impacts on forests [27]. In addition, rapid economic development has replaced wood, which is the main energy source for rural residents, with fossil fuels. Furthermore, imported timber from foreign countries lessened the pressure on newly planted forests [27].

Forest planting projects from 1946 to 2000 restored 5.3 m ha (average 97,000 ha $year^{-1}$) of forest, achieving approximately 83% forest coverage of total forestland [27]. Approximately 59% of total forestland was planted between 1961 and 1980. Artificial forestations have definitely had a great influence on the state of current forests. These efforts resulted in higher growing stock and more mature forests [27]. The growing stock in South Korea reached 125.6 $m^3 ha^{-1}$ in 2010 [27] and was composed of 52.8 $m^3 ha^{-1}$ of coniferous forest, 39.0 $m^3 ha^{-1}$ of mixed forest, and 33.8 $m^3 ha^{-1}$ of broad-leaved forest. The growing stock in South Korea would be expected to increase rapidly [27]. However, denser and even-aged forests resulted in continuous high fuel loading, which would lead to a high risk of large fires [4].

3. MODERN KOREAN FOREST FIRES

Fires can have significant effects on biodiversity, local economy, and aesthetic values [18,19,31,32,33]. From 2005 to 2014, 3842 forest fires were reported, and a total of 6307 ha of forests were burned [27]. Approximately 60% of these fires occurred in the dry and windy spring season. More than 80% of these forest fires were accidental anthropogenic fires caused by camping, memorial ceremonies, children's pranks, and levee fires [27].

The severity of fire depends greatly on local conditions such as fuel loading, fuel type, temperature, humidity, and topography [28]. Most of the recent large forest fires in Korea occurred in the eastern coastal region, including the Goseong forest fire in 1996 (3762 ha), the Donghaean forest fire in 2000 (23,794 ha), and the Yangyang forest fire in 2005 (973 ha), where fire-prone dense pine forests dominate. In addition, winds in this area are stronger and dryer than those in the western interior region due to Föhn winds that blow over the Taebaek Mountains [28,34]. High fuel loading in this region was mainly the result of inadequate forest management practices [4]. In particular, artificial forests, such as the Korean forests, require proper forest management. Without proper management, abundant ladder fuel is built up, and forests become prone to crown fires. Fire prevention requires continued management efforts, including pruning and thinning [20,35]. The pine mushroom (*Tricholoma matsutake* (S. Ito & S. Imai) Singer), an ectomycorrhizal basidiomycete associated with *Pinus* spp., grows extensively throughout the region. Local residents have traditionally protected and conserved pine forests in order to harvest pine mushrooms as a valuable source of income [18].

4. REVIEW ON THE CURRENT POST-FIRE FOREST RESTORATION PLANNING

Burned forests are subject to restoration by maximizing their economic, ecological, and social value (Figure 1). A restoration plan must be completed prior to conducting any restoration to circumvent the need for later modifications, which would otherwise incur higher costs with more technical difficulties. The following six fundamental functions should be considered in preparing a restoration plan: timber production, water conservation, disaster prevention, natural environment conservation, ecosystem conservation, and recreation [20]. Furthermore, a restoration plan after a large fire should follow the appropriate land use laws. If a fire does not allow the original land use designation to be maintained, a full plan should be developed that is suitable for the new land use classification. In contrast, a plan after a small fire should maximize the six forest functions. If burned areas are vulnerable to secondary damage such as soil erosion or landslide, which are very common post-fire natural hazards in Korea following heavy rain, burned forests should have a restoration plan that includes erosion control to prevent disaster. Previous studies [20] showed that a log strip terrace barrier (Figure 2a) is the most effective way to prevent soil erosion in the burned areas and it is a good way to utilize dead trees following a fire. Piling up burned trees horizontally (Figure 2b) is also an affordable option. In addition, a landslide forecasting program can be utilized to identify sites that are vulnerable to landslides and establish disaster-prevention forests beforehand. There are also several land use objective-specific rules to follow. Patch clearing and strip cutting are recommended for burned areas to regain timber production capability. The size of clear-cut patches should be less than 5 ha to minimize the soil disturbance [20]. The right tree species selection guide developed by the KFRI can be used to select tree species appropriate for timber production and forest establishment for the target area. Forests for water conservation should aim for mixed plant species restoration, with both deep- and shallow-rooted hardwood plants to make multi-layered root structures [20]. Forests with a natural environment conservation objective should adapt a passive restoration plan that includes the regeneration of remaining trees and sprouts to facilitate secondary growth forests [20]. Forests designated for ecosystem conservation and recreational use (e.g., near roads, cities, cultural landscape resource and tourist sites) can be restored through passive restoration, reforestation, and erosion control in a balanced manner. In addition, burned areas near former pine mushroom production sites should be reforested to reintroduce pine mushrooms for forest owners [4].

Passive and active restorations can be applied in a balanced way, depending on the site quality and vegetation conditions [20,36]. For active restoration, native and local tree seedlings can grow under similar weather conditions and site qualities. A plan for active restoration must seek the opinions of forest owners and residents to maximize the income for forest owners and/or benefits of residents [29].

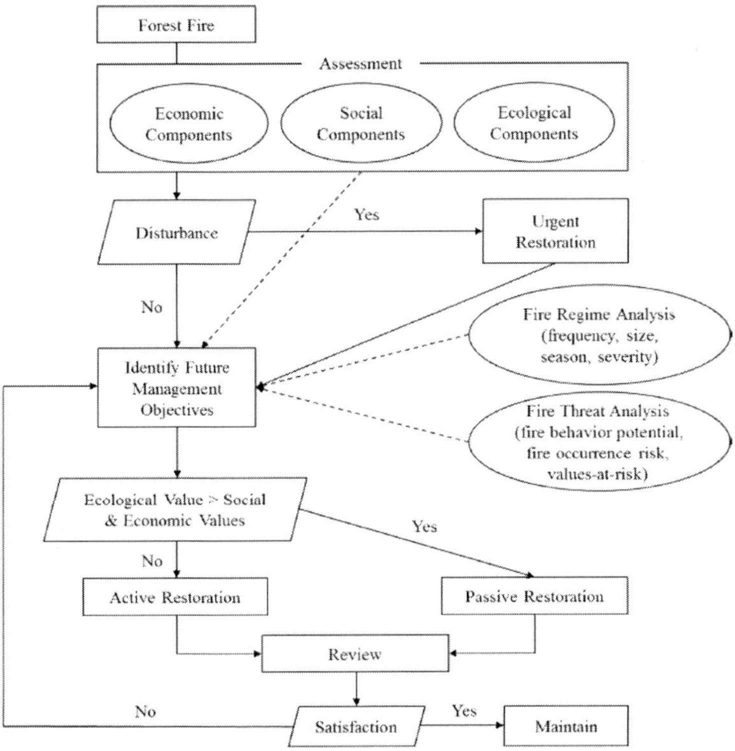

Figure 1. Flowchart of the development and execution of a plan to restore burned areas.

Figure 2. (**a**) Log strip terracing and (**b**) partially burned trees piling along a topographic contour in a burned area to prevent soil erosion (photos used with permission of the National Institute of Forest Science).

5. EXECUTING RESTORATION

The restoration of burned areas can use urgent and long-term restoration methods [20]. Some burned areas are vulnerable to secondary damage, such as landslide and debris flow, after heavy rainfall. Urgent restoration methods should be applied to these areas to prevent the loss of life and property. Long-term restoration methods can be used to maximize the economic, ecological, scenic, and environmental values. The opinions of residents should be considered in the long-term restoration process [4].

5.1. Urgent Restoration

Areas that are vulnerable to landslide and soil erosion or those near unstable streams require urgent restoration. Erosion control should be first implemented to prevent secondary post-fire damage. It is very important to stabilize slopes and install erosion control structures such as hillside stone masonry, sand bags, sod, and seed spraying (Figure 3). Hillside stone masonry is recommended for mountain slopes that have easy access and are steeper than 45°. Sod pitching is suitable for relatively fertile soil and a slope <45° [20]. Sand bag piling can be used for areas that are difficult to access or barren areas where sod is loosely attached to soil. Areas requiring quick vegetation establishment can be targeted for seed spraying after completing hillside stone masonry work or sand bag piling. The implementation must be completed before the rainy season and monitored by the "Erosion Control Regulation" of the Korea Forest Service [20]. Recently, field observations have shown that erosion control structures stabilize slopes and introduce vegetation [27]. Urgent restoration methods, such as hillside stone masonry, sand bags, sod, and seed-spraying treatments, in burned areas have been determined to be useful methods for preventing soil erosion and landslides.

5.2. Long-Term Restoration

Long-term restoration should consider the six forest functions and local priorities, such as the residents' preferences for pine mushroom production [4,29]. Long-term restoration can be divided into passive restoration and active restoration to maximize the economic, ecological, scenic, and environmental values [27]. Passive restoration can be used for areas that are minimally disrupted by forest fires and remnant forests. These areas are officially classified as environmental conservation forests and have the ability to naturally regenerate from the intact canopy. Passive restoration should avoid drastic changes to their forest structure [36]. Generally, passive restoration methods can be used for oak forests in Korea [4,20] because fire-damaged oaks can regenerate from their root sprouts and seeds. Shrubby bushclover (*Lespedeza bicolor* Turcz.) can also rapidly recover in burned areas. Oak seedlings and shrubby bushclover can gradually invade burned areas after forest fires (Figure 4). A typical commercial goal of oak management in Korea is to attain at least 350 harvestable trees per hectare with a minimum target diameter at breast height (DBH) of 30 cm at the end of a ca. 50-year rotation [27]. In burned areas, however, the goal for regenerated oak trees originating from root sprouts is at least 600 harvestable trees per hectare, with a target DBH of 20 cm and height of 6–8 m in a 30-year rotation (Figure 5a). This does not produce high-quality harvestable trees because of the damage to the roots from the forest fire [37]. The harvested oak trees in burned areas are mainly used for mushroom cultivation and pellet materials. Passive restoration can neglect the social and economic demands of local residents [4,38].

Figure 3. (**a**) Hillside stone masonry, (**b**) sand bags piling, (**c**) sod-pitching, and (**d**) seed spraying to control for soil erosion in an urgent restoration practice (photos used with permission of the National Institute of Forest Science).

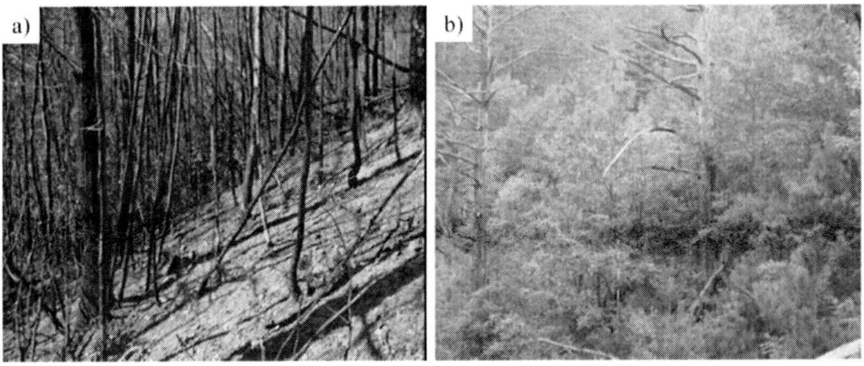

Figure 4. (**a**) Immediately after a fire (2000) and (**b**) passive restored site (2006) (photos used with permission of the National Institute of Forest Science).

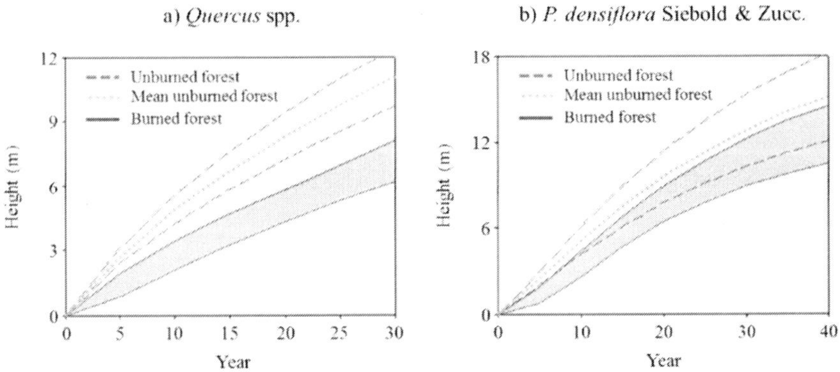

Figure 5. Growth curves of (**a**) *Quercus* spp. and (**b**) Japanese red pine (*Pinus densiflora*Siebold & Zucc. trees in unburned and burned areas (data modified from [37]).

The aim of active restoration is to improve the six major forest functions and local priorities such as pine mushroom production. It contains disaster prevention goals such as the construction of check dams and erosion dams [4,20]. Burned forests on gentle slopes (less than 30°) and fertile soil will offer the maximum timber yields. They can promote the sustainable production of high-quality timber by planting economic tree species to satisfy the national economic demands (Table 1). Pine forests have long been cultivated in Korea for the production of high-quality timber. Currently, a typical goal of pine management in Korea is to produce at least 600 harvestable trees per hectare [27]. The minimum target DBH is 25 cm with heights of 10–14 m at the end of a 40-year rotation period (Figure 5b). The recommended number of pine seedlings per hectare in row planting is approximately 5000–10,000 seedlings ha^{-1} (Figure 6c). The aim of row planting is to achieve a regular distribution of trees and structural homogeneity of the crown layer in the early stages of stand development to foster natural pruning and the development of stem quality [39]. Pine tree growth in burned areas also tends to be slower (Figure 5b) because of the poor nutrient conditions induced by erosion and leaching losses [4,20].

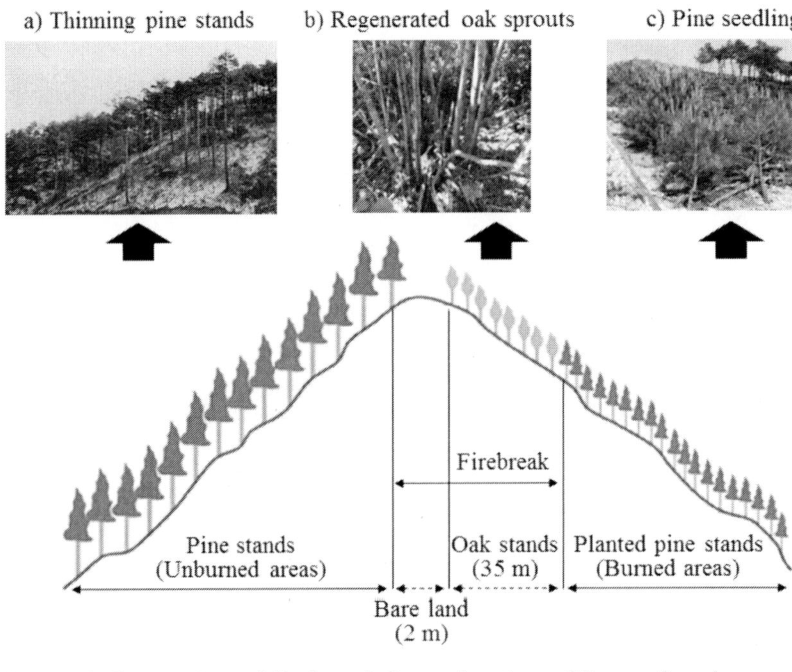

Figure 6. Formation of firebreak for reduction of forest fire damage near large planted pine stands (photos used with permission of the Korea Forest Service and National Institute of Forest Science).

Table 1. Major economic tree species by climatic zone of Korea [20].

Climate Zone	Main Species	Companion Species
Temperate	*Pinus koraiensis* Siebold & Zucc. *Larix kaempferi* (Lamb.) Carrière *P. densiflora* Siebold & Zucc. *Betula platyphylla* var. *japonica* (Miq.) H.Hara *Quercus* spp. *Liriodendron tulipifera* L.	*Pinus rigitaeda* S.K.Hyun & Ahn *Fraxinus rhynchophylla* Hance *Juglans mandshurica* Maxim.
Sub-Tropical	*Pinus thunbergii* Parl. *Chamaecyparis obtuse* (Siebold & Zucc.) Endl. *Liriodendron tulipifera* L.	*Pinus taeda* L. *Cryptomeria japonica* (L.f.) D.Don *Quercus myrsinifolia* Blume *Machilus thunbergii* Siebold & Zucc.

However, high-intensity fire damage in pine forests has prompted forest managers to convert pure pine stands established on large areas into mixed forests of pine and other broad-leaved trees [20,38]. Forests near large areas of pine trees must have firebreaks to reduce the risk of rapid fire spread [4,20,39]. Firebreaks can utilize fire-resistant tree species (Table 2), such as those with thick bark and leaves containing high levels of moisture. In addition, these trees must be able to sprout vigorously after a fire. Oak species are the dominant deciduous trees in Korea and have low heat yield, making them an effective natural fire break [4,20]. Based on high-intensity forest fires in 2000, Lim et al. [20] recommended that regenerated oak sprouts in belt-shaped configurations at least 35 m in width were necessary for safety considerations in forests near large areas of pine trees (Figure 6). Forest fires burn biomass in forests but do not destroy underground roots and seeds in plants. Oak in particular has the ability to regenerate from intact canopy naturally. Fast-growing oak sprouts develop in the upper layer (Figure 6b). Planting oak is a viable option when there are not enough sprouts and under conditions of high mortality [20]. The recommended oak restoration method is group planting, which was introduced to Europe in the 1980s and 1990s as an economic and ecological alternative to traditional row planting for reforestation of distributed areas [40,41,42]. Group planting uses larger seedlings or saplings (0.8–1.5 m tall) and a wider initial spacing (1 × 1 m). In group planting, both the total number of saplings per group (20–30) and the number of additional, shade-tolerant trainer tree saplings per group are designed. Trainer trees are commonly planted on the perimeter of groups to control ground vegetation and to shade oak stems, thus preventing development of epicormic sprouts. Irrespective of the group design, spacing between the group centers was kept at 10 × 10 m or 10 × 12 m, resulting in 80–100 groups ha^{-1}. Group planting resulted in greater stand-level tree species diversity than did row planting [40,41,42]. Oak group planting is a comparatively inexpensive option for the artificial regeneration of oak-dominated broad-leaved forests for a range of situations such as reforestation of disturbed areas. In addition, bare land installation (width 2 m) was recommended to reduce fire susceptibility (Figure 6). Fuel treatment outside the firebreak may contribute to their effectiveness. In particular, conifer forests that are not fire-resistant should be managed so they do not contact the crowns of other trees caused by pruning and thinning (Figure 6a). The suppression of forest fires requires ongoing management efforts, including branch pruning, thinning, and firebreak construction [4,20,38]. However, these initiatives mostly occur in government-owned forest areas because private forest owners are reluctant to undertake branch pruning, thinning, and firebreak construction, fearing their possible negative effects on pine mushroom growth [4]. Therefore, the effective planning of a systematic forest fire prevention system needs to encompass the stewardship of both government-owned and private forests.

Table 2. Fire-resistant tree species by climatic zone of Korea (data modified from [20,39]).

Climate Zone	Tree Species	Shrub Species
Temperate	*Ginkgo biloba* L. *Quercus variabilis* Blume *Quercus acutissima* Carruth. *Quercus dentata* Thunb. *F. rhynchophylla* Hance *Populus maximowiczii* Henry	*Picrasma quassioides* (D.Don) Benn. *Fraxinus sieboldiana* Blume *Sorbus commixta* Hedl.
Sub-Tropical	*Myrsinaefoliae* *Cinnamomum camphora* (L.) J.Presl *Cinnamomum japonicum* Siebold *Ilex rotunda* Thunb. *Machilus thunbergii* Siebold & Zucc. *Neolitsea sericea* (Blume) Koidz.	*Viburnum awabuki* K.Koch *Daphniphyllum macropodum* Miq. *Camellia japonica* L. *Illicium religiosum* Siebold & Zucc. *Cleyera japonica* Thunb. *Ternstroemia japonica* Thunb.

Fire burns biomass, and burned areas increase surface flow during rainfall events [4]. This may cause the collapse of soil aggregation and an increase in soil water repellency after fires, which results in a decline in the soil permeability, thereby reducing the water storage capacity in forests. Oak species are suitable for growth at sites with hydromorphic soil conditions that are prone to storm disturbance on bare slopes, where oaks can develop deep root systems (Figure 4b). To conserve the water supply within forests, deep-rooted tree species such as oaks should be planted [20]. Planting plays a major role in forests, especially in the reforestation of oak-dominated forests. In particular, artificial regeneration is the only way to establish oak stands in situations where acorn sources are lacking [20]. Supplementary shallow-rooted and intermediate-root species can be planted to increase water reservoirs in the soil through multi-layered root structures [20]. Therefore, a higher diversity of tree species in oak-dominated broad-leaved forests might provide more opportunity for the establishment of multilayered root structures. However, oak planting based on row planting in burned areas indicates that the tree species richness has not been achieved [40,41,42]. Based on other studies, group planting of oak is recommended as an alternative to traditional row planting for greater stand-level tree species diversity and natural regeneration, which may result from an increase in the size of the unplanted area between groups [40,41,42]. Recreational forests aim to improve recreational value to visitors and residents while preserving the ecological value. Landscaping and endemic tree species should be planted together to form a multilayered mixed forest with indigenous species. To reintroduce pine mushrooms, burned trees can be clear-cut, and locally grown Japanese red pine seedlings can be planted to benefit forest owners. The pine mushroom grows only in early mature pine stands but does not grow in the early stages of fire-regenerated stands.

The goal of long-term restoration is to maximize the economic, ecological, and environmental values through recovering forest functions. Restoration success can be measured by multilayered, mixed forest establishment and the recovery of major forest functions such as timber production, water conservation, ecosystem conservation, and recreation to satisfy residents [4,20]. Economic indices of success include the forest products and generated ecological services. Profit from recreation services and pine mushroom production can provide strong incentives to local residents to support restoration projects. Thus, a successful restoration site will undergo post-restoration management. The site should be monitored to evaluate the efficiency of management methods and provide direction for future management. Poorly restored areas are subject to having supplementary restoration work. In addition, collaborations among stakeholders such as local communities, government officials, and scientists are important to sustain natural forests.

6. CONCLUSIONS AND RECOMMENDATIONS

The impact of large-scale forest fires on human life has received attention in Korea since the Goseong forest fire in 1996 [29]. Restoration in burned areas is needed to rebuild their ecological, scenic, and economic values. Current diverse restoration methods need to be organized into a standard plan [4,20,29]. The Donghaean forest fire in 2000, which was the largest forest fire in modern Korean history, has triggered serious discussions about restoration methods, including controlling erosion to prevent damage to residential areas, planting fire-resistant trees to prevent the spread of fire, and planting pine trees to collect pine mushrooms in large-scale burned forests. To establish a standard post-fire restoration plan, the KFRI has studied burned areas extensively using the data about regional climate, vegetation, forest soils, hydrology, pests, wildlife, and pine mushrooms [29]. A standard restoration plan in burned areas has been developed to restore burned forests to full health (Figure 1). The proposed forest restoration method may not be perfect. Thus, it is crucial to monitor and evaluate the entire restoration, to obtain objective revisions. The concept of bioindication has been used to assess the restoration state in burned areas [4]. The abundance and species richness of microarthropod communities can be used to examine forest recovery in burned areas [4]. The dominant beetles in burned forests are members of the Curculionidae family, which attack trunks or branches of weakened trees, whereas detritivorous Scarabaeid beetles dominate unburned forests. The density of Collembola in burned forests declined because of the loss of aboveground vegetation and destruction of the litter layer [4]. Therefore, poorly restored areas may be dominated by the Curculionidae family, whereas successfully restored areas may be dominated by the detritivorous Scarabaeid beetles and Collembola. For cases with successful restoration, these sites will undergo post-restoration management. Poorly restored areas should be subjected to supplementary restoration work. The objectives of restoration are to return forest function and to provide incentives to local residents such as recreation services and pine mushroom production.

No species is 'fire adapted'; instead, they are all adapted to a particular fire regime, which can include fire frequency, fire intensity, and patterns of fuel consumption [43]. Oak trees have leaves containing high moisture levels and low heat yield, making them burn with less power [4,20]. Oak forests in Korea are resilient to periodic low-intensity fires at intervals of several decades. However, when the fire frequency increases, oak forests vigorously sprout epicormically along the length of the bole and replace the tree canopy within several years of such fires (Figure 4b). These traits are adaptive in fire-prone environments, and are key to resilience, to periodic low-intensity fires. The fire-tolerant oaks are likely to increase in importance as the climate warms in Korea. Hence, the oak cover in the Korea now constitutes up to 25% of the total forested area [27] as an adaptive management strategy against possible fire impacts. Plant species that exhibit traits that are adaptive under a particular fire regime can be threatened when that regime changes [43]. Pine forests can be established in areas where no forest fires occur, in which case the local residents will be protected and pine forest can be conserved to produce pine mushrooms as a valuable source of income [4]. However, pine forests are prone to large-scale forest fires. In pine forests that are subject to high-intensity fires, pine trees restrict seedling recruitment in the post-fire years, being replaced by shrubby bush cover [4]. If the fire frequency in pine forests increases, pine forests can be rapidly lost [4,20]. The forest restoration methods should be considered economic benefits to local residents and a source of ecological stability. Therefore, oak stands in belt-shaped configurations were constructed near large areas of pine trees to prevent the spread of fire and to improve the fire-adaptive traits in fire-prone environments and pine trees in areas where pine mushrooms are collected (Figure 6).

Restorations based on traditional row planting in burned areas of Korea have been positively evaluated by forest scientists and local residents [4,20]. The burned areas have been restored to produce green and lush forests, such as oak and pine stands, which has resulted in the rapid stabilization of soil erosion and landslides in the burned areas (Figure 2,Figure 4,Figure 6). Despite the gradual recovery of the forests in the burned areas, oak and pine do not produce high-quality harvestable trees [37]. In addition, the lack of natural regeneration of other tree species and the low species diversity reported for row planting are undesirable attributes of this classic type of crop cultivation [40,41,42]. Overall, these factors motivated this study to seek alternatives to the conventional row planting approach for establishing crop stands. Group planting represents a reforestation method that not only offers the opportunity to produce high-quality oak timber but also promotes the natural regeneration of additional tree species between the groups [40,41,42]. This increases woody species diversity and possible productivity and allows natural successional processes to occur. For burned areas, future research needs regarding the production of high-quality harvestable timber and adaptation to fire via group planting are highlighted.

ACKNOWLEDGMENTS

This research was supported by a grant (NRF-2013R1A1A1008790) of the Basic Science Research Program of the National Research Foundation of Korea (NRF) funded by the Ministry of Science, ICT, and Future Planning, Republic of Korea.

REFERENCES

1. Doblas-Miranda, E.; Martínez-Vilalta, J.; Lloret, F.; Álvarez, A.; Ávila, A.; Bonet, F.J.; Brotons, L.; Castro, J.; Curiel Yuste, J.; Díaz, M.; et al. Reassessing global change research priorities in mediterranean terrestrial ecosystems: How far have we come and where do we go from here? *Glob. Ecol. Biogeogr.* **2015**, *24*, 25–43.
2. Littell, J.S.; McKenzie, D.; Peterson, D.L.; Westerling, A.L. Climate and wildfire area burned in western U.S. ecoprovinces, 1916–2003. *Ecol. Appl.* **2009**, *19*, 1003–1021.
3. Williams, A.P.; Abatzoglou, J.T. Recent advances and remaining uncertainties in resolving past and future climate effects on global fire activity. *Curr. Clim. Chang. Rep.* **2016**, *2*, 1–14.
4. Ahn, Y.S.; Ryu, S.R.; Lim, J.; Lee, C.H.; Shin, J.H.; Choi, W.I.; Lee, B.; Jeong, J.H.; An, K.W.; Seo, J.I. Effects of forest fires on forest ecosystems in eastern coastal areas of Korea and an overview of restoration projects. *Landsc. Ecol. Eng.***2014**, *10*, 229–237.
5. Dennison, P.; Brewer, S.; Arnold, J.; Moritz, M. Large wildfire trends in the western United States, 1984–2011. *Geophys. Res. Lett.* **2014**, *41*, 2928–2933.
6. Westerling, A.L. Increasing western US forest wildfire activity: Sensitivity to changes in the timing of spring. *Philos. Trans. R. Soc. B* **2016**, *371*, 20150178.
7. Westerling, A.L.; Hidalgo, H.G.; Cayan, D.R.; Swetnam, T.W. Warming and earlier spring increase western U.S. forest wildfire activity. *Science* **2006**, *313*, 940–943.
8. Kasischke, E.S.; Turetsky, M.R. Recent changes in the fire regime across the North American boreal region-Spatial and temporal patterns of burning across Canada and Alaska. *Geophys. Res. Lett.* **2006**, *33*, L09703.
9. Kelly, R.; Chipmanb, M.L.; Higuerac, P.E.; Stefanovad, I.; Brubakere, L.B.; Hu, F.S. Recent burning of boreal forests exceeds fire regime limits of the past 10,000 years. *Proc. Natl. Acad. Sci. USA* **2011**, *110*, 13055–13060.
10. Batllori, E.; Parisien, M.A.; Krawchuk, M.A.; Moritz, M.A. Climate change-induced shifts in fire for Mediterranean ecosystems. *Glob. Ecol. Biogeogr.* **2013**, *22*, 1118–1129.
11. Lenihan, J.M.; Drapek, R.; Bachelet, D.; Neilson, R.P. Climate change effects on vegetation distribution, carbon, and fire in California. *Ecol. Appl.* **2003**, *13*, 1667–1681.
12. St.John, T.V.; Rundel, P.W. The role of fire as a mineralizing agent in a Sierran coniferous forest. *Oecologia* **1976**, *25*, 35–45.
13. Fulé, P.Z.; Covington, W.W.; Moore, M.M. Determining reference conditions for ecosystem management of southwestern ponderosa pine forests. *Ecol. Appl.* **1997**, *7*, 895–908.
14. Allen, C.D.; Savage, M.; Falk, D.A.; Suckling, K.F.; Swetnam, T.W.; Schulke, T.; Stacey, P.B.; Morgan, P.; Hoffman, M.; Klingel, J.T. Ecological restoration of southwestern ponderosa pine ecosystems: A broad perspective. *Ecol. Appl.* **2002**, *12*, 1418–1433.
15. Covington, W.W.; Moore, M.M. Post-settlement changes in natural fire regimes and forest structure: Ecological restoration of old-growth ponderosa pine forests. *J. Sustain. For.* **1994**, *2*, 153–181.
16. Hunter, M.E.; Omi, P.N.; Martinson, E.J.; Chong, G.W. Establishment of non- native species after wildfires: Effects of fuel treatments, abiotic and biotic factors, and post-fire grass seeding treatments. *Int. J. Wildland Fire* **2006**, *15*, 271–281.
17. Keeley, J.E. Fire management impacts on invasive plants in the western United States. *Conserv. Biol.* **2006**, *20*, 375–384.

18. Youn, Y.C. Assessment of social coasts of forest fire: Case of forest fire of spring 2000 in the east coast area of South Korea. *Korean For. Econ. Soc.* **2000**, *8*, 72–81, (In Korean with English abstract).
19. Ryu, S.R.; Chen, J.; Zheng, D.; Bresee, M.K.; Crow, T.R. Simulating the effects of prescribed burning on fuel loading and timber production (EcoFL) in managed northern Wisconsin forests. *Ecol. Model.* **2006**, *196*, 395–406.
20. Lim, J.; Lee, Y.; Jeong, J.; Lee, M.B. *Post-Fire restoration: To Establish a Healthy and Sustainable Forest Ecosystem*; Ukgo Press: Seoul, Korea, 2010; pp. 1–60.
21. Allingham, S.M.; Harvey, M. Effects of different fire regimes on amphibian communities in the Nyika National Park, Malawi. *Curr. Herpetol.* **2013**, *32*, 1–8.
22. Bury, R.B. Wildfire, fuel reduction, and herpetofaunas across diverse landscape mosaics in northwestern forests. *Conserv. Biol.* **2004**, *18*, 968–975.
23. Cano, P.D.; Leynaud, G.C. Effects of fire and cattle grazing on amphibians and lizards in northeastern Argentina (Humid Chaco). *Eur. J. Wildl. Res.* **2010**, *56*, 411–420.
24. Groves, C.R.; Steenhof, K. Responses of small mammals and vegetation to wildfire in shadscale communities of Southwestern Idaho. *Northwest Sci.* **1988**, *62*, 205–210.
25. Kelly, L.T.; Nimmo, D.G.; Spence-Bailey, L.M.; Taylor, R.S.; Watson, S.J.; Clarke, M.F.; Bennett, A.F. Managing fire mosaics for small mammal conservation: A landscape perspective. *J. Appl. Ecol.* **2012**, *49*, 412–421.
26. Law, B.S.; Dickman, C.R. The use of habitat mosaics by terrestrial vertebrate fauna: Implications for conservation and management. *Biodivers. Conserv.* **1998**, *7*, 323–333.
27. Korea Forest Service. Available online: http://www.forest.go.kr/ (accessed on 20 April 2017). (In Korean)
28. Lim, J. Forest fire and meteorology of eastern Korea. *Korean J. Agric. For. Meteorol.* **2000**, *2*, 62–67, (In Korean with English abstract).
29. Shin, J.H.; Lee, D.K. Strategies for restoration of forest ecosystems degraded by forest fire in Kangwon ecoregion of Korea. *For. Ecol. Manag.* **2004**, *201*, 43–56.
30. National Archives of Korea. In *The Department of Agriculture, Forestry, Ocean and Fishery*. Available online: http://www.archives.go.kr/ (accessed on 1 December 2016). (In Korean)
31. Fontaine, J.B.; Donato, D.C.; Robinson, W.D.; Law, B.E.; Kauffman, J.B. Bird communities following high-severity fire: Response to single and repeat fires in a mixed-evergreen forest, Oregon, USA. *For. Ecol. Manag.* **2009**, *257*, 1496–1504.
32. Kim, C.; Lee, W.K.; Byun, J.K.; Kim, Y.K.; Jeong, J.H. Short-term effects of fire on soil properties in *Pinus densiflora*stands. *J. For. Res.* **1999**, *4*, 23–25.
33. Neary, D.G.; Klopatek, C.C.; DeBano, L.F.; Ffolliott, P.F. Fire effects on belowground sustainability: A review and synthesis. *For. Ecol. Manag.* **1999**, *122*, 51–71.
34. Korea Meteorological Administration. Available online: http://www. kma.go.kr / (accessed on 1 December 2016). (In Korean)
35. Lee, S.Y.; An, S.H. Comparative analysis of mortality in species of tree after surface forest fire. *J. Korean Soc. Hazard Mitig.* **2009**, *9*, 39–43, (In Korean with English abstract).
36. Choung, Y.; Lee, B.C.; Cho, J.H.; Lee, K.S.; Jang, I.S.; Kim, S.H.; Hong, S.K.; Jung, H.C.; Choung, H.L. Forest responses to the large-scale east coast fires in Korea. *Ecol. Res.* **2004**, *19*, 43–54.
37. Lim, J.H; Kang, W.S.; Kwon, J.; Kim, J.; Cho, J.H.; Kwon, T.S.; Lee, Y.G.; Lee, W.S.; Kim, S.D.; Yoon, C.W.; et al. *Monitoring Forest Ecosystem Changes Following Large Scale Fires and Developing Forest Restoration Treatments*; National Institute of Forest Science: Seoul, Korea, 2016; pp. 1–191. (In Korean)
38. Lim, J.H.; Kwon, J.; Choi, H.; Park, K.H.; Kang, W.S.; Shin, M.H.; Kim, J.; Kwon, T.S.; Ka, K.H.; Lee, Y.G.; et al. *Forest Ecosystem Change since 1996 Wildfire in Korea*; Korea Forest Research Institute: Seoul, Korea, 2015; pp. 1–12.

39. Kang, Y.H.; Lim, J.; Shin, S.; Lee, M.B. A historical review on the forest fire prevention and establishment of fire line and firebreaks in Korea. *J. Korean For. Soc.* **2004**, *93*, 409–416, (In Korean with English abstract).

40. Saha, S.; Kuhne, C.; Bauhus, J. Lessons learned from oak cluster planting trials in central Europe. *Can. J. For. Res.***2017**, *47*, 139–148.

41. Saha, S.; Kuehne, C.; Bauhus, J. Tree Species Richness and Stand Productivity in Low-Density Cluster Plantings with Oaks (*Quercus robu*r L. and *Q. petraea* (Mattuschka) Liebl.). *Forests* **2013**, *4*, 650–665.

42. Saha, S.; Kuehne, C.; Kohnle, U.; Brang, P.; Ehring, A.; Geisl, J.; Leder, B.; Muth, M.; Petersen, R.; Peter, J.; et al. Growth and quality of young oaks (*Quercus robur* and *Quercus petraea*) grown in cluster plantings in central Europe: A weighted meta-analysis. *For. Ecol. Manag.* **2012**, *283*, 106–118.

43. Keeley, J.E.; Pausas, J.G.; Rundel, P.W.; Bond, W.J.; Bradstock, R.A. Fire as an evolutionary pressure shaping plant traits. *Trends Plant Sci.* **2011**, *16*, 406–411.

CHAPTER 14

Swamp Forest Use and Loss in the Niger Delta: Contextual and Underlying Issues

Nwabueze Ikenna Igu

Department of Geography and Meteorology, Nnamdi Azikiwe University, Awka, Nigeria

ABSTRACT

Forest ecosystems are vital not only for the ecosystem and biogeochemical processes, but also for the livelihood of forest dependent communities for which its continual existence is a necessity. This study explored the pattern of forest use in the hinterlands of the Niger Delta and sought to elucidate the drivers of forest loss and how the ownership and management of the forest plots influenced the changes in the forest ecosystem. Ecosystem services reduction and forest loss/degradation were found to be increasing over the years due to crude oil activities, urbanization/developments, population increase, agricultural activities and natural causes like flood. While each factor contributed to forest loss directly and indirectly, and varied from community to community, agricultural activities and population growth were responsible for most of the losses across the landscape. Even though agricultural activities were essential and thrived in the region, sustainable forest (land) uses could have reduced the associated implications of such land uses; but this was however hampered by the farming practises (shifting cultivation) across most of the communities. Lack of proper, effective and sustainable forest management structures, poor individual commitment and monitoring of forest activities were found to encourage forest loss at different spatial scales. Provision of alternative sources of livelihood and ensuring that suitable guidelines on forest abstraction and harvest are enforced across the region, are steps to promoting biodiversity conservation and resource management.

Keywords: Community Forest Management, Ecosystem Services, Forest Loss, Resilience, Sustainable Management

1. INTRODUCTION

Tropical forest ecosystems are very vital for the provision of ecosystem services, sustainability of livelihoods and global climatic processes. However, with increasing human impact, modification and consequential degradation of its biodiversity, its future has never been more uncertain (Kareiva et al., 2007; Gardner et al., 2009) . The underlying causes and proximate drivers of forest loss across the tropics are varied (Geist & Lambin, 2002) according to the patterns and processes of forest use in the different environments. Efforts to understand these patterns of loss across the tropics have been conducted at different spatial scales (De Fries et al., 2010; Asner et al., 2009; Hansen, 2013) , mainly with the use of satellite remote sensors, which has enabled the estimation of forest loss across different landscapes and ecosystems. This technique has been very useful in understanding the ecology and forest dynamics of forest ecosystems, especially the swamp forest ecosystems, whose poor terrain and marshy environments have acted as constraints to their accessibility and assessment (Harvey & Hill, 2001). Hence, it has both helped to facilitate a better understanding of the ongoing processes across the Niger Delta swamp forest ecosystems (James, 2008; Onojeghuo & Blackburn, 2011) , and will enable the modelling of its future scenarios.

Across the Niger Delta region, the causes of forest loss have mainly been linked or associated with crude oil exploration and pollution (Moffat & Linden, 1995; James, 2008; Onojeghuo & Blackburn, 2011), population increase (James, 2008; Mmom & Mbee, 2013), urbanization and development activities (James, 2008; Enaruvbe & Atafo, 2014; Kuenzer et al., 2014) and agricultural pressure(Enaruvbe & Atafo, 2014). While these factors of forest loss are normally viewed as the drivers of deforestation across the region as in most other tropical forest regions (Igu, 2016; Sassen et al., 2012), they are however not the main determinants underlying the spatial patterns of forest loss. A combination of institutional, economic, social factors as well as management frameworks have on the other hand contributed to the decline in forest cover through varied and inter-related ways across the region. Such information is not only vital in understanding the fundamental issues surrounding forest loss, but also useful for providing context specific and regional effective measures towards achieving an effective forest conservation and management.

To elucidate these issues, this study focuses on the associated effects of forest ownership and use across the ecosystem, the drivers and indices of forest loss, and the varied ways the peoples' attributes determined and influenced forest loss and degradation across the region.

2. MATERIALS AND METHODS

2.1. Study Region

The Niger Delta region is a rich sedimentary basin located in southern Nigeria (Figure 1). Located in the lower reaches of the Niger River, it is found between Latitude 4° and 6° north of the Equator and Longitudes 5° and 7° east of the Greenwich. It is a generally low landscape found in the lower reaches of Rivers Niger and Benue and criss-crossed by a network of rivers and creeks that drain into the Atlantic Ocean. The region is swampy and characterized by a tropical climate with long rainy season that last nearly throughout the year. Its rainy months are mainly from March to October, and a short dry season between December and February. Its annual precipitation increases from the North of the Delta (2500 mm) to the coastal areas, where it is as much as 4000 mm. Average monthly maximum and minimum temperatures across the region vary between 28°C to 33°C and 21°C to 23°C, respectively. The soils of the region are characterized by both seasonally and permanently waterlogged soils, which are broadly classified as hydromorphic (Areola, 1982). They

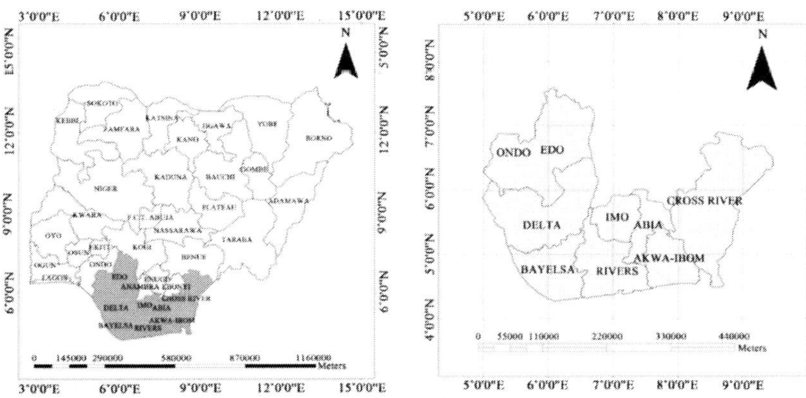

Figure 1. Map of Nigeria showing the Niger Delta region.

are whitish or greyish in colour due to the reduction of oxides in the soils (as a result of their poor drainage) and are mainly of fluvial origin. They are equally characterized by a succession of thinly bedded silts and clays that are inter-bedded with sands.

Administratively, the region encompasses the nine oil producing states across Nigeria, while it is delimited hydrologically to only comprise, Bayelsa, Delta and Rivers states. Ethnically, the region is made up of more than 40 ethnic groups that speak more than 250 dialects and characterized by a rich cultural heritage (Jike, 2004). The region is predominantly rural and composed of a populace that is mainly animist, and who that attach much importance and cultural values to the flora and fauna in their environment (Anwana et al., 2010; Adekola et al., 2015). Covering about 12% of Nigeria's total surface area, it is made up of a multi- cultural setting where 22% of Nigeria's population lives (UNDP, 2006; NPC, 2006). The Niger Delta has the most extensive mangrove and freshwater swamps in Africa (Ogon, 2006; Igu, 2013) and the third largest wetland in the world(Uluocha & Okeke, 2004). Both the forest resources and biodiversity of the region are features that contribute to the significance and importance of the area. This biodiversity importance is mainly from its vast mangrove forest ecosystem, of which three are endemic. Endemic mangrove families across the region includes the Rhizophoraceae family (red mangrove), Avicenniaceae family (white mangrove) and the Combretaceae family. These have been very useful for the provision of fuel, food, medicine, salt, timber, railway sleepers, tannins and dyes, across the region (Hamilton & Snedaker, 1984; Irikana, 2006; Igu, 2013; James et al., 2013). Freshwater swamp forests on the other hand are the region's major sources of timber and an important habitat for endangered and rare wildlife (UNDP, 2006).

2.2. Oil Resources and Social Concerns

The Niger Delta region is home to vast stores of hydrocarbons of reputable quality and the main source of Nigeria's wealth. With about 1182 exploration wells and about 400 oil and gas fields of varying capacities documented since the past fifty years (Obaje, 2009) and revenue generation of over $600 billion since the 1960's (Wurthmann, 2006), it has remained the main source of Nigeria's foreign exchange and government revenue. However, since its first commercial oil discovery in 1956, the region has steadily declined in its biodiversity potentials and suffered from "administrative neglect, crumbling social infrastructure and services, high unemployment, social deprivation, abject poverty, filth and squalor and endemic conflict" (UNDP, 2006). With about 36.4% of the inhabitants living below the poverty line (World Bank, 2003), only about 27% with access to safe drinking water, only 30% accessing electricity as at early 2000 (Ibeanu,

2002), an illiterate majority, a general lack of basic health care and employment, the region has not only been impoverished, but are a pathetic example of "resource curse". The region has been rife with aggression and violence following prolonged denials and poor human capital and regional impoverishment. Conflicts and insurgency have escalated across the region over the years, have affected the oil production adversely, shut down productions in some communities, led to the destruction of lives and property and created an endemic terror and insecurity across the region. While these have led to a drop in the oil production and consequent loss on the Nigerian economy in terms of oil revenue (Idemudia, 2009; Obi, 2011), it has not in reality improved the livelihood of the people who still depend on the forest ecosystem to meet their daily needs.

2.3. Methods

Data on the issues of forest ownership, management and use was elicited from the study communities through a household survey conducted between October 2013 and April 2014. The design of the questionnaire followed a literature review on forest use, drivers of forest loss and the ownership structure, across the region (NDES, 1997; Geist & Lambin, 2002; UNDP, 2006; James, 2008). The final draft of the questionnaire was arrived at after preliminary interviews were equally conducted during the reconnaissance survey and pilot study across the forest communities. This helped to ensure that the issues under consideration were both understood and applicable to the people.

The survey was conducted across 12 communities across the region (Delta and Bayelsa states). These communities were chosen at random across the region to capture differences in ethnicity, socio-economic backgrounds, remoteness, presence/absence of oil wells and forest uses. Due consultations with the community heads and leaders on the purpose and details of the survey were made before the selection of each of the communities. This helped to ensure that community ethics were not breached and that cooperation and openness among the community members were enhanced. To ensure that each of the communities was duly represented in the survey, they were first of all demarcated into four zones. Each of the quadrats (four zones) was then surveyed, starting from the first household and then every fourth household subsequently. This enabled the varied perceptions and status of the households in each of the communities to be captured and ensured that there were no biases in selection of the respondents. Being a household survey, it was targeted at the household heads across the settlements. This comprised mainly aged individuals, community leaders and individuals that understood the norms and day to day practices in the localities. This helped to ensure that the information elicited from the survey were a true picture of what applied to the people.

The data were analysed using rankings, frequencies and variance techniques with R statistical software version 3.1.0 (http://cran.r-project.org). Normality tests were first of all conducted on the data. Kruskal Wallis rank sum test was used to show how varied the drivers of forest loss were among the communities. Mann-Whitney tests were further used to show the ownership and usage patterns of the forests across the communities. Further details regarding the percentages of each of the variables were provided using descriptive statistics methods.

3. RESULTS

3.1. Overview of the Respondents

Most of the individuals that constituted the study had formal education (mainly primary and secondary education) and were found between less than and greater than 40 years of age (Table 1). They are divided into individuals whose full time occupation involved farming, civil service and other jobs (fishing, trading, craft making and collectors of non-timber forest products [NTFPs]) (Table 1). Though the people are divided into varying educational and occupation status, there are no wide gaps in the income levels of the populace.

3.2. Forest Ownership, Management and Loss

Forest ownership across the communities were varied (χ^2(U)= 3782.5, $p < 0.05$) between individual (family or households) and community ownership of forest lands (53.9% and 46.1%, respectively). As expected, these forests were used for different pur- poses (Kruskal-Wallis $\chi^2 = 58.6$, $p < 0.05$, D.F. = 11; Figure 2) by the resident commu- nities both for subsistence and for income generation. While these forests have no doubt sustained the people, its management structure has not been taken seriously by most of the people who are varied in their individual commitment to managing the forests (χ^2(U)= 4920, $p < 0.05$). Managing the forest or the forest resources has on other hand remained poor following a general lack of restriction or restraint to forest use across the communities (χ^2(U)= 4710, $p < 0.05$). Hence, forest loss and degradation has continued at different spatial scales across the forest communities. The direct and underlying causes of loss experienced across the communities are varied across the region (Kruskal-Wallis $\chi^2 = 33.72$, $p < 0.05$, D.F. = 11; Figure 3).

Table 1. Socio economic characteristics.

Variable	Value
Number of respondents	243
Highest level of education	Formal education 94.2% (21.2% primary, 37% secondary, 15% diploma, 19% bachelor degree, 2% masters degree) No formal education 5.8%
Age group	16 - 40 (≤40) years 53.9% 41-above 60 (>40) years 40.1%
Full time occupation	Farming 49.4% Civil service 35.4% Others 15.2%

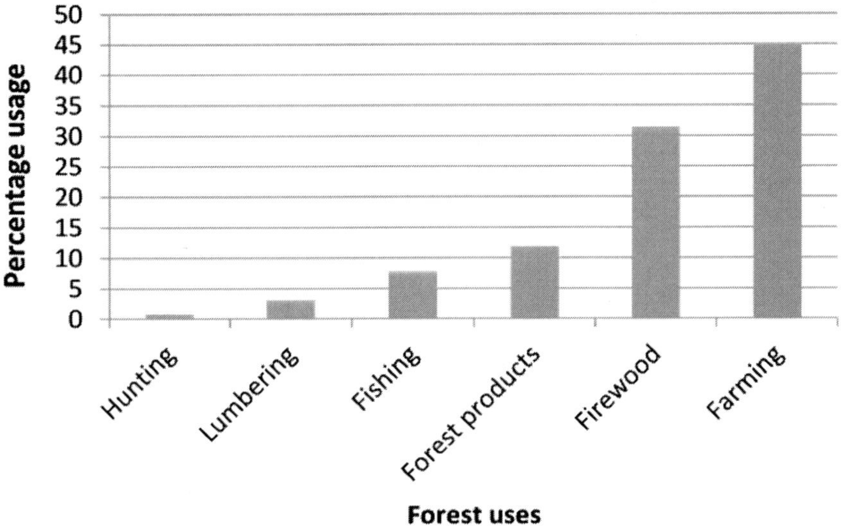

Figure 2. Forest uses across the communities.

4. DISCUSSION

The use of the forest across the region was influenced by the variations in the demographic variables of the people (Table 1). While the educational status of the people are expected to influence their income flows and in effect, their dependence on the forest for different resources(Godoy & Contreras, 2001; Mamo et al., 2007), they have however not followed such trends consistently. Since many of the educated people across the region are unemployed, the expected gap in forest extraction between those with formal and higher education and those without education are not wide. Hence, though those with higher education have more environmental and sustainable management knowledge, poor government intervention (as it concerns employment), hampers their quest to practise and promote forest conservation.

Total number or the population density of persons that reside around forested landscapes affect its rate of use and capacity to provide vital resources for the wider society. While this is important in understanding the extent of pressure such forest regions may be undergoing, the population structure of such landscapes show in greater detail the variations in use and dependence that may be inherent in such environments. Hence, for the Niger Delta region whose younger age class dominate the population, such that up to 62% of its populace are under 30 years of age (Francis et al., 2011), its pattern of forest utilization is expected to follow such dynamics. Since this younger populace do not have as much knowledge as the older people in the forest resources management and its utilization (Mamo et al., 2007), they are not likely to harness the resources optimally as would the older population. Furthermore, as some of the forest users in this younger age class use the forest resources as their only sources of livelihood to either fend for their young families or planning to start one, they tend to harvest the forest resources more and at rates that are far below the resilience of the ecosystem.

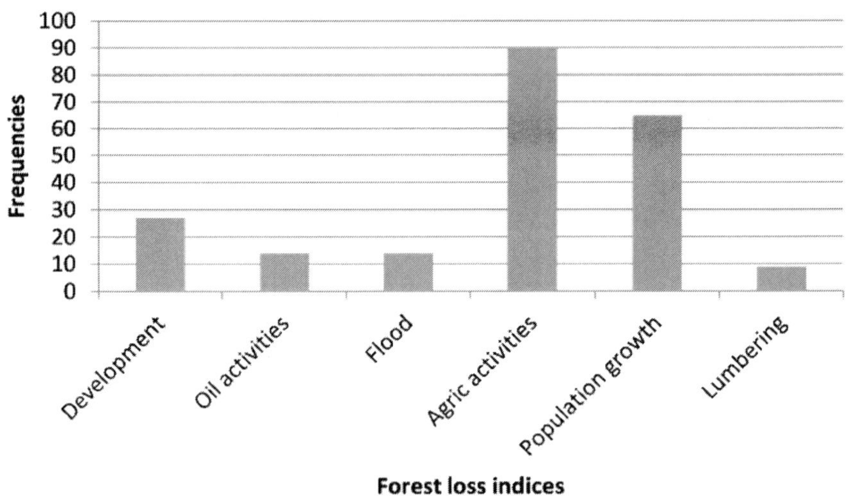

Figure 3. Direct and underlying causes of forest loss.

The varied occupation of the people is not only indicators to the heterogeneity seen even in rural forest landscapes but also indices to the livelihood variabilities and commitments inherent in such systems (environment). While this could be used to understand the income levels of the populace, it invariably depicts what each person spends the bulk of his time doing (Table 2). Since most of the people depend on agriculture for their sources of livelihood, there is then a higher tendency for such landscapes to experience an accelerated emission of greenhouse gases as well as a higher risk of forest degradation and deforestation. With the bulk of forest reductions and losses across the tropics dominated by the associated impacts of agricultural expansion and intensification (Angelsen, 2009; Hosonuma et al., 2012), the need to develop agricultural-forest friendly initiatives and guidelines are increasingly becoming a necessity.

The causes of forest loss and reduction of ecosystem services they provide vary across different landscapes mainly as a result of human induced factors. While lumbering is seen to contribute to the loss of forest cover across the communities, it is however a multi-faceted cause-effect factor which is induced by other underlying issues such as quest for financial gain, land clearing for agricultural activities and construction purposes. Crude oil exploration activities (seismic activities and pipeline constructions), as well as the effects from oil spills and gas flares is an endemic forest and biodiversity threat in the entire Niger Delta region (Moffat & Linden, 1995; NDES, 1997;James, 2008; Zabbey et al., 2010). However, the extent to which this leads to forest cover change at landscape or community levels differ widely. While some communities in the region do not have oil wells, and so are not expected to experience forest degradation occurrences as a result of oil activities, others that have oil wells or are criss-crossed by oil pipelines have had to grapple with the degradation and losses that such occurrences have caused. With the Land Use Act (of 1978) which neutralizes the traditional rights of the people over their lands and vests authority on the Nigerian state to revoke any rights of occupancy for an overriding public interest (which includes the requirement of land for all oil activities and other connected activities) (Emuedo & Abam, 2015)in operation, most forest lands are lost. This is seen to happen in instances where forest lands (where oil wells are found) are not only confiscated, but also experience biodiversity loss and degradation. Hence, while these (oil exploration) enriches the Nigerian state, the provisioning services such locations provide are not only degraded, but are eventually lost across most of the landscapes.

Table 2. Frequency of forest visit according to varying occupational status.

Full time occupation	Frequency of visit					
	Yearly	Seasonally	Monthly	Weekly	Daily	Total
Farming	3	8	13	38	58	120
Civil service	1	6	15	21	9	52
Teaching	4	9	3	10	5	31
Fishing	1	0	2	3	3	9
Trading	1	0	9	5	5	20
Craft making	0	0	0	0	4	4
Collector of NTFP	0	0	0	2	2	4
Labourer	0	2	0	0	1	3

Even though oil activities have impacted the forest ecosystem across the region, nonetheless, other anthropogenic activities have equally affected the forests adversely, both directly and through inter-related ways. While development activities or urbanization (which includes: roads, canals and construction activities for housing or industrial purposes) equally contributed to forest loss and reduction in the provisioning services, their impacts equally varied mainly on the basis of their proximity to urban centres. With the establishment of the Niger Delta Development Commission (which was laden with the responsibility of facilitating the rapid development of the region), most of the hinterlands started experiencing development activities (mostly in the form of roads, canals and rural electrification) for almost the first time. Such developments were beneficial to the entire region, but equally contributed to the modifications and loss of the marshy and fragile terrain that characterize the entire landscape. These dredging, landfilling, excavations and land reclamations (development activities) across the region, introduced a lot of physico-chemical changes to the ecosystem (Ohimain, 2001; Adekola & Mitchell, 2011) and exposed the biodiversity to fragmentation, physical and chemical stress and general degradation.

Even though the swampy zones of the Niger Delta experiences a yearly regime of flood, the ecosystems are not only adapted to it, but has suitable resilient capacities to cope with such occurrences. However, recent developments across the region have affected its natural coping mechanisms and altered its hydrological channels. Such events have contributed to the alteration of the region's normal flow cycle regime (which the ecosystem already has an adaptive and resilient capacities for) to more irregular-de- structive patterns. Initially, the flow/flood pattern of the Niger river (which flows down to the Niger Delta) was dependent on the volume of annual rainfall across the region. However, with the completion of the Kainji dam (on the Niger river) in 1968, the flow pattern has to a large extent been equally influenced by the opening and closing of the dam sluices(Brown & Thieme, 2005). Hence, in instances where the fluctuation in flow becomes higher than the normal capacity of the ecosystem (as seen in the year 2012), the flooding becomes so devastating that flora and fauna in the ecosystem are to a large extent destroyed, while the resilience for the ones that survived becomes affected.

Agriculture was seen as a main feature and determinant of forest decline and loss across the region. While this trend of forest loss (as a result of agricultural activities) seems to be similar with most other developing countries (Hosonuma et al., 2012), the extent to which it continues to contribute to forest loss is however a function of the population dynamics of the location. Hence, with an increasing population across the Niger Delta region (Watts, 2010), more forest lands are being converted to agricultural lands, so as to meet the needs of the teeming population. Furthermore, with high unemployment (at 8.8%) rates experienced across the region (Ukiwo, 2009) , most of the people have resorted to agricultural activities and harvesting of forest products as their main sources of livelihood. Such pressures on the forest land have encouraged forest degradation and loss across the region, especially because, most of the farmers (especially plantain farmers) would prefer to (plant on virgin soils) practise shifting cultivation rather than improving the fertility of their farmlands.

Generally, the forests across the region are found in community lands, sacred groves, family lands or in reserves. The forests found on community and family lands dominate the ownership structure in the area and are equally managed by the different communities or families that own them. Even though the pattern of management differs slightly from one community to the other, their effectiveness are mostly a function of the calibre of individuals that constitute that current administration of leaders. Since there is no model or structure of management, the consistency and effectiveness of the operational management have in most cases been short lived. Conversely, forests owned and managed by individuals and families have on the other hand been affected by land tenure, the economic status of the family members, their personal interests and attitudes. While the size or numbers of forest acres owned and managed by these families are not increasing in size, they are instead being threatened by parcelization, which not only increases the number of owners, but furthermore complicates the management and existence of the landscapes (Stein et al., 2005; D'Amato et al., 2010). Hence, with rising population growth across the region, forest plots are increasingly fragmented and equally subjected to different land uses and conversions by their multiple owners.

5. CONCLUSION

Forest ecosystems are continually lost and degraded across many tropical ecosystems mainly due to anthropogenic influences peculiar to such environment. In a bid to utilize the resources found in the forest ecosystems, most of the landscapes have been degraded, with the effect that some of its species are lost, while the ranges of others are reduced drastically. Achieving effective management and conservation of the swamp forests in the Niger Delta as well as other tropical ecosystems have been a big challenge because very minimal attention has been given to the underlying causes of forest loss in the region. Understanding such contextual and underlying issues has become more necessary than ever in order to suitably address such losses and provide efficient policies that could promote more environmental sustainability. This study has provided insights on these issues and recommends that suitable guidelines and monitoring of development activities, forest use and mineral exploitation be put in place, so that the swamp forests across the Niger Delta region could be protected and conserved.

ACKNOWLEDGEMENTS

This study forms part of the data collected during a doctoral training at the University of York. The Nigerian government that funded the study is hereby acknowledged. The host communities and those that participated in the study are equally appreciated.

REFERENCES

1. Adekola, O., & Mitchell, G. (2011). The Niger Delta Wetlands: Threats to Ecosystem Services, Their Importance to Dependent Communities and Possible Management Measures. International Journal of Biodiversity Science, Ecosystem Services and Management, 7, 50-68.
2. Adekola, O., Mitchell, G., & Grainger, A. (2015). Inequality and Ecosystem Services: The Value and Social Distribution of Niger Delta Wetland Services. Ecosystem Services, 12, 42-54.
3. Angelsen, A. (Eds.) (2009). Realizing REDD+: National Strategy and Policy Options. Bogor, Indonesia: Center for International Forestry Research (CIFOR).
4. Anwana, E. D., Martin, A. M., Cheke, A., Obireke, L., Ase, M., Otufu, P., & Otobotekere, D. (2010). The Crocodile Is Our Brother: Conservation Management of the Sacred Lakes of the Niger Delta, Nigeria. In B. Verschuuren, R. Wild, J. McNeely, & G. Oviedo (Eds.), Sacred Natural Sites Conserving Nature and Culture (pp. 129-138). Earthscan: Earthscan Publications Ltd.

5. Areola, O. (1982). Soils. In K. M. Barbour, J. S. Oguntoyinbo, J. O. C. Onyemelukwe, & J. C. Nwafor (Eds), Nigeria in Maps (pp. 22-23). USA: Africana Publishing Company.
6. Asner, G. P., Rudel, T. K., Aide, T. M., De Fries, R., & Emerson, R. (2009). A Contemporary Assessment of Change in Humid Tropical Forests. Conservation Biology, 23, 1386-1395. https://doi.org/10.1111/j.1523-1739.2009.01333.x
7. Brown, A., & Thieme, M. (2005). Niger Delta. In M. L. Thieme, R. Abell, M. L. J. Stiassny, J. Skelton et al. (Eds.), Freshwater Ecoregions of Africa and Madagascar: A Conservation Assessment (pp. 62-70). USA: World Wide Fund, Island Press.
8. D'Amato, A. W., Catanzaro, P. F., Damery, D. T., Kittredge, D. B., & Ferrare, K. A. (2010). Are Family Forest Owners Facing a Future in Which Forest Management Is Not Enough? Journal of Forestry, 108, 32-38.
9. De Fries, R. S., Rudel, T., Uriarte, M., & Hansen, M. (2010). Deforestation Driven by Urban Population Growth and Agricultural Trade in the Twenty-First Century. Nature Geoscience, 3, 178-181. https://doi.org/10.1038/ngeo756
10. Emuedo, C., & Abam, M. (2015). Oil, Land Alienation and Impoverishment in the Niger Delta, Nigeria. European Journal of Research in Social Sciences, 3, 8-23.
11. Enaruvbe, G. O., & Atafo, O. P. (2014). Analysis of Deforestation Pattern in the Niger Delta Region of Nigeria. Journal of Land Use Science, 11, 113-130.
12. Francis, P., Lapin, D., & Rossiasco, P. (2011). Securing Development and Peace in the Niger Delta: A Social and Conflict Analysis for Change. Washington DC: Woodrow Wilson International Centre for Scholars.
13. Gardner, T. A., Barlow, J., Chazdon, R., Ewers, R. M., Harvey, C. A., Peres, C. A., & Sodhi, N. S. (2009). Prospects for Tropical Forest Biodiversity in a Human-Modified World. Ecology Letters, 12, 561-582. https://doi.org/10.1111/j.1461-0248.2009.01294.x
14. Geist, H. J., & Lambin, E. F. (2002). Proximate Causes and Underlying Driving Forces of Tropical Deforestation. BioScience, 52, 143-150. https://doi.org/10.1641/0006-3568(2002)052[0143:PCAUDF]2.0.CO;2
15. Godoy, R., & Contreras, M. (2001). A Comparative Study of Education and Tropical Deforestation among Lowland Bolivian Amerindians: Forest Values, Environmental Externality, and School Subsidies. Economic Development and Cultural Change, 49, 555-574. https://doi.org/10.1086/452515
16. Hamilton, L. S., & Snedaker, S. C. (1984). Handbook for Mangrove Are Management. IUCN/ UNESCO/UNEP, Honolulu: East-West Center.
17. Hansen, M. C. et al. (2013). High-Resolution Global Maps of 21st-Century Forest Cover Change. Science, 324, 850-853. https://doi.org/10.1126/science.1244693
18. Harvey, K. R., & Hill, G. J. E. (2001). Vegetation Mapping of a Tropical Freshwater Swamp in the Northern Territory, Australia: A Comparison of Aerial Photography, Landsat TM and SPOT Satellite Imagery. International Journal of Remote Sensing, 22, 2911-2925. https://doi.org/10.1080/01431160119174
19. Hosonuma, N., Herold, M., Sy, V. D., Fries, R. S. D., Brockhaus, M., Verchot, L., Angelsen, A., & Romijn, E. (2012). An Assessment of Deforestation and Forest Degradation Drivers in Developing Countries. Environmental Research Letters, 7, Article ID: 044009. https://doi.org/10.1088/1748-9326/7/4/044009
20. Ibeanu, O. (2002). Janus Unbound: Petro-Business and Petro-Politics in the Niger Delta. Review of African Political Economy, 29, 163-167. https://doi.org/10.1080/03056240208704599
21. Idemudia, U. (2009). Oil Extraction and Poverty Reduction in the Niger Delta: A Critical Examination of Partnership Initiatives. Journal of Business Ethics, 90, 91-116. https://doi.org/10.1007/s10551-008-9916-8
22. Igu, N. I. (2013). Mangrove Forest Loss in Nigeria: Implications for Climate Change. International Journal of Ecosystem, 3, 55-63.
23. Igu, N. I. (2016). Freshwater Swamp Forest Ecosystem in the Niger Delta: Ecology, Disturbance and Ecosystem Services. PhD Thesis, UK: University of York.
24. Irikana, G. J. (2006). Mangrove Utilization and Poverty Alleviation in the Niger Delta: A Case Study of Andoni Women. PhD Thesis, Port Harcourt: University of Port Harcourt.

25. James, G. K. (2008). Assessment of Environmental Change and Its Socio-Economic Impacts in the Mangrove Ecological Zone of the Niger Delta, Nigeria. PhD Thesis, Kansas City: University of Missouri-Kansas City.
26. James, G. K., Adegoke, J. O., Osagie, S., Ekechukwu, S., Nwilo, P., & Akinyede, J. (2013). Social Valuation of Mangroves in the Niger Delta Region of Nigeria. International Journal of Biodiversity Science, Ecosystem Services and Management, 9, 311-323. https://doi.org/10.1080/21513732.2013.842611
27. Jike, T. V. (2004). Environmental Degradation, Social Disequilibrium and the Dilemma of Sustainable Development in the Niger Delta of Nigeria. Journal of Black Studies, 34, 686-701. https://doi.org/10.1177/0021934703261934
28. Kareiva, P., Watts, S., McDonald, R., & Boucher, T. (2007). Domesticated Nature: Shaping Landscapes and Ecosystems for Human Welfare. Science, 316, 1866-1869. https://doi.org/10.1126/science.1140170
29. Kuenzer, C., van Beijma, S., Gessner, U., & Dech, S. (2014). Land Surface Dynamics and Environmental Challenges of the Niger Delta, Africa: Remote Sensing-Based Analyses Spanning Three Decades (1986-2013). Applied Geography, 53, 354-368. https://doi.org/10.1016/j.apgeog.2014.07.002
30. Mamo, G., Sjaastad, E., & Vedeld, P. (2007). Economic Dependence on Forest Resources: A Case from Dendi District, Ethiopia. Forest Policy and Economics, 9, 916-927. https://doi.org/10.1016/j.forpol.2006.08.001
31. Mmom, P. C., & Mbee, M. D. (2013). Population Pressure and Forest Resources Depletion in Gele-Gele Forest Reserve of Edo State, Nigeria. International Journal of Physical and Human Geography, 1, 31-42.
32. Moffat, D., & Linden, O. (1995). Perception and Reality: Assessing Priorities for Sustainable Development in the Niger Delta. Ambio, 24, 7-8.
33. NDES (Niger Delta Environmental Survey) (1997). Environmental and Socio-Economic Characteristics. Lagos: Environmental Resources Managers Limited.
34. NPC (2006). Population and Housing Census of the Federal Republic of Nigeria 2006 Census. Abuja: National Population Commission.
35. Obaje, N. G. (2009). Geology and Mineral Resources of Nigeria (221 p). Berlin: Springer. https://doi.org/10.1007/978-3-540-92685-6
36. Obi, C. I. (2011). Oil Extraction, Dispossession, Resistance and Conflict in Nigeria's Oil-Rich Niger Delta. Canadian Journal of Development, 30, 219-236.
37. Ogon, P. (2006). Land and Forest Resource Use in the Niger Delta: Issues in Regulation and Sustainable Management. Green Governance Project, Berkeley, CA: Institute of International Studies.
38. Ohimain, E. I. (2001). Bioremediation of Heavy Metal Contaminated Dredged Spoil from a Mangrove Ecosystem in the Niger Delta. Unpublished PhD Thesis, Port Harcourt: University of Port Harcourt.
39. Onojeghuo, A. O., & Blackburn, G. A. (2011). Forest Transition in an Ecologically Important Region: Patterns and Causes for Landscape Dynamics in the Niger Delta. Ecological Indicators, 11, 1437-1446. https://doi.org/10.1016 /j.ecolind. 2011.03.017
40. Sassen, M., Sheil, D., Giller, K. E., & Braak, C. J. F. (2012). Complex Contexts and Dynamic Drivers: Understanding Four Decades of Forest Loss and Recovery in an East African Protected Area. Biological Conservation, 159, 257-268. https://doi.org/10.1016/j.biocon.2012.12.003
41. Stein, S. M., McRoberts, R. E., Alig, M. D., Nelson, D. M., Theobald, M., Eley, M., Dechter, M., & Carr, M. (2005). Forests on the Edge: Housing Development on America's Private Forests. General Technical Report PNW-GTR-636, Portland, OR: US Department of Agriculture (USDA) Forest Service. https://doi.org/10.2737/pnw-gtr-636
42. Ukiwo, U. (2009) Causes and Cures of Oil-Related Niger Delta Conflicts. Policy Notes, Nordiska Afrika Institutet, the Nordic Africa Institute.

43. Uluocha, N., & Okeke, I. (2004). Implications of Wetlands Degradation for Water Resources Management: Lessons from Nigeria. GeoJournal, 61, 151-154. https://doi.org/10.1007/s10708-004-2868-3
44. UNDP (2006). Human Development Report 2006. Lagos: United Nations Development Program. https://doi.org/10.18356/334c604b-en
45. Watts, M. (2010). Resource Curse? Governmentality, Oil and Power in the Niger Delta, Nigeria. Geopolitics, 9, 50-80. https://doi.org/ 10.1080/ 14650040412331307832
46. World Bank (2003). Global Problems and Local Concern. Pathways to a Sustainable Future: Sustainable Development in a Dynamic World, Transforming Institution's Growth and Quality of Life. World Development Report, Washington DC: The World Bank.
47. Wurthmann, G. (2006). Ways of Using the African Oil Boom for Sustainable Development. Economic Research Working Research Paper Series No. 84, African Development Bank.
48. Zabbey, N., Erondu, E. S., & Hart, A. I. (2010). Nigeria and the Prospect of Shrimp Farming: Critical Issues. Livestock Research for Rural Development, 22, 198.

CHAPTER 15

Floristic Diversity of Forests of the Northwest Republic of the Congo

Félix Koubouana[1], Suspense Averti Ifo[2], Jean-Marie Moutsambote[1], Romarick Mondzali-Lenguiya[1]

[1]*ENSAF, BP 69 Université Marien Ngouabi, Brazzaville, Congo*

[2]*ENS, BP 69 Université Marien Ngouabi, Brazzaville, Congo*

ABSTRACT

The study of structural and tree flora biodiversity of the forest of the Mbomo-Kellé UFA was done on an area of 88.5 ha (0.014% of the total UFA area). The inventory was carried out in plots of 200 m × 25 m (0.5 ha), arranged in a contiguous way along the opened trails in the forest with an equidistance interval of 2 km. 11,012 trees of DBH ≥20 cm were counted, divided into 107 genera, 31 families and 133 species including 121 completely identified and 12 defined only in terms of genus. Data analysis consisted of calculating relative and weighted frequencies, the Shannon and Weaver diversity indexes, Pielou's evenness index, and the Jaccard and Sorensen similarity indices. Relative frequencies showed that the most representative family in terms of species is the Fabaceae-Caesal piniodeae with 18.05% of species and the most representative in terms of individuals is the Burseraceae with 19.17% of inventoried trees. The principal components analysis allowed to show inter- and intra- block heterogeneities.

Keywords: Floral Diversity, Management, Floristic Composition, Floristic Heterogeneity, Mbomo-Kellé UFA, Congo

1. INTRODUCTION

Tropical forests are the subject of several studies to better understand the role they could play in sustainable development, climate change, and floristic biodiversity. Tropical forests are characterized by high biodiversity, this diversity is an indicator that allows to appreciate links between the richness and the abundance of individuals; it reflects the degree of heterogeneity or stability of vegetation (Basnet, 1992; Trichon, 1997) . Plant diversity can therefore integrate several parameters such as the vegetation structure and dynamics (Moutsamboté, 1985;Koubouana, 1992; Koubouana, 1993) , In the Republic of Congo, according to the current national definition of the forest, forests cover 69% (CNIAF, 2015) of the territory. Sustainable management of these forests requires a sound knowledge of all the natural forest resource; this knowledge can be relied upon only by field studies of the forest environment. Vegetation studies lead to either conducting physiognomic research of the architectural type, either to identifying a number of representative vegetation parameters, allowing to define it simply, in order to compare it to others vegetations (Lescure et al., 1988) . For the present study, the second approach was used, that of the floristic and structural parameters. Long-term biodiversity conservation depends basically on knowledge of the structure, the species richness and the ecological characteristics of the vegetation.

Many studies have been made in Republic of Congo on improving the knowledge of the plant resource. But this knowledge remains generally piecemeal and predominantly localized in protected areas and logging forest concessions. These previous studies are related to ethnobotanical characteristics and general knowledge of the flora (Cusset, 1980; Adjanohoun et al., 1988; Cusset, 1989; Hecketsweiler, 1990; Hecketsweiler & Mokoko-Ikonga, 1991; Dowsett & Dowsett, 1991; Dowsett, 1991; Doumenge, 1992; Kimpouni et al., 2007; Banzouzi, 2008) ; for the massif of Mayombe (Koubouana & Moutsamboté, 2006) in the massif of Chaillu (Mondzali-Lenguiya, 2008;Moutsamboté, 2012) , for the vegetation of the Northern Congo. As far as concerns us, there are not enough available data on the forest of the central western parts of Republic of Congo. These forests are poorly known in Republic of Congo.

This work contributes to the knowledge of this part of the Congolese forest. Mbomo- Kellé UFA , by its geographical location, could be a transition zone between the flora of southern UFA and the UFA of Northern Congo. This study has as its main objective, to improve the knowledge of the floristic resource of the Mbomo-Kellé UFA, and as specific objectives to: (i) conduct a floristic analysis of the vegetation of the Mbomo-Kellé UFA from three blocks of inventory data; (ii) characterize the UFA's flora, through the calculation and analysis of diversity indexes (iii) make an analysis of floristic heterogeneity intra- and inter-blocks of inventory from inventory track data.

2. MATERIAL AND METHODS

2.1. Study Site

Experimental plots were installed within the Mbomo-Kéllé forest management unit (UFA) in the Department of Cuvette Ouest between 14'00" E and 15°00"E; 00°27'35.9" North and 00°22'03" South. It is limited to the North by the Republic of Gabon, to the Northeast by the Odzala-Kokoua National Park, to the South by the Tsama-Bamana UFA, to the West and South-West by the Republic of Gabon, and to the East by the Tsama-Bamana UFA (Figure 1). The Mbomo-Kéllé UFA is crossed by the Equator and under the influence of the Congolese equatorial climate. The average annual rainfall is about 1900 mm/year. The average temperature is about 25°C, the warmest month is March (with 25.9°C) and the coldest month is May (with 20.26°C). Annual average relative humidity is 84%, with a monthly average maximum relative humidity of 97.33% observed in June and a minimum average monthly relative humidity of 59% observed in September.

The study area is characterized by high hills whose altitude varies between 350 and 680 m, which extend a large area of the border of the Congo and Gabon. This relief extends continuously from the Mbomo-Kellé UFA to the Mbinda area in the Chaillu massif. The river network of the Mbomo-Kellé UFA is organized around two large rivers including the Lebango tributary North of the UFA and the Lekona tributary to the South. All the rivers that are in the UFA are either the Lebango or the Lekona tributaries. The two rivers flow into the Likouala-Mossaka in Itoumbi.

2.2. Data Collection

For a better inventory of field data, Mbombo-Kellé UFA was subdivided into 13 blocks of inventory in which plots were installed to inventory all trees o DBH ≥20 cm. For this work, we used data obtained from three blocks of inventory traversing South UFA (blocks 9, 4 and 8) (Figure 2). The size of each plot was of 200 m × 25 m, i.e. 0.5 ha. A total of 88.5 ha were inventoried, distributed unevenly in the three blocks with 61.5 ha in block 4, 18 ha in block 8 and 9 ha in block 9. All trees of DBH ≥ 20 cm were identified and their diameter at 1.30 m measured.

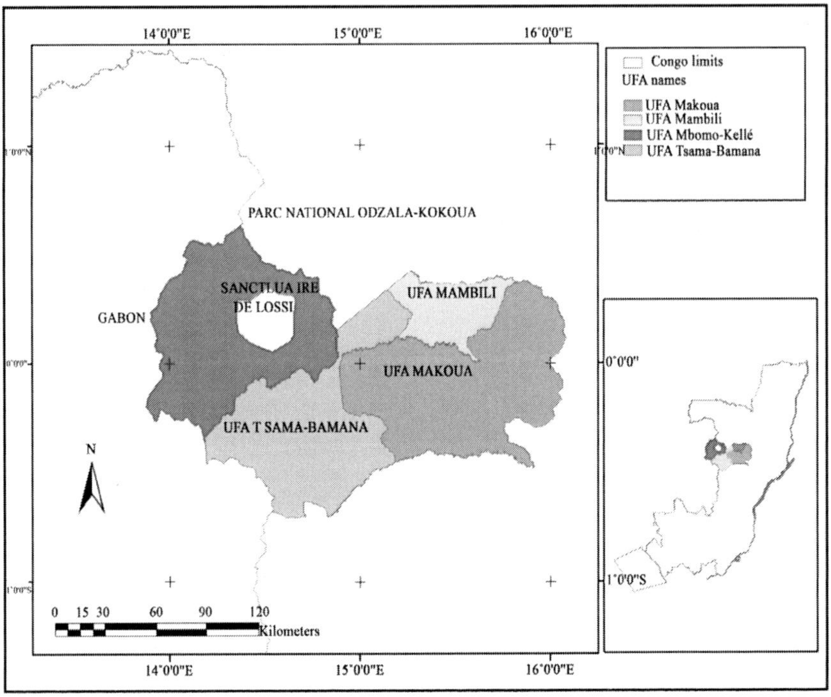

Figure 1. Map of limits and geography situation of Mbomo-Kellé UFA.

2.3. Data Analysis

Floristic diversity

Diversity was determined by using the Shannon Index where pi is the frequency of species I (Ni /N), Ni the number of individuals of species i, and N the number of individuals of all the species.

$$H' = -\sum pi \log 2 pi$$

The Pielou or evenness index

Is the ratio of the Shannon and Weaver index of diversity on the maximum diversity index. It marks the level of diversity achieved by the system compared to the maximum possible (Kimpouni et al., 2013) .

$$J = H'/H' \max$$

$$H' \max = \log 2S$$

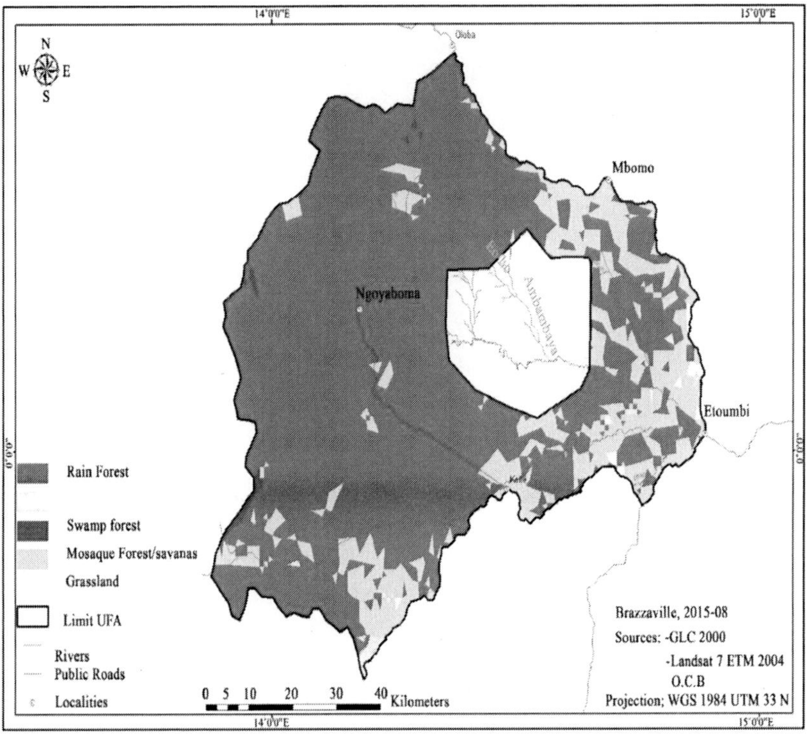

Figure 2. Map of the vegetation of UFA Mbomo-Kellé.

H' max is the logarithm of the total number of species (*S*):

S = total number of species; *NI* = number of individuals of a species in the sample;

When the Index is low, a phenomenon of dominance is revealed, while when it is high, it marks regular compensation of individuals among species (Kimpouni et al., 2013) .

Principal components analysis

This multivariate analysis is used to discriminate between the blocks and the inventoried trails according to edaphic affinities.

2.4. Similarity

Similarity is determined by the coefficients of similarity. These coefficients are used to measure the association between objects. The similarity of two samples (floristic surveys) is based on the presence or absence of certain species in the two samples (Legendre & Legendre, 1984a) . To study the similarity of our different floristic surveys, we used two binary factors excluding the double zeros; i.e. the coefficient of Sorensen (K) and the coefficient of Jaccard (S).

The SØrensen coefficient provides a two times higher weight to double presence, we can consider the presence as more informative than its absence (Legendre & Legendre, 1984b) .

$$S(\%)=(ax100)/(a+b+c)$$

$$K(\%)=(2ax100)/(2a+b+c)$$

a = number of common presence for both floristic surveys

b = number of presence for the first floristic survey

c = number of presence for the second floristic survey

d = number of species absent in both floristic surveys

According to Legendre & Legendre (1984b) , the SØrensen coefficient is dull compared with the Jaccard coefficient, that is, if the similarity of a pair of objects computed by the Jaccard coefficient is higher than the similarity of another pair of objects, it will be also higher if using the coefficient of Sorensen for the calculation of similarity.

3. RESULTS

Floristic richness

The floristic inventory carried out in blocks4, 8 and 9 of the Mbomo-Kellé UFA allowed to record11012 trees of DBH ≥20 cm, divided into 34 families, 107 genus and 133 species, including 12 species defined only as shown in Table 1.

The calculation of gross and weighted frequencies allowed us to appreciate the wealth specific (in number of species) and weight (in number of trees of DBH ≥20 cm) per family (Figure 2).

It appears in Figure 3 that the Fabaceae-Caesalpinioideae family (18.05%) is the most abundantly represented in number of species in the surveyed zone, followed by the Meliaceae (7.52%), Fabaceae Mimosoideae (6.02%), Eupho-rbiaceae (5.26%), Annonaceae (4.51%) and the \Myristicaceae (3.76%).

Table 1. Floristic composition of the study area.

Family	Species	Number of species	Number of trees
Anacardiaceae	*Antrocaryon klaineanum Pierre* *Pseudospondias microcarpa (A. Rich.) Engl.* *Trichoscypha oddonii De Wild.*	3	77
Annonaceae	*Anonidium mannii (Oliv.) Engl. & Diels* *Cleistopholis patens (Benth.) Engl. & Diels* *Duguetia spp* *Greenwayodendron suaveolens Engl. & Diels* *Xylopia staudtii Engl. & Diels* *Xylopia aethiopica (Dunal) A. Rich.*	6	1054
Apocynaceae	*Alstonia boonei De Wild.* *Alstonia congensis Engl.* *Funtumia africana (Benth.) Stapf* *Tetrorchidium didymostemon (Baill.) Pax & K. Hoffm.*	4	87
Malvaceae-Bombacoideae	*Tetrorchidium didymostemon (Baill.) Pax & K. Hoffm.* *Ceiba pentandra (L.) Gaertn.* *Rhodognaphalon brevicuspe (Sprague) Roberty*	2	8
Burseraceae	*Canarium schweinfurthii Engl.* *Dacryodes buettneri (Engl.) H.J. Lam* *Dacryodes edulis (G. Don) H.J. Lam* *Dacryodes pubescens (Vermoesen) H.J. Lam* *Santiria trimera (Oliv.) Aubrév.*	5	2111
Cannabaceae	*Celtis adolfi-friderici Engl.* *Celtis tessmannii Rendle*	2	192
Urticaceae	*Musanga cecropioides R. Br.*	1	69

Chrysobalanaceae	*Maranthes glabra (Oliv.) Prance*	1	4
Ebenaceae	*Diospyros crassiflora Hiern*	2	13
	Diospyros spp		
Euphorbiaceae	*Croton sylvaticus Hochst.*	7	924
	Dichostemma glaucescens Pierre		
	Croton spp		
	Macaranga spp		
	Maesobotrya ssp		
	Margaritaria discoidea (Baill.) G.L. Webster		
	Plagiostyles africana (Müll. Arg.) Prain		
	Ricinodendron heudelotii (Baill.) Pierre ex Heckel		
	Afzelia bipindensis Harms		
	Afzelia pachyloba Harms		
	Amphimas ferrugineas Pierre ex Pellegr.		
	Anthonotha spp		
	Berlinia bracteosa Benth.		
	Daniellia klainei (Pierre) De Wild.		
	Daniellia soyauxii (Harms) Rolfe		
	Dialium bipindense Harms		
	Dialium dinklagei Harms		
	Dialium pachyphyllum Harms		
	Distemonanthus benthamianus Baill.		
	Erythrophleum ivorense A. Chev.		

Fabaceae-Fabioideae	Pachyelasma tessmannii (Harms) Harms	3	134
	Pterocarpus soyauxii Taub.		
Fabaceae-Mimosoideae	Albizia ferruginea (Guill. & Perr.) Benth.	8	813
	Albizia Spp		
	Cylicodiscus gabunensis Harms		
	Parkia bicolor A. Chev.		
	Pentaclethra macrophylla Benth.		
	Piptadeniastrum africanum (Hook. f.) Brenan		
	Tetrapleura tetraptera (Schumach. & Thonn.) Taub.		
	Pentaclethra eetveldeana De Wild. & T. Durand		
Achariaceae	Oncoba glauca (P. Beauv.) Planch.	2	7
	Oncoba spp		
Guttiferae	Allanblackia floribunda Oliv.	3	107
	Mammea africana Sabine		
	Symphonia globulifera L. f.		
Hypericaceae	Harungana madagascariensis Lam. ex Poir.	1	12
Irvingiaceae	Irvingia gabonensis (Aubry-LeComte ex O'Rorke) Baill.	3	129
	Klainedoxa gabonensis Pierre ex Engl.		
	Irvingia spp		

Irvingiaceae	*Klainedoxa gabonensis Pierre ex Engl.*	3	129
	Irvingia spp		
Lauraceae	*Beilschmiedia obscura (Stapf) Engl. ex A. Chev.*	1	7
Lecythidaceae	*Petersianthus macrocarpus (P.Beauv.) Liben*	1	138
Gentianaceae	*Anthocleista schweinfurthii Gilg*	2	7
	Anthocleista vogelii Planch.		
Meliaceae	*Carapa procoera DC.var. procera*	10	262
	Entandrophragma candollei Harms		
	Entandrophragma congoense (Pierre ex De Wild.) A. Chev.		
	Entandrophragma cylindricum (Sprague) Sprague		
	Guarea thompsonii Sprague & Hutch.		
	Khaya ivorensis A. Chev.		
	Lovoa trichilioides Harms		
	Trichilia tessmannii Harms		
	Turraeanthus africanus (Welw. ex C. DC.) Pellegr.		
Moraceae	*Milicia excelsa (Welw.) C.C. Berg*	2	82
	Morus mesozygia Stapf		
Myristicaceae	*Coelocaryon botryoides Vermoesen*	5	1463
	Coelocaryon preussii Warb		
	Pycnanthus angolensis (Welw.) Warb.		
	Scyphocephalium mannii (Benth.) Warb.		
	Staudtia kamerunensis var. gabonensis (Warb.) Fouilloy		
Ochnaceae	*Lophira alata Banks ex C. F. Gaertn.*	1	122

	Ongokea gore (Hua) Pierre		
	Strombosia grandifolia Hook. f.		
Olacaceae	*Strombosiopsis tetrandra Engl.*	5	554
	Strombosia pustulata Oliv.		
	Strombosia zenkeri Engl.		
Pandapoceae	*Panda oleosa Pierre*	1	103
	Uapaca guineensis		
Phyllanthaceae	*Uapaca sp*	3	149
	Maesobotrya spp		
	Maesobotrya spp		
Putranjivaceae	*Drypetes gossweileri S. Moore*	1	86
Rhizophoraceae	*Anopyxis klaineana (Pierre) Engl.*	1	15
	Brenania brieyi (De Wild.) E.M.A. Petit		
	Hallea stipulosa (DC) Leroy		
	Nauclea diderrichii De Wild. & T. Durand		
Rubiaceae	*Pausinystalia johimbe (K. Schum.) Pierre*	6	317
	Pausinystalia macroceras (K. Schum.) Pierre		
	Sarcocephalus pobeguinii Pobeg.		
	Vepris louisii G.C.C. Gilbert		

Rutacaeae	*Vepris louisii G.C.C. Gilbert* *Zanthoxylum heitzii (Aubrév. & Pellegr.) P. G. Waterman* *Zanthoxylum gilletii (De Wild.) P. G. Waterman*	3	35
Sapotaceae	*Baillonella toxisperma Pierre* *Chrysophyllum spp.* *Chrysophyllum africanum A. DC.* *Chrysophyllum lacourtianum De Wild.* *Letestua durissima (A. Chev.) Lecomte* *Tieghemella africana Pierre*	6	75
Malvaceae Sterculioideae	*Cola acuminata (P. Beauv.) Schott & Endl.* *Cola spp* *Eribroma oblongum (Mast.) Pierre ex A. Chev.* *Pterygota bequaertii De Wild.*	4	20
Malvaceae-Tilioideae	*Duboscia macrocarpa Bocq.* *Duboscia viridiflora (K. Schum.) Mildbr.*	2	153
Ulmaceae	*Holoptelea grandis (Hutch.) Mildbr.*	1	1
	TOTAL	133	11102

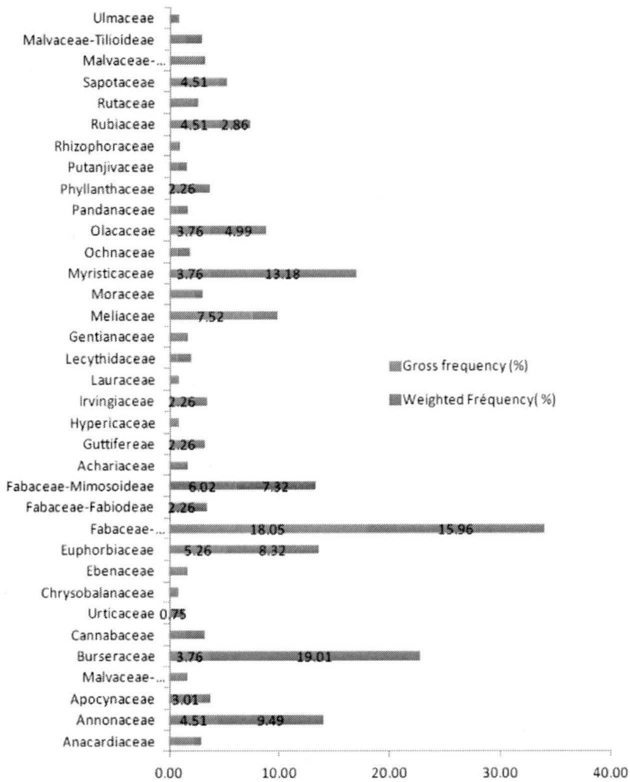

Figure 3. Floristic richness and weight of different inventoried families.

If we consider the quantitative aspect, it is the Burseraceae family (19.01%) which is the most represented, followed by the Fabaceae-Caesalpinioideae (15.96%) (16.09), Myristicaceae (13.18%), Annonaceae (9.49%), Euphorbiaceae (8.32%), and Fabaceae Mimosoideae (7.32%).

The quantitative abundance of Burseraceae, Myristicaceae and Annonaceae is explained by the strong representation respectively of species Santiria trimera (14.57%), *Scyphocephaluim mannii* (7.55%) and *Greenwayodendron suaveolens* (4.30%).

The Shannon-Weaver and Pielou evenness indexes were calculated in each block considering the number of individuals per species and the total number of individuals of each block. The results showed that the index of Shannon and Weaver varies between 5 and 6 bits, and evenness between 0.4 and 0.6 (Figure 4).

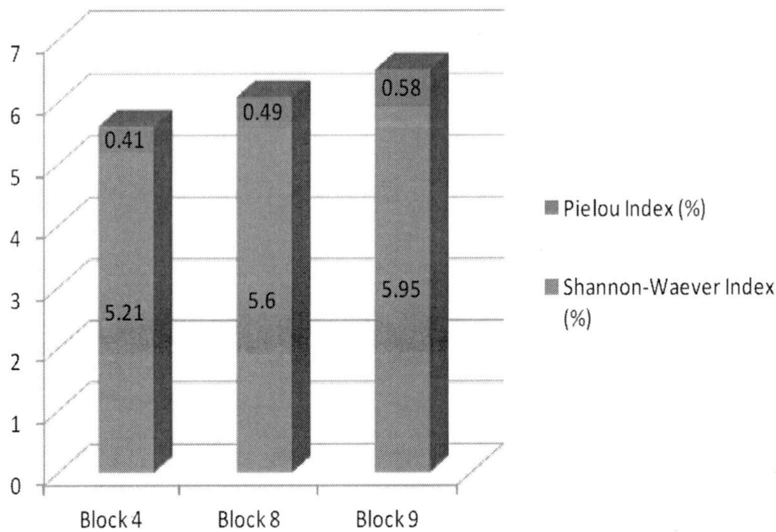

Figure 4. Values for the Shannon and Weaver index and the Pielou evenness index.

These indexes did not vary significantly between inventory blocks. It should be noted that through both the relative frequencies that index calculated, neither of these settings show inter and intra-bloc heterogeneity.

Heterogeneity between blocks

Inter-block heterogeneity was studied from a principal component analysis made on a presence-absence basis, by observing encountered species and as variable or column represented by the different inventory blocks Table 2 shows the contribution of the variables on the factor axes.

This table shows that the essential information on the distribution of the variables in the correlation circle (Figure 5) is explained by F1 and F2 axes.

Figure 5 shows that all the variables, i.e. all the blocks, are arranged in the positive side of the F1 axis. This could explain the association of the inventory blocks with a same forest type. This result is confirmed by the calculated similarity index values (Figure 6).

Indeed, the values of the Jaccard and Sorensen indexes are greater than 60%, indicating a similarity between inventory blocks. However, this similarity cannot hide the local variations of flora. So we note that on the F2 axis, there are two groups: the Group made of block 4, and group 2 made by blocks 8 and 9. Block 4 is opposed to block 9. Indeed, block 4 was located in a mostly surface area or barely marshy, while in block 9, many trails cross swampy areas or follow the banks of the rivers where characteristic wetland species are present, such as *Sarcocephalus pobeguinii* Pobeg., *Coelocaryon botryoides* Vermoesen, *Xylopia staudtii* Engl. & Diels and *Alstonia congensis* Engl. Many trails of block 8 cross fallow areas, where species include *Dacryodes edulis* (G. Don.) H.J. Lam, and many shoots of *Musanga cecropioides* R. Br. The factorial axis F2 discriminates against our blocks according to not only the local substrate variability, but also according to the degree of human activity in the forest.

Intra-block heterogeneity

Theoretically, inter-block heterogeneity would arise from intra-block heterogeneity. To test this hypothesis, we performed a principal component analysis (PCA) on the basis of an array of presence-absence of species on the distance of each block. Obviously, it is interesting to a PCA in the blocks where there were more than two past inventoried i.e. blocks 4 and 8.

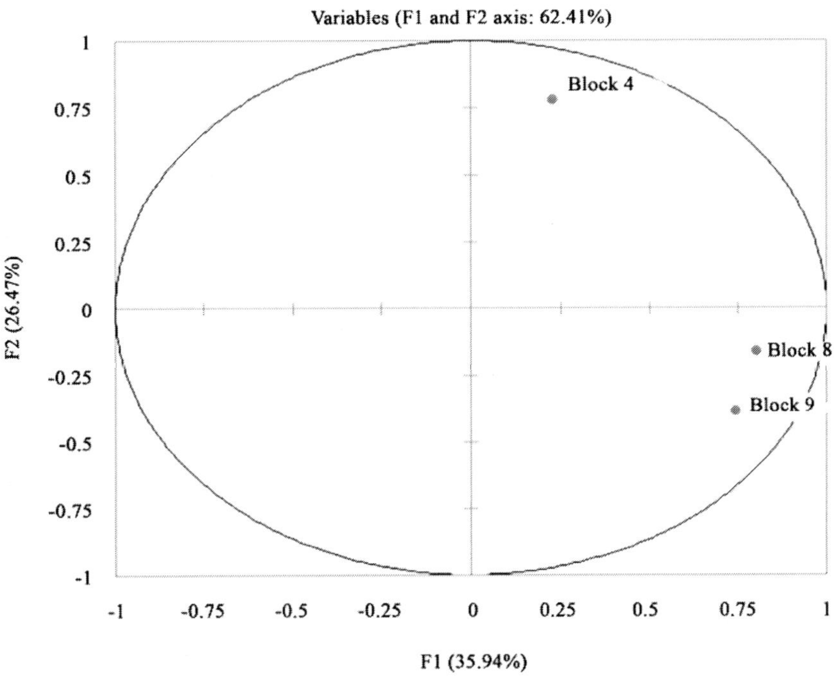

Figure 5. Distribution of variables (blocks) in the correlation circle.

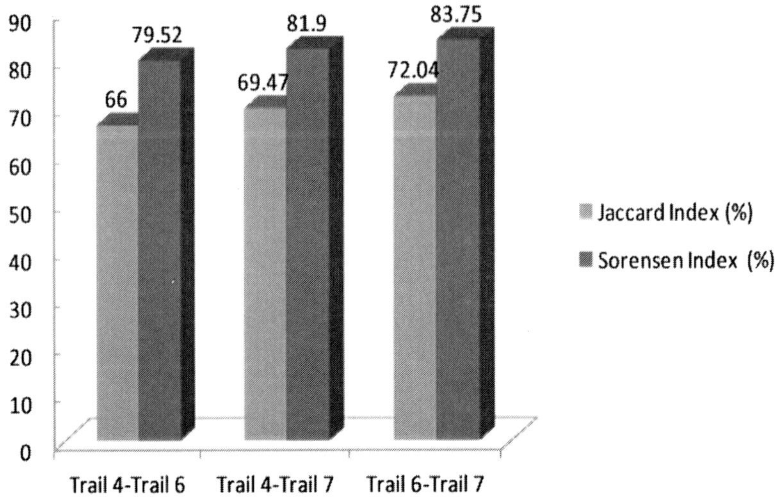

Figure 6. Jaccard and Sorensen similarity Indexes.

Table 2. Contribution of variables (in %) to the total inertia.

	F1	F2	F3	F4
BLOCK 4	3.619	57.627	37.622	1.133
BLOCK 8	44.519	2.417	1.220	51.845
BLOCK 9	37.954	14.015	2.403	45.628

Heterogeneity in block 4

With regard to block 4, Table 3 indicating the contribution of the variables represented by the trails in the different factorial axes, shows that the essential information is represented in the F1 - F2 (Figure 7) plan below.

This figure shows considering axis F1, all the trails are the positive side of the axis indicating that they belong to a single block of inventory. However on the F2 axis, two groups of Trails are distinguished: the trails 5- 6 Group and the trails 2-3-4 Group. Indeed, the F2 axis discriminates between trails depending on the nature of the substrate. The distance of the first group have not gone into a swamp of hydromorphic soils while those in the second group cut the marshy areas. This is explained by the presence of species such as *Sarcocephalus pobeguinii*. Pobeg, *Xylopia staudtii* Engl. & Diels, *Alstonia congensis* Engl., *Hallea stipulosa* (DC.) Leroy. These results are confirmed by the values of the index of similarity (Figure 8).

It appears on this figure that the values of the index of similarity are low between two different trails of different groups whilst the values are high for trails of the same group.

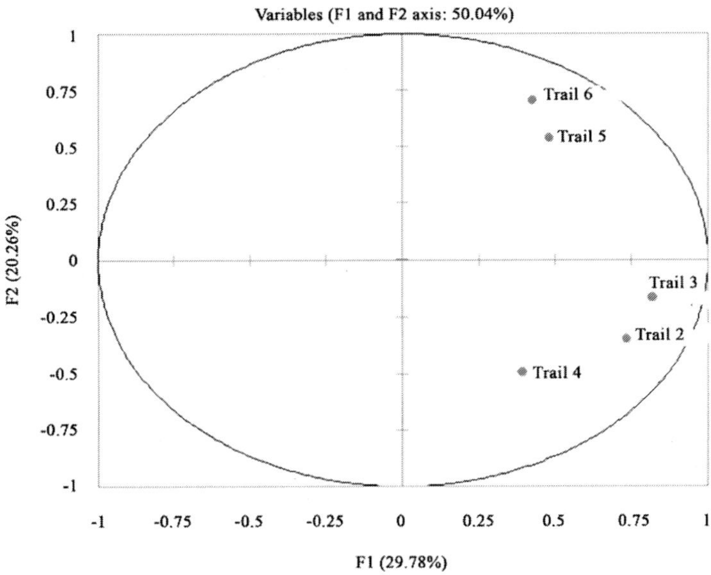

Figure 7. Distribution of variables (trails) in the circle of correlation in block 4.

Table 3. Contribution of the variables (in %) to the total inertia.

	F1	F2	F3	F4	F5	F6
N°	0.432	2.832	89.166	2.369	4.870	0.332
Trail 2	30.105	9.723	1.422	10.523	13.248	34.978
Trail 3	37.665	2.153	0.108	7.775	2.779	49.520
Trail 4	8.758	19.708	2.631	49.220	17.030	2.653
Trail 5	12.935	24.242	0.288	27.047	33.665	1.824
Trail 6	10.106	41.342	6.384	3.066	28.409	10.694

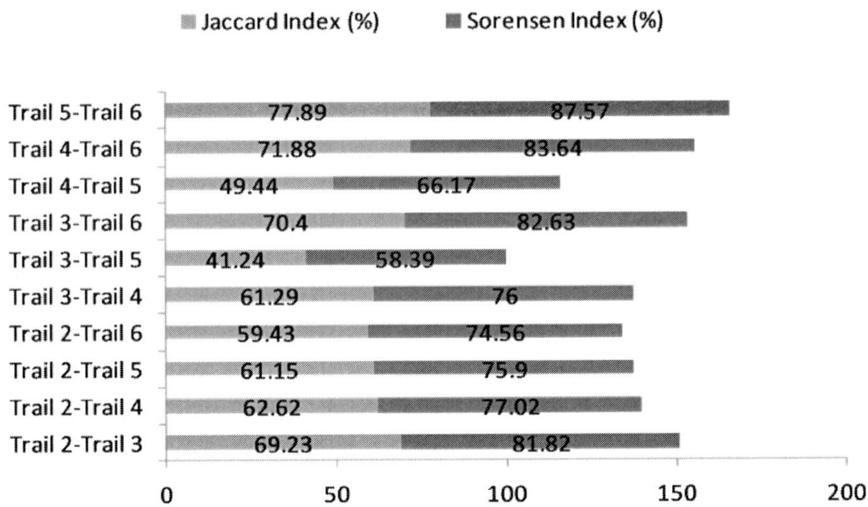

Figure 8. Index of similarity of Jaccard and Sorensen's for block 4.

Heterogeneity in block 8

Figure 9 shows the Trail distribution in block 8.

The point made in block 4 is valid for block 8. Indeed, all trails are arranged in the positive side of the F1 axis, indicating their belonging to a single block. In the F2 axis, trails 4 and 7 are floristically close because they are situated on the positive side of the F2 axis. Trail 6, located on the negative side of the F2 axis, contains species that are characteristic of marshy areas.

4. DISCUSSION

By comparing the results obtained in the study's zone and those obtained in the UFA of the North of the Republic of Congo, in particular the UFA Pokola and Missa, it appear to be more higher than those of the other UFA of the North of Congo. The forest of the UFA Mbomo-Kellé is floristiquement richer than of the UFA Missa.

The six most representative families in terms of species in the forest of the Mbomo-Kellé UFA are those of: Fabaceae-Caesalpinioideae with 18.05%, Meliaceae 8.27%, Euphorbiaceae 6.02%, Fabaceae-Mimosoideae 6.02%, Annonaceae 4.51% and 4.51% Sapotaceae. These results are similar with those of Mondzali-Lenguiya (2008) observed in the forest of the Missa UFA where three families have a high percentage of representation (Table 4).

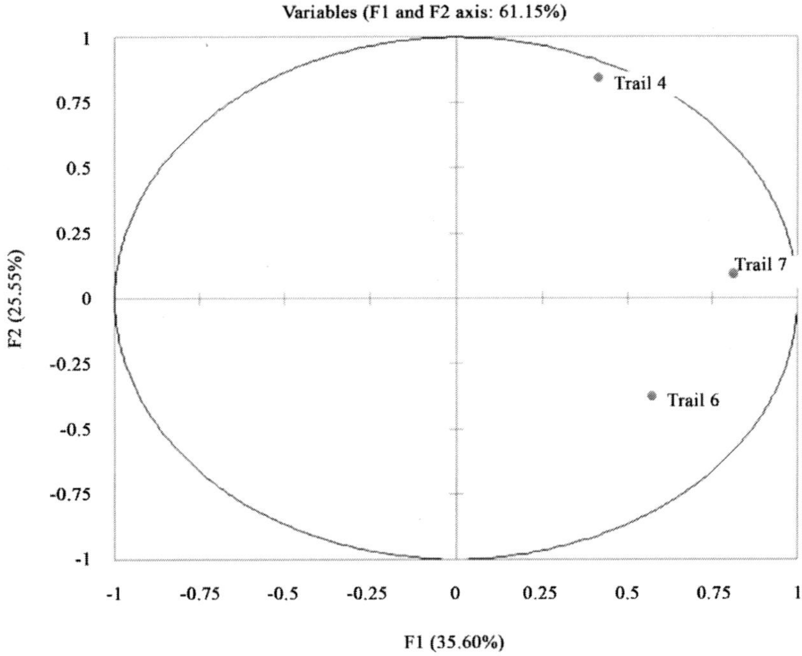

Figure 9. Distribution of variables (trails) in the circle of correlation in block 8.

Table 4. Comparison of the six richest families of the Mbomo-Kellé UFA and the Missa UFA.

Family	% per family at the Mbomo-Kellé UFA	% per family at the Missa UFA
Annonaceae	4.51	5.55
Fabaceae-Caesalpinioideae	18.05	8.51
Euphorbiaceae	6.02	8.88
Meliaceae	8.27	5.92
Fabaceae-Mimosoideae	6.02	5.30
Sapotaceae	4.51	7.77
Total	48.88	41.93

Fabaceae-Caesalpinioideae species richness appears to be a commonality of Congolese forests because Kimpouni et al. (2007) also note a gross frequency of 18.05% in the forest near Tobin on the Congolese coast.

The forest species density studied varies between 0.01 tree·ha^{-1} to 18.12 tree·ha^{-1} and the overall density of the forest is 124.43 tree·ha^{-1}. This result is similar to that obtained in the forest of the Ngombe UFA by IFO company (2007). Indeed, the logging concession company IFO found a density of 124.13 tree·ha^{-1} for the Ngombe UFA. However this result is slightly lower than the density in the forest species of the Missa UFA which is of 189.001 tree·ha^{-1} calculated in relation to the data of Mondzali-Lenguiya (2008) .

Quantitatively, the forest of the Mbomo-Kellé UFA showed a preponderance of individuals from the family Burseraceae, Fabaceae-Caesalpinoideae, Euphorbiaceae and Myristicaceae. These results are similar to those observed in the Boubissi UFE in the Mayombe (located in South of Republic of Congo) by reflecting its character of transition between the forests of the South and North Republic of Congo.

Weighted spectrum shows that in addition to the Fabaceae-Caesalpinioideae, the families of the Burseraceae (19.17%), of the Myristicaceae (13.29%) and of Euphorbiaceae (9%) are the most represented in the number of trees regardless of the inventoried block. These high values are dependent on the abundance of the species respectively *Santiria trimera, Scyphocephalium mannii* and *Plagiostyles africana* as we found Kimpouni et al. (2007) in the forests of the Bambama UFA (in the South-Congo) thereby reinforcing the transition forest characteristic of the Mbombo-Kellé UFA, between the forests of the South and North Congo forests.

The floristic variability or the floristic heterogeneity inter-block was studied from the factorial analysis of correspondence. The results showed on the one hand that the sampled blocks of inventory belong to the same forest type and on the other hand which there are floristic variabilities due to the variation of the substrate, like had underlined it (Basnet, 1992; Gartlan et al., 1986) . The membership of the same forest type is confirmed by the values of indices of similarity. This heterogeneity inter-block is also observed in the tailboards of the same block. The variability of the edaphic factors (composition of the grounds, grounds dewatered and hydromorphic grounds) explains the richness of the forest of the UFA Mbomo-Kellé.

The floristic variability or the floristic heterogeneity between-block was studied starting from the factorial analysis of correspondence. The results showed on the one hand that the sampled blocks of inventory belong to the same forest type and in the other hand which there are floristic variabilities due to the variation of the substrate, like had underlined it (Basnet,

1992; Gartlan et al., 1986) . The various blocks belong to the same type of plot highlighted by the indices of similarities. This heterogeneity inter-block is also observed in the tailboards of the same block. The variability of the edaphic factors (composition of the grounds, grounds dewatered and hydromorphic grounds) could explained the richness of the forest of the UFA Mbomo-Kellé.

5. CONCLUSION

The floristic inventory of the Mbomo-Kéllé UFA forest allowed us to appreciate its floristic composition. Qualitatively, all inventoried blocks were rich in Fabaceae- Caesalpinoideae, Meliaceae, Myristicaceae, Euphorbiaceae, Fabaceae Mimosoideae, Burseraceae and Annonaceae. Shannon-Weather calculated index values are high, around 6-bit indicating a recovery of the same importance for most of the inventoried species. However, the Pielou evenness index was relatively low in block 4 indicating a low abundance of some species; these include the *Santiria trimera* (Oliv.) Aubrév. *Santiria trimera* (Burseraceae), *Plagiostyles africana* (Mull. Arg.) *Prain Plagiostyles africana* (Euphorbiaceae), *Scyphocephalium mannii* (Beth.) Warb. *Scyphocephalium mannii* (Myristicaceae). The results of the principal components (PCA) analysis revealed floristic heterogeneities within blocks. Regarding the inter-block variance, the results of the PCA have shown that three blocks belong to a same ecosystem. What was confirmed by the indices of similarity of Jaccard and Sorensen. The PCA also indicated the floristic differences between blocks due to the local variability of the substrate. As regards inter-block heterogeneity, the factorial axes of the PCA have discriminated between Trails of blocks 4 and 8 depending on the nature of the substrate. Some past crossed swampy areas of hydromorphic soils. *Scyphocephalium mannii* is present in the forests of Southern Congo in the Chaillu massif and is reported in the forests of Northern Congo in the Mbomo-Kellé UFA in Odzala National Park and in parts of the Western Sangha. This floristic fact shows that the Mbomo-Kéllé UFA is a transition zone between the South-Congo forests and the forests of Northern Congo. This observation is confirmed by the similarity of relief between the Chaillu massif and the Mbomo-Kellé UFA.

Structurally, the density of trees in the surveyed area is 124.42 trees/ha. Weight frequency showed a preponderance of trees from the families Burseraceae, Myristicaceae, and Euphorbiaceae.

REFERENCES

1. Adjanohoun, E. J., Ahyi, A. M. R., Ake Asi, L., Baniakina, J., Chibon, P., Cusset, G., Doulou, V., Enzanza, A., Eyeme, J., Goudote, E., Keita, E., Mbemba, C., Mollet, J., Moutsamboté, J. M., Mpati, J., & Sita, P. (1988). Contribution aux études ethnobotaniques et floristiques en République populaire du Congo : médecine traditionnelle et pharmacopée (605 p). Paris: ACCT
2. Banzouzi, F. (2008). Inventaire d'aménagement de l'UFE Boubissi: Végétation et flore du Bloc B (46 p). Rapp. de fin de formation, IDR, Brazzaville: Univ. Marien Ngouabi
3. Basnet, K. (1992). Effect of Topography on the Pattern of Trees in Tabanuco (Dacryodes Excels) Dominated Rain Forest or Puerto-Rico. Biotropica, 24, 31-42
4. CNIAF (2015) Carte de changement de couverture forestière en République du Congo pour la période 2010-2012
5. Cusset, G. (1980). Etude phytoécologique préliminaire de la forêt Yombé (78 p). Université paris VI et Eq. Biol. For. Trop. Dimonika
6. Cusset, G. (1989). La flore et la végétation du Mayombe congolais, état des connaissances. In J. Sénéchal et al. (Eds.), Revue de connaissances sur le Mayombe (pp. 103-136). Paris: Unesco
7. Doumenge, C. (1992). La réserve de Conkouati, Congo: le secteur sud-ouest. UICN, Chevron, Gland, Suisse, 213 p
8. Dowsett, R. J., & Dowsett-Lemaire, F. (1991). Flore et faune du Bassin du Kouilou (Congo) et leur exploitation. Tauraco Research Report (CONOCO), 4, 340 p
9. Dowsett-Lemaire, F. (1991). The Vegetation of the Kouilou basin in Congo. In R. J. Dowsett, & F. Dowsett-Lemaire (Eds.), Flore et faune du Bassin du Kouilou (Congo) et leur exploitation. Tauraco Research Report, 4, 17-51
10. Gartlan, J. S., Newbery, M., Thomas, D. W., & Waterman, P. G. (1986). The Influence of Topography and Soil Phosphorus on the Vegetation of Korup Forest Reserve, Cameroun. Vegetation, 65, 131-148. http://dx.doi.org/10.1007/BF00044814
11. Hecketsweiler, P., & Mokoko-Ikonga, J. (1991). La réserve de Conkouati, Congo: Le secteur sud-est (323 p). Gland: UICN
12. Hecketsweiler, P. (1990). La conservation des écosystèmes forestiers du Congo. Gland: UICN
13. Kimpouni, V., Loumeto, J., & Mizingou, J. (2007). Caractérisation de l'okoumé (Aucoumea klaineana) dans la forêt du littoral congolais. DGRST, Actes des premières journées nationales sur la recherche forestière au Congo, Brazzaville, 26-29 juin 2006
14. Kimpouni, V., Mbou, P., Gakosso, G., & Motom, M. (2013). Diversité floristique du sous-bois et regeneration naturelle de la forêt de la Patte d'Oie de Brazzaville, Congo. International Journal of Biological and Chemical Sciences, 7, 1255-1270. http://dx.doi.org/10.4314/ijbcs.v7i3.31
15. Koubouana, F., & Moutsambote, J.-M. (2006). Etude préliminaire de la végétation des UFE Tsinguidi et Bambama (42 p). Brazzaville: Rapp. D'étude
16. Koubouana, F. (1992). Etude comparée de plusieurs types d'analyses spatiales sur les forêts semi-décidues de la Vallée du Niari (CONGO). Bulletin d'écologie, 23, 163-174
17. Koubouana, F. (1993). Les forêts de la Vallée du Niari (Ongo): Etudes floristique et structurales (265 p). Thèse doct, Paris: University of Paris VI
18. Legendre, L., & Legendre, P. (1984a). Ecologie Numérique, tome 1: Le traitement multiple des données écologiques (2rd éd., 260 p). Paris: Masson
19. Legendre, L., & Legendre P. (1984b). Ecologie Numérique, tome 2: La structure des données écologiques (2rd éd., 335 p). Paris: Masson

20. Lescure, J. P., & Boulet, R. (1985). Relationship beetwen Soil and Vegetation in a Tropical Rain Forest in Franch Guiana. Biotropica, 17, 155-160. http://dx.doi.org/10.2307/2388508
21. Mondzali-Lenguiya, R. (2008). Inventaire d'aménagement de l'UFA Missa: aspect floristique des zones 1,2,3 and 4 (43 p). Rapport de stage de fin de formation en vue de l'obtention du diplome d'ingénieur des travaux de développement rural. Brazzaville: technique forestière, Institut de Développement Rural (IDR), Université Marien Ngouabi
22. Moutsamboté, J.-M. (1985). Dynamique de reconstitution de la forêt Yombe (Dimonika, R.P. du Congo) (301 p). Bordeau: These 3 cycle University of Bordeaux III
23. Moutsamboté, J.-M. (2012). Etude écologique, phytogéographique et phytosociologique du Congo septentrional (Plateaux, Cuvettes, Likouala et Sangha) (632 p). Thèse de Doctorat d'Etat, Brazzaville: University of Marien Ngouabi
24. Trichon, V. (1997). Hétérogénéité spatiale des forêts d'une forêt tropicale humide de Sumatra: Effet de la topographie sur la structure floristique. Annals of Forest Science, 54, 431-446. http://dx.doi.org/10.1051/ forest:19970502

CHAPTER 16

Silvopasture Practices and Perspectives in the Northeastern United States

Joseph Orefice, John Carroll, Drew Conroy, Leanne Ketner

Paul Smith's College, Routes 86 & 30, Post Office Box 265, Paul Smiths, NY 12970, USA

Natural Resources, University of New Hampshire, 105 Main St, Durham, NH 03824, USA

Applied Animal Science, Thompson School of Applied Science, University of New Hampshire, 105 Main St, Durham, NH 03824, USA

Integrative Studies, Paul Smith's College, Routes 86 & 30, Post Office Box 265, Paul Smiths, NY 12970, USA

ABSTRACT

The use of silvopasture systems on farms in the Northeastern United States has never been documented. Our objective was to gather baseline data to describe silvopasture practices and perspectives in the Northeastern United States. To accomplish this, we investigated the structure, management of, and reasons for use of silvopastures in New York state and New England through a series of interviews and inventories on 20 farms purposefully chosen as practicing silvopasture. Thematic content analysis was conducted to summarize interview results and identify trends related to silvopasture practices. Three farmers in this study had been practicing silvopasture on their farms over 30 years; the rest were new to silvopasture in the past 10 years. Only three of 20 farmers interviewed in this study had experience practicing silvopasture prior to implementing it on their farms. Forest conversion to silvopasture was the primary starting point for silvopastures observed on regional farms. Orchard, open field edge, outdoor living barn, and plantation silvopastures were also documented on multiple farms. Shade and a desire to maximize use of farm woodlands were primary reasons for silvopasture utilization. This research provides evidence that silvopastures are being used to diversify regional farms. For the practice to be advanced in the region further research is needed on the topic.

Keywords: Agroforestry, Pastured woodlands, Woodland grazing, Permaculture, Livestock, Woods

INTRODUCTION

Silvopasture in the Northeastern United States has never been formally documented in the academic literature. Two syntheses on agroforestry and silvopasture science in North America describe silvopasture systems in all regions of the United States except the Northeast (Clason and Sharrow 2000; Garrett et al. 2004). While it is clear that some regions of the United States have a strong history with silvopasture, the occurrence of this agroforestry practice in the Northeast is relatively unknown. Recent publications have called for adoption of silvopasture by farmers in the Northeast and the topic has been highlighted during regional workshops and conferences over the last 5 years (Carroll 2011; Chedzoy and Smallidge 2011a, b).

A stumbling block in the adoption of silvopasture systems in the Northeastern United States may be that there are few publicly known examples of silvopasture in the region. A 2011 publication on silvopasture in the Northeast describes the benefits and general components of silvopasture systems, but few specific examples are provided (Chedzoy and Smallidge 2011b). It is risky for a farmer to adopt a new system without an understanding of its benefits and tradeoffs in the form of established regional examples.

In other areas of the world the adoption of agroforestry practices has been slow due to farmer bias against trees (Neumann et al. 2007) and limited knowledge of landowners on agroforestry practices (Barbieri and Valdivia 2010). Semi-structured interviews with farmers in Colombia indicate that the primary barriers to silvopasture adoption were high establishment costs and lack of knowledge and resources available to farmers about the practice (Calle 2008). In Argentina, researchers using semi-structured interviews found that 84 % of farmers practicing silvopasture would increase the amount of land they have in silvopasture if given the opportunity (Frey et al. 2012).

Agroforestry research in the United States provides insight into attitudes toward unconventional farming and forest management practices. In Missouri, for instance, many farm landowners had little knowledge of agroforestry practices yet they had interest but little knowledge in how to practice agroforestry, including silvopasture (Arbuckle et al. 2009). Another study found that family farmers in Missouri had little understanding of agroforestry practices (Barbieri and Valdivia 2010). A survey of woodland owners and farmers in Pennsylvania found the barriers to agroforestry adoption to be a lack of ability to experiment, expenses of additional management, and unknown markets for products (Strong and Jacobson 2005). Four of the top five ways farmers prefer to learn about new practices are demonstration, farm visits, field days, and discussions (Franz et al. 2010).

The path to ensure the sustainable management of regional silvopasture systems starts by providing land managers with documented experiences of others to learn from and consider. Therefore, our objective was to gather baseline data to describe silvopasture practices and perspectives in the Northeastern United States. These data act as a reference point for future scientific inquiry and advancement of silvopasture. To accomplish this, we investigated the structure, management of, and reasons for use of silvopastures in New York state and New England through a series of interviews and inventories on farms practicing silvopasture.

METHODS

Interviews

Twenty-two semi-structured interviews were conducted by phone and on-farm with silvopasture practitioners to document the details of, and reasons for the current use of silvopasture within New York state and New England. Interview questions were reviewed by multiple researchers at the University of New Hampshire and Cornell Cooperative Extension for clarity and comprehensiveness. These questions were then pilot-tested for clarity with three silvopasture practitioners prior to implementation. Phone interview questions were consistent with on-farm interviews but included an additional question asking the practitioner to describe the tree and understory species compositions of their silvopastures. In addition, we conducted quantitative inventories of silvopasture systems on selected farms. Interviews and inventories occurred in 2014.

A snowball sampling technique was used to identify and purposefully sample farms practicing silvopasture. Snowball sampling allows for the identification of practitioners from other practitioners and people in the broader field (Patton 2002). Professionals in the field of silvopasture, cooperative extension agents, and professional farming and forestry organizations were used to locate self-identifying practitioners of silvopasture. Additionally, attendees of the 2011 and 2014 Northeast Silvopasture Conferences who identified as farmers were solicited for interviews. Fifty-two farms were identified through this process as potentially practicing silvopasture. Silvopasture was defined as having intentional and sustainable management of tree crops, livestock, and forages on the same unit of land. Farms for interviews and site visits were selected by the following ranked measures: willingness to offer an interview, number of years practicing silvopasture, multiple types of silvopasture systems integrated in the same farm, and number of hectares in silvopasture. Preference for an interview was given to three farms that were practicing silvopasture in states which were under-represented via the above measures, enabling the scope of this research to encompass all states in New England and New York. Of 52 farms identified, 20 farms were selected for an interview and 15 of these interviews were conducted on-farm while the remaining five were over the telephone. Three farms were not selected for interviews or site visits due to unwilling participants and the remaining 29 farms not selected for interviews were in their first year or planning stages of silvopasture development. Two foresters were interviewed on farms where they were consulting on silvopastures, this was in addition to interviews with the farmers on those two farms.

Interviews lasted between 30 and 120 min and interviewees had the opportunity to answer and expand on many questions regarding their perspectives toward silvopasture, farm demographics, and the management of on-farm silvopastures. To ensure consistency, the primary author conducted all interviews. With the permission of the interviewee, interviews were tape-recorded and detailed notes were taken. Upon completion, recorded interviews were transcribed and reviewed by multiple researchers (investigator triangulation) to account for interpretation bias (Denzin 1978; Patton 2002). Thematic content analysis was conducted to summarize interview results and identify trends related to silvopasture practices (Patton 2002). Interview results were coded into the following broad categories: demographics, reasons for silvopasture use, silvopasture management (trees, livestock, and forages), and challenges of silvopasture use. Additional coding of results were completed within each main category to quantify similar responses.

Inventories

In addition to interviews, an inventory was conducted in silvopastures on each farm visited to determine overstory conditions and forage species composition. Twenty-three unique silvopastures at various stages of establishment were inventoried on 15 farms visited. The sampling design was a nested plot design using variable radius sampling for overstory and fixed area plots for understory plants. Sampling intensity varied between (but not within) silvopastures due to time constraints and overstory conditions. Silvopastures with low variability in tree spacing and recently established silvopastures were sampled less intensively than silvopastures which had been established for multiple years and had high variability in tree spacing. Data recorded in the overstory sample included tree species, diameter at 1.37 m off the ground, and product height. Understory sampling consisted of a percent cover of dominant forage species present, and a tally of non-forage plant species present. Qualitative notes were taken on tree vigor, tree root exposure, and bare soil exposure. Qualitative notes were also taken regarding pasture conditions and management on farms visited.

Data analysis

Inventory data of trees, stand relative density, and forages were summarized and analyzed using Microsoft Excel and NED-2, a forest inventory and analysis program developed and available from the US Forest Service (Twery et al. 2005). Inventory data were compiled to include a summary of overstory tree stocking, health, financial value, understory forage species composition, existent non-forage plants, and photosynthetically active radiation. Inventory data and interview transcriptions were used to categorize regional silvopastures into the following groups: uniform spacing with forest origin, patch systems with forest origin, variable tree density systems with forest origin, open field edge silvopastures, plantation silvopastures, orchard silvopastures, outdoor living barns, and a silvopasture maple (*Acer* spp.) sugarbush. Systems that farmers perceived as silvopasture but were missing intentional management of tree crops, livestock, and forages were categorized as wooded livestock paddocks instead of silvopasture. Wooded livestock paddocks were found on some farms even though the sampling design purposefully sought out silvopasture practitioners. The exception to this was outdoor living barns which were considered silvopastures, but may have been missing the forage component, but had direct management of tree health through active livestock rotations. Areas where livestock were kept for multiple months without rotation were also classified as wooded livestock paddocks instead of silvopasture.

RESULTS

Farm demographics

Ten of 20 farms had at least one full time farmer with no off-farm employment and the remaining 10 had farmers with off-farm jobs in addition to their farm business. Off-farm jobs were diverse and included professionals in the medical field, law, forestry, and agricultural extension. Farmer experience was also diverse. Farmers had been the principal operator of a farm for an average of 13 years (standard deviation = 11 years) with a maximum of 42 years, and a minimum of 2 years. Tenure on their current farm, regardless of being a principal operator, averaged 14 years (standard deviation = 13 years) with a maximum of 44 years, and a minimum of 2 years. Farmers interviewed included both owner/operators and hired workers.

Primary farm products were diverse between and within farms, although a primary farm product on 16 of 20 farms was some type of livestock for meat. Two farms had primary farm products of dairy cattle and the other two farms' primary products were tree crops, including a tree nursery. Timber sales were cited as additional primary farm product on six of the farms. The size of farms practicing silvopasture varied in both land holdings (12–486 ha), percent of land in silvopasture (1–32 %), and number of livestock.

Farmer experience with silvopasture

Only three of 20 farmers interviewed in this study had experience practicing silvopasture prior to implementing it on their farms. All of these three farmers' experiences with silvopasture prior to implementing it occurred in Europe or Central America. Four additional farmers claimed to have some knowledge of silvopasture prior to implementing it on their farm. The remaining 14 farmers had no, or extremely limited, prior knowledge and experience with silvopasture before implementing it on their farms. Three farmers had been practicing silvopasture on their farms over 30 years in the region and the rest were new to silvopasture in the past 10 years. The longest silvopasture documented had been in production for 42 years (in this case a forest was converted to silvopasture and trees were over 42 years old), although the median age of land managed as silvopasture was 4 years.

Silvopasture was a fairly new concept to most farmers, becoming familiar with the practice over the last decade. However, seven farms had been practicing silvopasture prior to finding out it was an agroforestry practice.

For example, one farmer who had been utilizing silvopastures for 30 years had first heard the term when an extension professional in his region suggested him to be a part of this study. A misconception that any integration of livestock in a wooded area would be silvopasture was held by four farmers in this study. Three farms were continuously pasturing pigs in woodlands. One farm was continuously grazing dairy cattle and horses in wooded areas. Two of the four farms utilizing wooded livestock paddocks also had well managed silvopastures being grazed by other species. Additionally, a misconception was found to exist among practitioners in the region that any use of livestock to actively eliminate or manage woody vegetation could be called silvopasture.

Reasons for, timing, and challenges of silvopasture utilization

Farmers were utilizing silvopasture for a variety of reasons (Table 1). Shade for livestock was the most commonly stated reason for incorporating silvopastures into farms with 16 of 20 farmers independently citing this as a reason for use. Expanding pasture acreage and diversity was also greatly cited by farmers, 14 of 20. Utilizing and incorporating woodlands into primary farming ventures was a reason for silvopasture adoption by 12 of 20 farmers.

Table 1 Reasons for, and challenges of, silvopasture utilization by 20 farmers practicing silvopasture in New York and New England. Farmers were interviewed in 2014 and may have provided more than one reason for or challenge of silvopasture utilization

Incorporation of silvopasture into farm management systems was also diverse. Farms were primarily utilizing silvopastures during the grazing periods of late-spring, summer, and fall. All farmers used silvopastures during the hot periods of the summer. All farms, except one, utilized both silvopastures and open pastures in their grazing operations, but not necessarily in the same paddock. Some farms reserved silvopastures for certain times of the year, such as hot periods in the summer or inclement winter weather, while others kept them as a patchwork within on-farm livestock rotations. Farmers identified the early spring (mud season) as a time when livestock were excluded from silvopastures (although one farm utilized silvopastures year round). During mid-summer and times of droughts, farmers were utilizing silvopastures because they perceived that silvopastures had greater forage availability during these periods (Table 1).

Fencing establishment, such as type to utilize and methods of construction, and maintenance of fencing of paddocks was stated as a challenge by 9 of

20 farmers interviewed when asked what their major challenges were when managing silvopastures (Table 1). Lack of knowledge toward silvopasture and lack of time for silvopasture management were cited as challenges by six and five farmers of 20 interviewed, respectively. Forage management and unknown forage quality was another area farmers expressed as a challenge toward managing silvopastures. One of 20 farmers interviewed was not planning to continue practicing silvopasture in the future. This farmer intended to phase out practicing silvopasture in order to create a better view through tree removal, increase options for his agricultural land, and make fencing more time efficient. However, the 19 other farmers interviewed were pleased with the practice, and 14 of these farmers intended to increase the amount of land on their farm in silvopasture.

Silvopasture characteristics

The amount of silvopastures on farms ranged from less than 1 ha to 73 ha, with a median of 5 ha per farm. Sizes of individual silvopastures on farms were typically less than 1–2 ha. Forest conversion to silvopasture was the primary starting point for silvopastures observed on regional farms (Table 2). The most common of these was a conversion to uniform tree spacing. In these systems mature hardwood, softwood, and mixedwood forests were heavily thinned from the lower canopy classes, leaving well-formed co-dominant and dominant stems as residuals. Oak, maple, and eastern white pine were the most common species favored as residuals in silvopasture converted from forests. Farmer goals for these species were primarily timber, but in the case of oak, acorns were also favored by many farmers as a livestock supplement. Residual basal area of forests converted to uniformly spaced silvopastures ranged from 6 $m^2\ ha^{-1}$ (546 trees ha^{-1}) to 30 $m^2\ ha^{-1}$ (282 trees ha^{-1}), with an average basal area of 17 $m^2\ ha^{-1}$(343 trees ha^{-1}).

Five silvopasture systems converted from forestland utilized patched grouping of residual trees. Patch sizes were small, <0.25 ha, and greatly variable in shape. Multiple patches of trees were interspersed within similarly sized patches of open pasture in these systems. Farmers' reasons for choosing grouped tree retention in silvopastures included working with unequal distribution of quality trees in the pre-silvopasture forest, ease of creation, and ease of management. Tree spacing was so heterogeneous in three hardwood silvopastures that these were classified as irregular tree density due to their difference from both uniformly spaced systems and patch systems.

Table 2 Type of silvopasture systems found on 20 farms in New York in New England purposefully identified. In some cases, multiple types of silvopasture existed on the same farm. Farms were inventoried in 2014

Silvopasture type	Number of farms
Forest conversion to uniform tree spacing	13
Open field edges	7
Orchards	6
Forest conversion to patch tree spacing	5
Outdoor living barns	4
Forest conversion to irregular tree spacing	3
Hardwood plantations	2
Conifer plantations	1
Maple sugarbush	1

Seven farms in this study were incorporating silvopastures into the edges of open pastures (Table 2). To benefit from forest edge encroachment into open pastures, some farmers had converted overgrown field edges into silvopastures by thinning trees. Others had converted portions of (non-encroaching) forest adjacent to open pastures to silvopastures in an effort to diversify the shade conditions of regularly used open pastures. Grazing partitioning of these edge silvopastures contained both open areas and wooded areas, as opposed to grazing the silvopasture edge separately from the open pasture area. Thinned and managed trees in the field border was also valued for aesthetic reasons by practitioners.

Three types of plantation silvopastures were observed in this study. One, conifer plantations, were being utilized as outdoor living barns and will be discussed in a following paragraph. The other two types were hardwood plantations. Two farms had established black walnut (*Juglans nigra*) in open pastures and were utilizing the plantations as a silvopasture. Black locust was the other hardwood silvopasture plantation documented. These plantations had basal areas of 17 $m^2\ ha^{-1}$ (734 trees ha^{-1}) and 21 $m^2\ ha^{-1}$ (974 trees ha^{-1}). Trees were being grown primarily for use as fence posts, and harvested through commercial thinnings at years 15, 20, and 25. Initial establishment of trees in these systems was through seedlings, but on one farm the next cohort was being established through a coppice system.

Four farms in this study were utilizing outdoor living barns (Table 2). Outdoor living barns are silvopasture systems in which tree density is maintained at an abundant level to maximize the amount of shelter that trees provide to livestock. Forages in all of these outdoor living barns documented were very sparse or non-existent. Outdoor living barns consisted of areas of dense conifers. Farmers stated the purpose of their outdoor living barns were to produce timber or fence posts, in the case of northern white cedar (*Thuja occidentalis*), while also providing shelter for livestock during exceptionally cold periods of the year. Farmers stated that outdoor living barns were not utilized as permanent winter paddocks nor did these areas experience livestock pressure during the spring thaw of frozen ground. One farm maintained an outdoor living barn to provide shelter and biting fly relief for livestock during the summer grazing period.

Six farms in this study incorporated livestock into orchards as a form of silvopasture. Farmers stated the value of these systems were fertilization to trees, grass management, livestock nutrition, and reduction in rodent habitat. Orchards were primarily comprised of apple trees (*Malus* spp.) and, in some cases, with lesser components of other fruit or nut trees. Farmers were using fruit products from orchards for on-farm consumption, direct marketing to consumers, livestock feed, and scion wood. Spacing between trees in orchard silvopastures were typically uniform in pattern with space between tree crowns, but variable in terms of actual distance between trees.

Primarily sheep or cattle were incorporated into orchard silvopastures by grazing in the summer months, and in the fall after excess fruit has dropped. Farmers grazing sheep in orchard systems did not express a fear of damage to fruit trees from livestock while farmers utilizing cattle stated the importance of short grazing periods to avoid tree damage in orchards. Regeneration of new fruit trees was accomplished through individual tree protection mechanisms. Farmers did not mention any food safety concerns regarding livestock and orchard integration.

One farm in this study was utilizing a maple sugarbush as a silvopasture for beef cattle to keep brush down in the sugarbush and provide shade for livestock. This farmer was periodically grazing a herd of around 90 beef cattle in a 6 ha production sugarbush for over 25 years. Cattle were only introduced to the sugarbush during dry periods of the summer, such as late July and early August. The farmer stated that he intentionally installed sap lines as high as possible to avoid livestock damage. The farmer attributed low forage availability in the sugarbush to high tree density.

Livestock management in silvopastures

Livestock type raised on farms in this study was diverse and ranged from 1 to 6 varieties of livestock on each farm and incorporated into silvopastures. Livestock incorporated into silvopastures included beef (12 farms) and dairy cattle (2 farms), sheep for meat and/or fiber (6 farms), meat (3 farms) and dairy goats (1 farm), chickens for meat or eggs (4 farms), turkeys (2 farms), and horses (2 farms). Pigs were used in the establishment phase of silvopastures on 4 farms. Four other farms were raising pigs in wooded areas for a month or longer without pasture rotation, these areas were not considered silvopastures due to a lack of forage and tree management. Number of livestock was diverse between farms with maximum values of 130 beef cattle, 8900 poultry, or 200 dairy goats. The minimum end of the range for livestock on farms included two dairy cattle, 30 poultry, or nine sheep.

Only one of 20 farmers interviewed was not using rotational grazing techniques when managing livestock, although this farm was planning to transition to a rotational grazing system. Some farmers considered moving animals once a year as rotational grazing, while others understood rotational grazing to mean moving animals at least every few days. Rotations lengths used for cattle, sheep, goats, and horses in silvopastures ranged from less than one day to a maximum of 21 days. Rotation lengths utilized by farms for pigs ranged from two days to 365 days. Poultry were integrated into silvopastures on six farms, but four of these farms simply allowed poultry to free-range into silvopastures. The other two farms integrating poultry into silvopastures rotated them on a one to three day rotation.

With the exception for pigs, farmers were using forage height and availability as a measure of when to move livestock into and out of silvopasture paddocks. All farmers pasturing pigs utilized signs of site or tree damage as indicators for when to move them. Site damage included muddy ground, visible soil erosion, and visible soil compaction. Challenges of moving pig fencing, housing, and watering systems were referenced by farmers as a primary reason for long (or no) rotations of pigs in wooded areas.

The primary reason for pig incorporation into wooded areas was for the welfare of the pig from the shade of trees and as a form of vegetation management. Farmers saw the forest as a foraging area for pigs to consume roots, nuts, and insects. Where pigs were utilized in the first year of silvopasture establishment and in wooded paddock situations, between 32 and 100 % of inventoried trees had some form of basal damage or root exposure due to livestock. Three farms had damage on 100 % of the trees in pastures where pigs were incorporated. Additionally, bare mineral soil exposure ranged from 20 to 100 % within areas actively pastured with pigs.

When asked about animal health concerns in silvopastures compared with open pastures, nine of 20 farmers explicitly stated that they felt animal health was improved in silvopastures, primarily because of shelter and a diversified diet. Farmers expressed the following concerns regarding animal health in silvopastures: predators (3 farms), falling tree branches (2 farms), hunters (2 farms), parasites (2 farms), toxic plants (2 farms), physical injuries to livestock (1 farm), and limited visibility and access to livestock (1 farm). Only two farmers had livestock injury actually occur in treed pastures; one farmer stated hoof injury to pigs and the other farmer stated two cows' tails being caught and torn-off by woody vegetation.

Forage management in silvopastures

Orchardgrass (*Dactylis glomerata*), bentgrasses (*Agrostis* spp.), bluegrasses (*Poa* spp.), red clover (*Trifolium pretense*), white clover (*Trifolium repens*), timothy (*Phleum pretense*), and fescues (*Festuca* spp.) were commonly observed in silvopastures (Table 3). In newly established silvopastures converted from forests bentgrasses and poverty oatgrass (*Danthonia spicata*) were common volunteer grasses inventoried. Eleven of 20 farms interviewed were actively seeding forages into silvopastures. Broadcast seeding in the spring and fall and out-feeding of hay in silvopastures were being used to establish forages. Across the region, forage species observed in silvopastures were similar to those commonly found in open pastures on similar quality soils. Three farms were specifically managing woody browse as a component of forage in their silvopastures, but these farms did not provide specifics in terms of preferred woody browse species. Twelve of 20 farmers considered woody invasive alien shrubs as undesirable plants in their silvopastures. Eight of these 12 farms specifically named multiflora rose (*Rosa multiflora*) as challenging weed in their silvopastures. Other undesirable plants, native and introduced, that were mentioned by more than one farm are listed in Table 4.

Tree management in silvopastures

Farmers were primarily managing trees in silvopastures for sawtimber, firewood, and nuts/fruit (Table 5). Other management goals for trees in silvopastures are listed in Table 5. Trees species/groups stated as favorable by multiple silvopasture practitioners included oak, maple, fruit trees, eastern white pine, and others (Table 5).

Table 3 Forage and non-woody understory plants occurring in more than 5 silvopasture inventories on 20 farms in New York state and New England in 2014

Common forages	Common non-woody plants	Forages actively managed for
Red clover (*Trifolium pratense*)	Sedges (*Carex*spp.)	Red clover (*Trifolium pratense*)
White clover (*Trifolium repens*)	Ferns	White clover (*Trifolium repens*)
Orchardgrass (*Dactylis glomerata*)	Brambles (*Rhubus* spp.)	Timothy (*Phleum pratense*)
Bentgrasses (*Agrostis* spp.)	wood-sorrel (*Oxalis acetosella*)	Orchardgrass (*Dactylis glomerata*)
Bluegrasses (*Poa* spp.)	Dandelion (*Taraxacum officinale*)	Ryegrasses (*Lolium* spp.)
Fescues (*Festuca* spp.)		Diversified woody browse
Timothy (*Phleum pratense*)		

Table 4 Undesirable plants stated by more than one silvopasture practitioner in New York and New England in 2014

Plant species	Scientific name
Multiflora rose	*Rosa multiflora*
Japanese barberry	*Berberis thunbergii*
Japanese knotweed	*Fallopia japonica*
Oriental bittersweet	*Celastrus orbiculatus*
Buckthorn	*Rhamnus* spp.
Honeysuckle	*Lonicera* spp.
Privet	*Ligustrum* spp.
Thistle	*Cirsium* spp.
Sedges	*Carex* spp.
Mountain laurel	*Kalmia latifolia*
Ferns	spp.

Ten of 20 farms in this study did not receive direct financial benefit from the trees in their pastures. Four farms had only received financial benefits from trees in their silvopastures when timber was sold as part of the initial timber harvest converting a forest to silvopasture. Five farms received direct financial income from trees in their silvopastures; these were a commercial tree nursery, farm with black locust thinned for fence posts, fruit from orchards, and maple sap.

Table 5 Tree composition and uses of silvopastures on 20 farms in New York and New England in 2014

Dominant tree species/groups (scientific name)	Number of farms
Oaks (*Quercus* spp.)	11
Maples (*Acer* spp.)	10
Fruit trees, primarily apples (*Malus* spp.)	8
Eastern White Pine (*Pinus strobus*)	4
Hickories (*Carya* spp.)	4
Eastern Hemlock (*Tsuga Canadensis*)	3
Commercial nut trees, primarily walnuts (*Juglans* spp.)	2
Black locust (*Robinia pseudoacacia*)	2
Goals for trees in silvopastures	–
Sawtimber	12
Firewood	12
Fruit or nuts	11
Maple sugar potential	4
Wildlife habitat	3
Fence posts	2
Scion wood	1

Six farms were actively regenerating trees in their silvopastures, the remaining 14 farms stated that they were not actively regenerating trees at that time. Individual tree fencing was being utilized by six farms to regenerate trees in silvopastures, one farm was also using a coppice system for black locust, and another farm was allowing hardwood sprouts to regenerate in the piles of slash left over from the initial conversion of forest to silvopasture.

When asked about concerns regarding tree health in silvopastures, nine of 20 farmers stated concerns related to invasive alien forest pests, such as emerald ash borer (*Agrilus planipennis*) and hemlock woolly adelgid (*Adelges tsugae*). Four farmers also found the springtime to be a high risk for tree damage from livestock perceived to be caused by sap flow at this time. Stripping of bark was of concern, and observed, by farmers with goats and pigs, although not all farmers were taking action to prevent it. Pig farmers recognized that their pigs may be damaging tree roots, but they were uncertain as to how much, if any, damage was being done. A cattle, goat, and sheep farmer expressed the importance of management: "We've never really seen any debarking, or girdling by livestock, at least in areas that are being managed."

Twelve of 20 farms had worked directly with a forester when developing silvopastures, and eight of those 12 found the forester to be supportive of silvopasture. One farmer actively avoided working with a forester in silvopastures, stating that foresters did not know much about silvopasture. Three farmers had switched the foresters they were working with and hired new foresters who were more open to the practice of silvopasture. The two foresters interviewed in this study stated their involvement with silvopasture was due to demand for the practice from clients.

Farmer needs for silvopasture optimization

Three farmers expressed lack of support from agricultural extension agencies for silvopasture as a major challenge they faced in adopting the practice (Table 1). Farmers were especially frustrated when extension personnel confused their silvopastures with poorly managed wooded livestock paddocks. However, the converse confusion also occurred by farmers practicing continuous pasturing of woodlands with pigs and calling it silvopasture.

Table 6 Areas of silvopasture research requested by two or more farmers during interviews with 20 farmers practicing silvopasture in New York and New England in 2014

Requested areas of silvopasture research
Forage/browse quality, selection, and management
Tree care, regeneration strategies, and management
Overall silvopasture management
Soil properties and management
Best management practices for pasturing pigs
Vegetation management using livestock
Fencing systems
Quantification of animal health and production
Environmental benefits
Management of orchard silvopastures
Air temperature dynamics
Economics

Areas of research desired by farmers on silvopasture were diverse (Table 6). Eight farmers requested visuals and commercially viable case studies of regional silvopastures. When asked what resources they would utilize to learn about silvopasture, farmers varied greatly in their responses. Farm tours were cited as important educational opportunities by 12 farmers, but timing of these tours was cited as a challenge. Farmers were split between desiring online resources such as webinars and web pages while others preferred printed material. Extension personnel, conferences, and other farmers were cited as educational resources farmers would utilize in obtaining information about silvopastures.

DISCUSSION

Silvopasture systems utilized in New York state and New England were diverse in terms of structure and reasons for use. Tree density and spacing differed between silvopastures, and in some cases within silvopastures. Coupled with this, farmers' goals for the trees in their silvopastures were multiple and on-farm use was often one component. Forages in these systems were also variable and seemed to be highly dependent on multiple site conditions. Despite this heterogeneity, four key issues emerged that warrant follow-up: the need to distinguish between silvopasture and livestock in wooded paddocks, the complexity of integrating livestock and tree management in silvopastures, the unique challenges of managing pigs in silvopasture, and the opportunity and demand for silvopasture outreach in this region.

The most pressing concern is the confusion around what is and is not silvopasture. Specifically, confusion existed between silvopasture and any incorporation of livestock into areas with trees or woody vegetation, regardless of tree health or livestock management. Of significant concern were some farmers mislabeling degraded wooded livestock paddocks as silvopasture. This confusion poses a severe risk to the successful adoption of the practice in the region as it furthers the confusion between farmers, extension professionals, and foresters as to what silvopasture really is. Worse yet was the degradation happening to woodlands and not realized by farmers who might have believed they were doing the "right" thing. A clear and consistent message toward what makes successful silvopastures coupled with best management practices needs to be developed for silvopasture in the Northeastern United States. Little knowledge of landowners about agroforestry practices is not unique to the Northeastern United States (Barbieri and Valdivia 2010), and farmer education has been shown to lead to successful agroforestry adoption (Frey et al. 2007). While farmers interviewed in this study favored trees, many of them were unaware of how to manage trees.

Farmers were challenged by the complexity of integrating livestock and tree management in silvopastures. Farmers typically prioritized the care they give to each silvopasture and woodland component based on their primary economic crop. For example, farmers whose primary farm income was cattle-based would respond to questions about silvopasture management in terms of cattle production and wellbeing and not mention tree crops or health. The reverse was true when speaking to farmers about silvopasture orchards where the primary economic crop was fruit. In the cases of wooded livestock paddocks, this difference could be extreme; for example, one farm with wooded livestock paddocks received $20,000 per year net income from pork, but utilized woodland trees only for heating a small home with firewood. This farm recognized the damage being done to forest soils by pasturing pigs on long rotations, but the short-term (annual) economics did not dictate caring for their trees.

An opportunity exists for agricultural extension professionals and foresters to aid farmers in managing and optimizing these complex systems. Resources need to be developed to assist farmers in managing silvopastures. Best management practices regarding livestock, trees, and forages coupled with case studies and silvopasture demonstration areas would go a long way in ensuring that farmers are integrating functional silvopastures into the regional landscape. Additionally, identifying the benefits and tradeoffs of silvopasture to livestock, the environment, and farming economy is an important regional need.

Managing pigs in a silvopasture system is uniquely challenging due to their rooting and destructive nature. The fundamental problem with pasturing of pigs was that the movement of pigs out of a paddock was reactionary and driven by indicators of site damage, such as heavy soil compaction or damage to trees, and/or persistent breakouts by pigs to find new areas. Movement of any livestock to a new paddock should be proactive and before damage occurs.

Ironically, on farms which pastured both pigs and other livestock, the other livestock were moved based on signs of reduced forage availability, such as forage height, while pigs were still moved in reaction to site degradation. This was also the case on farms where pigs were the primary source of income; cattle were moved daily yet pigs were moved monthly. This difference in species management may have been the result of farmers being required to bring feed to pigs regardless of site conditions, whereas feed was an extra cost to farmers when their grazing animals were on pastures without available forage. Additionally, farmers were pasturing pigs by trial and error, but were aware of recommended management practices for other livestock species.

In Europe the use of pigs in areas with trees has been going on for centuries, but it is often only in the fall to allow pigs to glean fallen nuts or fruit (Rigueiro-Rodriguez et al. 2009). Lessons from Europe would suggest that the incorporation of pigs into forestland necessitates active management. According to the German law, pigs are banned from forests unless natural regeneration of beech and oak trees is guaranteed (Rigueiro-Rodriguez et al. 2009). Pigs in European systems are primarily consuming mast, whereas what was witnessed in this research were pigs consuming actual components of trees, primarily roots and lower bark. The grazing of pigs to forage fruit and nuts is much less destructive to tree health than allowing pigs to browse tree roots. Grazing implies forage availability and management, therefore the systems documented here would be best termed rooting or neglect.

There was a desire among farmers to be doing the right thing, despite the destructive nature of pasturing pigs documented here. The challenge was those farmers did not have the resources to determine what the right thing was: "Was I really doing silvopasture or was I just running pigs in the woods?" Timing was a major factor in this degradation and a simple recommendation may be for farmers to set up multiple paddocks for pigs prior to their introduction into silvopastures. Development and maintenance of a sod layer in silvopastures may also help to buffer soil degradation and rooting from pigs.

Twelve farmers interviewed in this study had been utilizing silvopastures for less than 5 years, suggesting that silvopasture is a budding regional practice. Agricultural extension professionals and researchers are in a unique position to influence the development of silvopasture practices at the beginning stages of their adoption in the region. Farmers were well aware of public education and extension efforts toward invasive alien forest pests, suggesting that similar efforts toward silvopasture management would reach the right people. One challenge faced by farmers in this study was an inconsistent message being put out by extension professionals within and between states. For example, three farms were actively working with extension professionals in development of silvopastures, while other farms were very frustrated with the lack of support, and in some cases clear mistrust for silvopasture by extension professionals. Additionally, four farms in this study practicing silvopasture were actually owned and operated by agricultural extension professionals in differing states. In a region as small as the Northeastern United States, it is important that messages toward agricultural practices are consistent between states as regional farms commonly cross-pollinate information.

CONCLUSIONS

This study was limited by a small sample size, time constraints for silvopasture inventories, and an intentionally biased participant identification. A limitation to use of purposeful sampling is that it intentionally favors sampling of a single group. This study did not seek out non-adopters of silvopasture who were familiar with the practice. Follow-up studies should consider addressing this group to identify concerns regarding silvopasture that inhibited adoption. A regional assessment should be conducted in the Northeastern United States to address the extent and full diversity of silvopasture and differentiate it from livestock simply being kept in wooded paddocks. This study was also limited in being able to fully assess how well silvopastures were being managed, as no management recommendations exist for silvopastures in the region.

Examples exist of decades old, well managed silvopastures in New York state and New England, although the majority of silvopasture identified in this study were in the first few years of establishment. Farmers practicing silvopasture found it to be a functional and desirable component of their farm landscape. Confusion between silvopasture and degraded wooded livestock paddocks exists in the region and poses a significant threat to the success of silvopasture. Specifically, the use of pigs in wooded areas needs to be addressed as farmers in the region are causing severe damage to woodlands through poorly managed pasturing of pigs.

Regardless of livestock species or silvopasture type, this study provides evidence that silvopastures are being used to diversify regional farms. If best management practices regarding silvopasture are developed, it is likely that they will reach and be considered by farmers. Farmers' knowledge about silvopastures was homegrown or based on systems from other parts of the world. For the practice to be advanced in the region, further research is needed on the topic. Farmers in this study have identified numerous areas of applied research which would help them improve their silvopasture management. Research into the integration of farm woodland into agricultural ventures, benefits of shade to livestock, potential for increased forage availability during mid-summer and droughts, and diversified livestock diets would serve the region well. Additionally, outreach should occur on fencing strategies, vegetation management, and forage establishment in silvopasture systems.

ACKNOWLEDGEMENTS

We would like to thank all of the farmers who contributed to this research by sharing their experiences and pastures with us. Thanks is due to Richard Smith, Heidi Asbjornsen, and Ted Howard at the University of New Hampshire for their guidance and assistance with this study. We would also like to thank Brett Chedzoy, Peter Smallidge, and Jeff Jourdain for their contributions in the development of this project. This project was supported by the Northeastern States Research Cooperative through funding made available by the USDA Forest Service. The conclusions and opinions in this paper are those of the authors and not of the NSRC, the Forest Service, or the USDA.

REFERENCES

1. Arbuckle JG, Valdivia C, Raedeke A, Green J, Rikoon JS (2009) Non-operator landowner interest in agroforestry practices in two Missouri watersheds. Agrofor Syst 75:73–82
2. Barbieri C, Valdivia C (2010) Recreation and agroforestry: examining new dimensions of multifunctionality in family farms. J Rural Stud 26:465–473
3. Calle (2008) Using PES and techincal assistance to promote silvopastoral systems in Quindio, Colombia: attitude change as a key to permanent adoption. Yale University
4. Carroll JE (2011) The real dirt. New Hampshire Agricultural Experiment Station, Durham
5. Chedzoy BJ, Smallidge PJ (2011a) Silvopasture. Cornell Cooperative Extension. http ://silvopasture.ning.com/. 2015
6. Chedzoy BJ, Smallidge PJ (2011b) Silvopasturing in the Northeast an introduction to opportunities and strategies for integrating livestock in private woodlands. Cornell Cooperative Extension, New York
7. Clason TR, Sharrow SH (2000) Silvopastoral practices. In: Garrett HE, Rietveld WJ, Fisher RF (eds) North American agroforestry: an integrated science and practice. American Society of Agronomy, Wisconsin, pp 119–147
8. Denzin NK (1978) The research act: a theoretical introduction to sociological methods, 2nd edn. McGraw-Hill, New York
9. Franz NK, Piercy F, Donaldson J, Westbrook J, Richard R (2010) Farmer, agent, and specialist perspectives on preferences for learning among today's farmers. J Ext 48:3RIB1
10. Frey GE, Fassola HE, Pachas A, Colcombet L, Lacorte SM, Cubbage FW, Perez O (2007) Adoption and economics of silvopasture systems by farm size in Northeastern Argentina. Paper presented at the Proceedings of the 10th North American Agroforestry Conference, Quebec City
11. Frey GE et al (2012) Perceptions of silvopasture systems among adopters in northeast Argentina. Agric Syst 105:21–32
12. Garrett HE, Kerley MS, Ladyman KP, Walter WD, Godsey LD, Van Sambeek JW, Brauer DK (2004) Hardwood silvopasture management in North America. Agrofor Syst 61–2:21–33. doi:10.1023/B:AGFO.0000028987.09206.6b
13. Neumann PD, Krogman NT, Krahn HJ, Thomas BR (2007) 'My grandfather would roll over in his grave': family farming and tree plantations on farmland. Rural Sociol 72:111–135

14. Patton MC (2002) Qualitative research & evaluation methods, 3rd edn. Sage, London
15. Rigueiro-Rodriguez A, McAdam JH, Mosquera Losada MR (2009) Agroforestry in Europe: current status and future prospects. Springer, New York
16. Strong NA, Jacobson MG (2005) Assessing agroforestry adoption potential utilising market segmentation: a case study in Pennsylvania small-scale. Forestry 4:215–228
17. Twery MJ et al (2005) NED-2: a decision support system for integrated forest ecosystem management. Comput Electron Agric 49:24–43

CHAPTER 17

Substrate Chemistry and Rainfall Regime Regulate Elemental Composition of Tree Leaves in Karst Forests

Ernesto Medina[1,2,3,*], Elvira Cuevas[3] and Ariel E. Lugo[1]

[1]*International Institute of Tropical Forestry (IITF), USDA Forest Service, Rio Piedras 00926-1115, Puerto Rico*

[2]*Centro de Ecología, Instituto Venezolano de Investigaciones Científicas, Altos de Pipe 1020A, Venezuela*

[3]*Departmento de Biología, Universidad de Puerto Rico, San Juan 00926; Puerto Rico*

ABSTRACT

Forests on calcareous substrates constitute a large fraction of the vegetation in Puerto Rico. Plant growth on these substrates may be affected by nutrient deficiencies, mainly P and Fe, resulting from high pH and formation of insoluble compounds of these elements. The occurrence of these forests in humid and dry areas provides an opportunity to compare nutrient relations, water use efficiency, and N dynamics, using biogeochemical parameters. We selected sites under humid climate in the north, and dry climate in the southwest of Puerto Rico. Adult, healthy leaves of species with high importance values were collected at each site and analyzed for their elemental composition and the natural abundance of C and N isotopes. Calcium was the dominant cation in leaf tissues, explaining over 70% of the ash content variation, and Al and Ca concentration were positively correlated, excepting only two Al-accumulating species. Karst vegetation consistently showed high N/P ratios comparable to forests on P-poor soils. Dry karst sites had significantly higher $\delta^{13}C$ and $\delta^{15}N$ ratios. We conclude that forests on karst are mainly limited by P availability, and that mechanisms of nutrient uptake in the rhizosphere lead to linear correlations in the uptake of Ca and Al. Isotope ratios indicate higher water use efficiency, and predominant denitrification in dry karst forest sites.

Keywords: tropical karst; element concentration; N/P ratios; Ca/Al relationship; $\delta^{13}C$; $\delta^{15}N$

1. INTRODUCTION

Limestone areas are widespread in the Caribbean under dry to subhumid climatic conditions. Those areas developed mostly during the middle Oligocene to the middle Pliocene and have been documented extensively in Jamaica, Cuba, Santo Domingo, and Puerto Rico [1,2,3,4,5,6]. Several forest types, from dry to moist, covering large areas of those geologic formations, have been described in detail in many locations [4,7,8,9].

In Puerto Rico, karst areas constitute about 27% of the territory and are separated as moist karst forming a northern belt crossing the island from east to west, and a dry karst formation in the southwest, under semiarid climate [4]. These formations have a similar geological age (Pliocene-Oligocene) but are under different rainfall regimes [10]. The limestone formations are covered throughout the island by sediments constituted by non-calcareous material that, in the northern karst belt, accumulate in depressions between haystack (*mogotes*) hills [10,11]. On top of the *mogotes* in the karst belt, or on the calcareous hills of the southern karst, this material may accumulate within crevices and fractures of the calcareous strata. Surface soils in karstic areas, when present, are highly heterogeneous, frequently shallow, and, in many cases, difficult to sample. Soils derived from the calcareous substrate itself are infrequent. On ridge tops and slopes, soil is almost inexistent, and substrate is a rocky, partially fragmented surface, covered with organic residues produced by the vegetation. Fragments of the carbonaceous rock and organic matter slide downhill and accumulate in the valleys [4,11,12]. On the valleys, true soils develop constituted by calcareous rocks fragments and the remaining material transported from volcaniclastic areas during the time when the carbonate rocks were under water (see [11] for an explanation of the composition of blanket sands). Assessment of nutrient availability for plants on the ridge top sites using soils as reference is therefore not a practical approach. The alternative is the analysis of plant organs, particularly leaves [13].

Nutrient availability for plant growth in soils derived from carbonate rocks is usually limited, particularly in P, as it tends to be immobilized as Ca compounds of low solubility [14]. High soil pH determines lower mobility of elements such as Mn and Fe, resulting in potential limitation for plant growth (Ca induced iron chlorosis). The nutrient relationships of plants from calcareous substrates have been mainly studied in temperate climates [15] and little is known from tropical vegetation [16].

Typical vegetation of moist karst forests has been described by Alvarez et al. [17] and Acevedo-Rodríguez and Axelrod [18] and that of the dry karst forest was studied by Murphy and Lugo [19], and documented in detail by Monsegur [20].

Our study took advantage of detailed phytosociological analysis of the northern karst vegetation undertaken by Chinea [21] and Aukema et al. [22]. In the latter, the authors quantified the composition of forest units according to species dominance relationships and geomorphological positions on the *mogotes* (top, slopes, valleys). This study was critical to the identification of tree assemblages on the top of *mogotes*. In addition, we used the vegetation analysis of Murphy and Lugo [19] and Molina and Lugo [23] for the Guánica dry forest in southern Puerto Rico. The dominant tree species in typical sites of northern and southern karts areas, as defined by their importance value, reveals the occurrence of common and restricted species. For example, *Coccoloba diversifolia* (Polygonaceae) and *Gymnanthes lucida* (Euphorbiaceae) occur as dominant species in both moist and dry karst, whereas *Pisonia albida* (Nyctaginaceae), *Thouinia portoricensis* (Sapindaceae), and *Pictetia aculeata* (Fabaceae) seem to be restricted to dry karst forests, and *Lonchocarpus lancifolius* (Fabaceae), and *Prunus myrtifolia* (Rosaceae) are restricted to the northern moist karst belt [24].

The objective of the present study was to provide a geochemical characterization of the karst areas in the island of Puerto Rico through the elemental analysis of substrate and plant leaves. Leaf analyses provide information on both the availability of specific elements in the soil solution, and reveal physiological properties such as exclusion or facilitation of element uptake at the root level [13]. We tested hypotheses related to the leaf elemental composition, substrate geochemistry, and environmental humidity: (1) Ca is the main component of leaf ash due to its expected high availability in the calcareous substrate; (2) Lower total P concentrations and high N to P ratios reflect relative limited availability of P, probably associated with the formation of insoluble Ca compounds; (3) Concentrations of heavy metals (Fe and Mn) and Al, are probably low compared to species from forests on non-calcareous substrates; (4) Environmental humidity determined by seasonality and magnitude of rainfall affect plant water use efficiency and rate of organic N mineralization in the substrate.

2. MATERIALS AND METHODS

For the study of the nutritional relationships of tree species strictly associated with karstic substrates, we selected sites in the humid karst belt and in the semiarid south west karst area in Puerto Rico. The sampling was restricted to plants growing on the top of haystack hills in the northern karst belt, and on ridge sites and coastal carbonate pavement in the southwestern karst area.

We selected six *mogote* tops sites in the moist northern karst belt and two sites in the dry Guánica forest in southwestern Puerto Rico (Table 1). The former sites were located along a west-east line on the northern karst belt, on top of the Aymamon limestone type, whereas those on the dry karst areas are on the Ponce limestone type [10]. The northern karst belt receives, on average, annual rainfall amounts of 1000–1500 mm, whereas in the southern karst areas long-term annual rainfall averages lie below 1000 mm. This contrasting rainfall pattern is clearly exemplified by two stations, Utuado (158 m), located in the middle of the Aymamon karst [10] in the northern karst belt, and Ensenada (46 m), on the southwestern coast, representative for the rainfall regime of the southern karst belt (Figure 1).

Table 1. Karst site locations (Lat, Long) and approximate altitude.

	Karst Site	North Lat (°)	West Long (°)	Altitude (m)
Moist sites	Guajataca	18.42	66.83	230
	El Tallonal	18.41	66.73	160
	Cambalache	18.45	66.59	45
	Río Lajas	18.40	66.26	50
	Nevarez	18.41	66.25	35
	Hato Tejas	18.40	66.19	100
Dry sites	Guánica ridge	17.97	66.87	160
	Guánica Dwarf	17.95	66.83	13

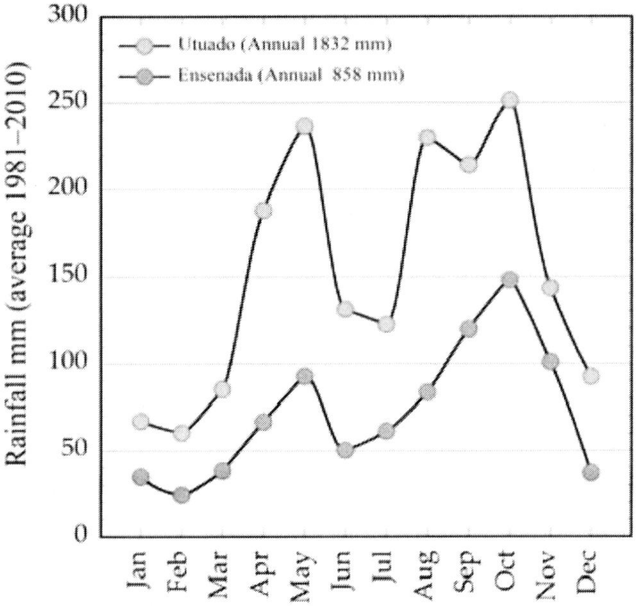

Figure 1. Average rainfall pattern of humid and dry karst sites.

The species sampled at each site were those more common, with high importance values [16,19,22,23] (Table 2). The number of species varied from site to site due to different areal extension, and the number of tree species growing on the *mogote* top. In total, we sampled 30 species from the northern, moist sites, and 26 species from the dry sites, representing 29 families and 18 orders. Only five species were sampled from both humidity sites.

For elemental analyses, we sampled at each site healthy mature leaves, without visible herbivore consumption, of trees from the upper forest canopy. Leaves were dried in the laboratory in a ventilated oven at 65 °C (3–7 days) and ground to pass an 18-mesh screen. Rock fragments on the forest floor were collected from each site. In the laboratory, rock samples were dissolved in 2N HCl for elemental analysis.

Elemental composition of samples was determined (C, N, S, P, K, Ca, Mg, Al, Mn, Fe) using standard methodology of the laboratory at the International Institute of Tropical Forestry [25]. Briefly, in acid digested samples, C, N, and S were determined using a LECO elemental analyzer, whereas P and cations were measured using Inductively coupled plasma atomic emission spectroscopy (ICP-AES) techniques.

Table 2. Species sampled at each site designated by name and municipality.

Guajataca, Isabela	El Tallonal, Arecibo	Cambalache, Barceloneta
Calophyllum antillanum	*Amyris elemifera*	*Coccoloba diversifolia*
Clusia rosea	*Coccoloba diversifolia*	*Crossopetalum rhacoma*
Coccoloba pubescens	*Exothea paniculata*	*Gymnanthes lucida*
Comocladia glabra	*Gymnanthes lucida*	*Krugiodendron ferreum*
Eugenia monticola	*Pimenta racemosa* var. *grisea*	*Sideroxylon salicifolium*
Lonchocarpus glaucifolius	*Prunus myrtifolia*	
Neolaugeria resinosa	*Tabebuia karsensis*	
Tetrazygia elaeagnoides		
Nevarez, Toa Baja	**Rio Lajas, Dorado**	**Hato Tejas, Bayamon**
Ardisia obovata	*Ardisia obovata*	*Coccoloba diversifolia*
Coccoloba diversifolia	*Calyptranthes pallens*	*Eugenia monticola*
Drypetes alba	*Coccoloba diversifolia*	*Ficus citrifolia*
Exothea paniculata	*Cojoba arborea.*	*Phyllanthus epiphyllanthu*
Garcinia portoricensis	*Exothea paniculata*	*Sideroxylon salicifolium*
Gaussia attenuata	*Krugiodendron ferreum*	
Guapira fragrans	*Licaria salicifolia*	
Licaria salicifolia	*Ottoschultzia rodoxylon*	
Ottoschultzia rodoxylon	*Sideroxylon salicifolium*	
Picramnia pentandra		
Guánica Ridge F., Guánica	**Guánica Dwarf F., Guánica**	
Amyris elemifera	*Amyris elemifera*	
Bourreria succulenta	*Bourreria virgata*	
Bucida buceras	*Canella winterana*	
Elaeodendrum xylocarpa	*Coccoloba diversifolia*	
Colubrina arborescens	*Colubrina arborescens*	
Erythalis fruticosa	*Crossopetalum rhacoma*	
Erythroxylum areolatum	*Erithalis fruticosa*	
Exostemma caribaeum	*Eugenia foetida*	
Gymnanthes lucida	*Ficus citrifolia*	
Pisonia albida	*Jacquinia arborea*	
Tabebuia heterophylla	*Jacquinia berteroi*	
Thouinia striata var. *portoricensis*	*Pisonia albida*	
	Quadrella cynophallophora	
	Reynosia uncinata	
	Sideroxylon salicifolium	
	Stenostomum acutatum	
	Strumpfia maritima	
	Tabebuia heterophylla	
	Thrinax morrisii	

Plant names after [24].

In addition, the natural abundance of C ($\delta^{13}C$) and N isotopes ($\delta^{15}N$) in leaves was determined at the Stable Isotope Laboratory of the University of Miami, Coral Gables, using mass spectrometric techniques. The $\delta^{13}C$ values are used as indicators of long-term water use efficiency, because they are related to lower leaf conductance [26]. The leaf $\delta^{15}N$ values are related to the natural abundance of N isotopes in the substrate. More positive values point to higher rates of mineralization of organic N in the soil [27].

Element concentrations and C and N isotopic ratios of all species per site were submitted to a one-way analysis of variance and post-hoc tests for comparison of site averages. Inter-element correlations were estimated for the whole set of species to establish commonalities among sites. All statistical analyses were performed using the JMP statistical program.

3. RESULTS

3.1. Elemental Composition of the Substrate in Karst Areas in Puerto Rico

The analysis of the elemental composition of rocks from different karst types of the northern karst belt published by Monroe [12] indicate that nearly 83% is constituted by $CaCO_3$, and that they contain small but significant amounts of Mg, Al, and Fe. The elements P and particularly K are found at concentrations below 1 mmol/kg. Rock fragments from the sites selected for vegetation sampling showed important differences in the proportion of minor elements. The sequence of elemental concentrations in the northern karst site confirmed the sequence Ca >> Mg > Al > Fe, corresponding to the average values from Monroe [12] (Figure 2). The samples from the southern karst had a similar elemental distribution, but Na was present in the Guánica samples at higher concentrations than any other site. Phosphorus concentration was at the same level in all samples (≈1 mmol/kg). Potassium reached concentrations around 10 mmol/kg in the southern samples, whereas it was always well below the 1 mmol/kg level in the northern karst samples.

The presence of non-calcareous sediments in cracks, holes, and crevices of the karstic substrate raised the question about their potential as a source of nutrients for the vegetation. They have been characterized in the Dominican Republic as lateritic in nature, with high concentrations of Al and Fe [1]. We do not have information on the elemental composition of those sediments in Puerto Rico. However, we calculated sediment concentrations from a nutrient inventory study in the northern karst belt [28]. Both A and C horizons indicate much higher concentrations of Fe, Mn, and Al, and lower concentrations of Ca and Mg compared to those measured in calcareous rocks (Figure 3). The higher concentrations of Ca and Mg of the A horizon reveal the influence of in situ weathering of karst rocks. It appears then that these sediments may be a source of nutrients for the vegetation, and due to their clayey texture, they have much higher water retention and cation exchange capacities than the calcareous substrates.

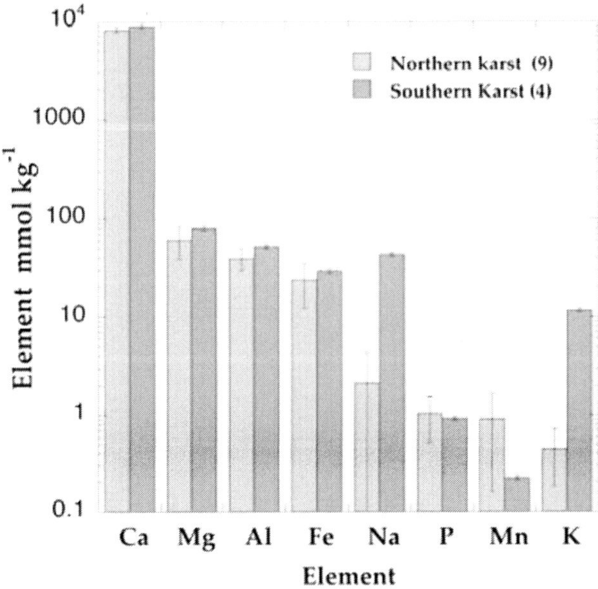

Figure 2. Average elemental composition of karst rocks from northern and southern sites in Puerto Rico.

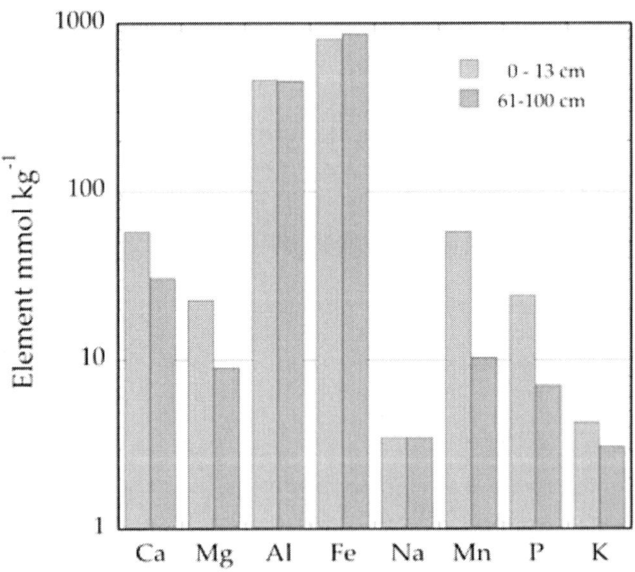

Figure 3. Elemental composition of clay soils on karst in El Tallonal, Arecibo, Puerto Rico. Values for the A horizon (0–13 cm) and the C horizon (61–100 cm) were generated from [28].

3.2. Leaf Elemental Composition

The concentration of the main organic matter elements did not differ significantly between sites. The coefficients of variation were small for C (<10%), intermediate for N and P (10 to 100%), and large for S (>100%) (Table 3). The concentration of metallic elements differed between sites for K, Ca, Fe, Mn, and Na (Table 4). The coefficients of variation were intermediate for K, Mg, Ca, and Fe, large for Na, and very large for Al and Mn (>250%). In the southern karst, average K concentration was higher, whereas for Ca and Mn concentration averages were lower than the overall mean. The average Na concentration in the Guánica Dwarf forest site was much larger than the overall mean, probably because this site is located near the coast line, under the influence of sea salt spray, and Na may be present in the rooting substrate, as indicated by the composition of the limestone from Guánica (Figure 2).

Table 3. Mean (±SE) concentrations of main elements of organic matter C in mol/kg, the rest in mmol/kg.

System	*n*	C	N	S	P
Guajataca	8	43.7 (1.2)	1247 (163)	60 (12)	18.7 (1.9)
Tallonal	7	43.7 (1.0)	1193 (100)	58 (11)	26.7 (3.0)
Cambalache	5	41.7 (1.1)	1315 (217)	102 (37)	22.8 (3.1)
Nevarez	10	41.7 (0.7)	1226 (179)	128 (34)	21.4 (1.8)
Río Lajas	10	43.2 (0.6)	1375 (179)	60 (8)	21.9 (3.1)
Hato Tejas	5	42.1 (1.2)	1144 (103)	77 (18)	23.1 (3.7)
Guánica Ridge	13	41.5 (0.7)	1361 (168)	75 (10)	25.4 (3.7)
Guánica Dwarf	26	42.3 (0.5)	991 (65)	94 (26)	20.5 (2.5)
Overall mean	84	42.4 (0.3)	1192 (51)	84 (10)	22.2 (1.1)
Coefficient of Variation (CV)		6	39	105	46
Analysis of Variance (ANOVA)		1.1 (ns)	1.3 (ns)	0.7 (ns)	0.6 (ns)

ns, not significant.

Table 4. Mean (± SE) concentrations of ash (%) and metallic elements in leaves per site in mmol/kg.

System	*n*	Ash	K	Mg	Ca	Al	Fe	Mn	Na
Guajataca	8	6.8 (1.3)	132 (20)	110 (17)	450 (95)	91.6 (86.7)	1.3 (0.4)	3.6 (1.7)	61 (21)
Tallonal	7	7.7 (1.4)	162 (29)	114 (17)	583 (145)	5.1 (1.1)	0.7 (0.1)	0.3 (0.1)	25 (8)
Cambalache	5	10.9 (1.6)	102 (32)	148 (22)	872 (178)	6.3 (1.0)	0.7 (0.1)	0.3 (0.1)	77 (19)
Nevarez	10	9.9 (1.0)	129 (20)	154 (36)	708 (75)	7.1 (0.5)	1.4 (0.1)	0.5 (0.1)	92 (21)
Rio Lajas	10	9.8 (1.4)	231 (60)	100 (15)	574 (93)	5.6 (0.8)	1.2 (0.2)	0.4 (0.1)	83 (28)
Hato Tejas	5	9.4 (1.6)	206 (55)	127 (28)	733 (148)	7.1 (1.1)	1.4 (0.2)	0.3 (0.1)	90 (22)
Guánica Ridge	13	8.4 (0.7)	315 (41)	162 (36)	409 (45)	4.1 (0.4)	0.6 (0.1)	0.3 (0.1)	40 (12)
Guánica Dwarf	26	7.8 (0.5)	212 (22)	138 (17)	340 (40)	4.6 (0.4)	0.6 (0.0)	0.5 (0.1)	221 (34)
Overall mean	84	8.5 (0.4)	202 (14)	132 (9)	509 (33)	13.6 (8.4)	0.9 (0.1)	0.7 (0.2)	112 (14)
CV		37	65	65	59	560	62	250	115
ANOVA (F; P)		1.3 (ns)	3.3 < 0.01	0.6 (ns)	4.9 < 0.01	1.4 (ns)	6.3 < 0.01	4.4 < 0.01	5.4 < 0.01

ns, not significant.

Some species within the sampled set stand out, with concentrations of particular elements 2–3 times larger or smaller than the overall average (Table 5). These differences may be attributed to variations in element availability between sites, to differences in leaf age, or to actual physiological differences between species. Leaf age between samples of the same species at different sites cannot be discarded, as leaf sampling was conducted during several months; however, leaves collected were always fully expanded, healthy, and without visual signals of senescence or herbivory.

Table 5. Maximum and minimum concentrations of elements (mmol/kg).

Element	Species	Site	Max	Mean	Min	Species	Site
N	*Erythroxylum areolatum*	Grf	2731	1197	639	*Cassine xylocarpa*	Grf
	Guapira fragrans	N	2621		565	*Erithalis fruticosa* Grf, Gdf	
S	*Quadrella cynophallophora*	Gdf	707	83	30	*Coccoloba pubescens*	G
	Gaussia attenuata	N	367		19	*Prunus myrtifolia*	T
P	*Erythroxylum areolatum*	Grf	62	22	9	*Erithalis fruticosa* Gdf, Grf	
	Coccoloba diversifolia	Gdf	48		7	*Jacquinia berteroi*	Gdf
K	*Bourreria succulenta*	Grf	615	202	51	*Sideroxylon salicifolium*	C
	Canella winterana	Gdf	524		50	*Strumpfia maritima*	Gdf
Mg	*Pisonia albida*	Grf, Gdf	553	132	38	*Thrinax morrisii*	Gdf
	Guapira fragrans	N	415		22	*Licaria salicifolia* N, RL	
Ca	*Crossopetalum rhacoma*	C	1478	509	133	*Calophyllum brasiliense*	G
	Coccoloba diversifolia	T, HT, Gdf	1281		75	*Thrinax morrisii*	Gdf
Al	*Tetrazygia elaeagnoides*	G	699	14	1.7	*Tabebuia karsensis*	T
	Lonchocarpus lancifolius	G	13		1.4	*Calophyllum brasiliense*	G
Fe	*Tetrazygia elaeagnoides*	G	3.7	0.9	0.4	*Prunus myrtifolia*	T
	Krugiodendron ferreum	RL	2.7		0.3	*Clusia rosea*	G
Mn	*Tetrazygia elaeagnoides*	G	13.2	0.7	0.04	*Prunus myrtifolia*	T
	Clusia rosea	G	9.3		0.04	*Jacquinia berteroi*	Gdf

C, Cambalache; G, Guajataca; Gdf, Guánica dwarf forest; Grf, Guánica ridge forest; HT, Hato Tejas; N, Nevárez; RL, Rio Lajas; T, Tallonal.

The design of the research was not oriented to differentiate species occurring at different sites; therefore, the data set is insufficient to conduct within species comparisons.

Elemental ratios between organic matter forming elements (C, N, S, P) do not differ significantly between sites (Table 6). However, K to Ca ratios were larger and the Ca to Mg ratios lower for the two sites on the southern karst in correspondence with the relative elemental concentration of those elements measured in the calcareous rocks. The other sites showed K to Ca ratios ranging from 0.14 to 0.47, revealing the high availability of Ca in the soil solution.

Table 6. Molar elemental ratios of leaves from trees on calcareous substrates.

Site	n	C/N	C/P	C/S	N/P	K/Ca	Ca/Mg
Guajataca	8	39	2507	900	67	0.40	4.08
Tallonal	7	39	1831	996	46	0.47	5.09
Cambalache	5	35	1992	654	58	0.14	6.43
Nevarez	10	39	2086	505	56	0.20	7.35
Río Lajas	9	39	2514	838	67	0.34	7.34
Hato Tejas	5	38	2006	717	53	0.31	6.10
Guánica Ridge forest	13	37	2082	639	56	0.91	3.29
Guánica Dwarf forest	25	47	2819	744	59	0.87	2.90
Overall mean	84	41	2328	732	58	0.59	4.63
ANOVA (F; P)		1.0 (ns)	1.0 (ns)	1.5 (ns)	1.3 (ns)	3.4 <0.01	4.2 <0.01

ns, not significant.

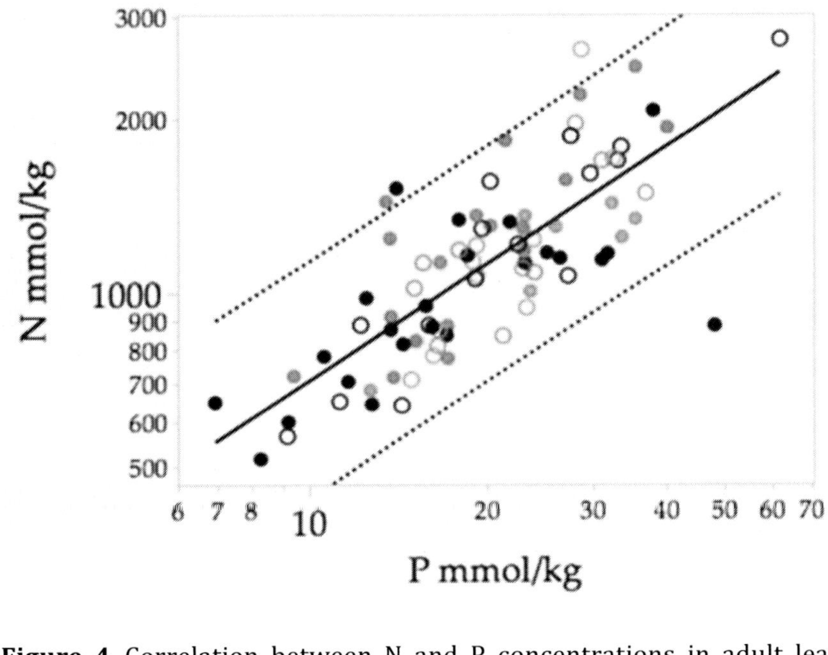

Figure 4. Correlation between N and P concentrations in adult leaves from different karst forest species. Guánica forests: dwarf, Black; ridge, empty black; Guajataca, solid blue; Cambalache, empty blue; Tallonal solid green; Nevarez, empty green; Río Lajas, solid red; Hato Tejas, empty red. Log (N mmol/kg) = 5.011 + 0.672 × Log (P); Fratio = 120; F < p < 0.001; R^2 adj = 0.592; n = 83.

Inter-Element Correlations

The correlation between leaf N and P was positive and significant, as usually reported for forest foliage throughout the world [29] (Figure 4). As expected, the Ash% was inversely correlated with C concentration (R^2 adj = 0.57), but positively correlated with Ca (R^2 adj = 0.72), Mg (R^2 adj = 0.24), and Mn (R^2 adj = 0.08). Clearly Ca and Mg were the main constituents of ash, explaining more than 80% of the variance.

We found an unexpected highly significant correlation between Ca and Al concentrations for all sites (Figure 5). Only two species departed strongly from this relationship, *T. elaeagnoides* and *L. glaucifolius*, both from the Guajataca site. This relationship deserves special attention because of its implications regarding the mechanisms of nutrient uptake from karstic complex substrates. Considering the low concentrations of Al, Fe, and Mn in the calcareous substrate in karst areas, it is not surprising that these elements are only a minor component of leaf ash of trees growing there. If clayey sediment accumulated in cracks and crevices of the rocky substrates on *mogote* tops was a source of those metallic elements, we expected higher concentrations of them in tree foliage, but that was not the case in our data set, with the exception of *T. elaeagnoides*, identified as a metal accumulator (Table 5).

3.3. Isotopic Ratios

The $\delta^{13}C$ values did not differ significantly in the northern karst forest locations (Figure 6). The only outlier was represented by *C. rosea*, a tree with crassulacean acid metabolism [30] commonly occurring in the northern karst belt forests. It was excluded from average calculations. As expected, the two sites on the southern karst had significantly higher $\delta^{13}C$ values, because of the drier conditions under which this vegetation grows. Average $\delta^{13}C$ of leaves from humid karst was −30.83 ± 1.85‰ (n = 52), and from dry karst was −27.37 ± 1.16‰, the difference was highly significant (Tukey Honest Significant Difference test $p = 0.01$).

The distribution of $\delta^{15}N$ values varied significantly among sites, showing a more complex pattern than that of the carbon isotopes (Figure 6). The sites on southern karst and the Nevarez site showed positive values, whereas the rest of the sites had negative $\delta^{15}N$ average values. Overall averages were positive in the dry karst, 3.49 ± 2.20‰ (n = 40), and slightly negative, −0.28 ± 1.82‰ (n = 51), in the moist karst forests. The degree of overlap was higher in the case of ^{15}N, revealing the more complex regulation of this factor under natural conditions.

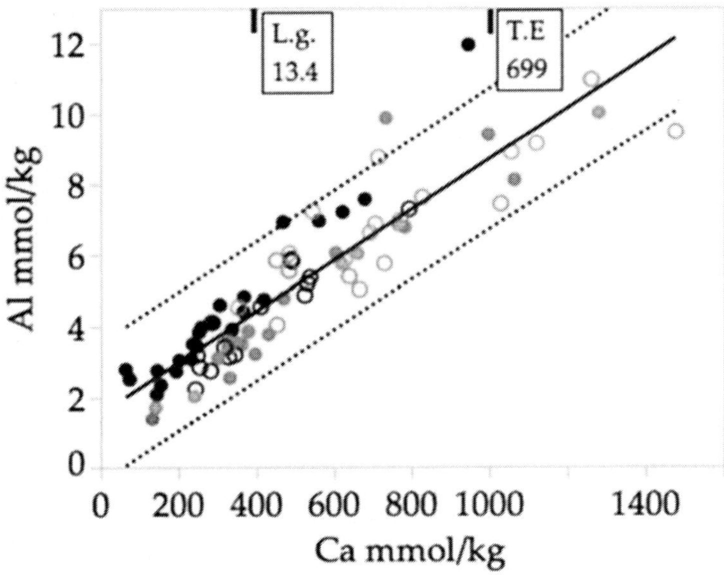

Figure 5. Linear correlation between the concentrations of Ca and Al in adult leaves of karst tree species. The arrows indicate the values of Al accumulators which were not included in the regression. Al (mmol/kg) = 1.55 + 0.007 × Ca (mmol/kg). R^2 adj = 0.825; F = 379, $p < 0.0001$, $n = 81$. Symbols are the same as in Figure 4.

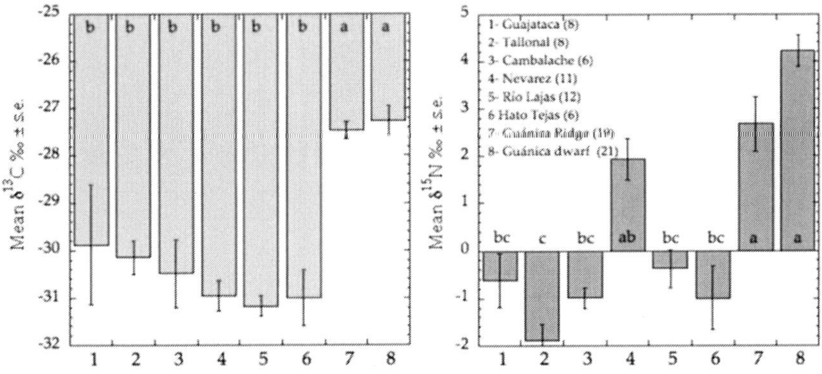

Figure 6. Average ± SE. values of N and C isotopes in adult leaves from different karst forests in Puerto Rico. Columns with the same letter at the base indicate statistical similarity (Tukey Honest Significant Difference, $p = 0.01$).

4. DISCUSSION

4.1. Comparison between Humid and Dry Tropical Forests on Contrasting Soils

Comparison of the dry/moist karst leaf data with other assessments of leaf nutrient concentration in wet and seasonally dry tropical forests shows, in general, that karst forests in Puerto Rico are more limited by P than N, judging by their high N/P ratios, similar to those reported for P-poor forests in the Amazon basin (Table 7). A lower montane rain forest on volcanic soils in Puerto Rico (El Verde, Luquillo Experimental Forest) [31] has much higher concentrations of K and P, and lower N/P ratio than both the humid and dry karst leaves. Moreover, dry and humid karst leaves have quite similar average concentrations of Mg, P, and N, and large N/P ratios. Remarkably, dry karst leaves are richer in K, but poorer in Ca compared to humid karst leaves. Possible explanation for the high K may be the coastal location of the dry karst sites, whereas the higher Ca concentration in moist karst is probably associated with the prevailing humidity of the substrate, which leads to higher availability of soluble Ca.

Table 7. Average leaf nutrient concentrations in a range of humid and dry tropical forests (mmol/kg).

Site	*n*	N	P	K	Ca	Mg	N/P	Reference
Rainforests								
Oxisols	462	1371	26	-	-	-	58	[32]
Other	110	1307	37	-	-	-	38	[32]
French Guyana	45	1143	23	-	-	-	54	[33]
El Verde PR	41	1156	26	291	257	163	46	[31]
Seasonally dry forest								
Chamela, Mexico	19	1281	33	-	-	-	40	[34]
Goias, Brazil	13	1680	46	293	389	89	43	[35]
Karst region southwestern China								
	121	1224	47	210	517	125	26	[36]
Karst moist forest	44	1243	22	150	640	123	59	Present study
Karst dry forest	38	1123	21	248	360	141	58	Present study

The average values of C, N, and P from our data set are similar to those reported by Hättenschwiler et al. [33] for 45 tree species in a lowland wet neotropical forest. The range of variation of the species from karst forests (Table 3) is the same for C but much larger for N and P than that of the neotropical forest in French Guiana. The average N/P ratio of their data set amounts to 54, slightly lower than the average of our data set (58), but the C/N and C/P ratios are similar.

Absolute concentrations of N and P in karst forest species are also lower than those reported for many species from forests on P-poor oxisols in Amazonas by Townsend et al. [32], but the N/P ratio is quite similar, amounting to 58 on a molar basis. It may be concluded from these comparisons with moist and wet tropical forest sites that the karts forests from Puerto Rico are relatively limited by P.

Most seasonally dry forest soils in the neotropics are comparatively rich in Ca, but are as low in P as the karst forest reported here. Leaf P concentrations in seasonally dry forest in Chamela, Mexico [34], and a deciduous forest on calcareous outcrops in the cerrado of Goias in Brazil [35], are much higher than the values reported here (Table 7). Therefore, their average leaf N/P ratios are much lower than those measured in our karst forests.

A recent study from vegetation on karst in China slightly above 26° N reports surprisingly much higher P concentrations in leaves, and lower N/P values than those reported here for Puerto Rico [36]. Concentrations of the other elements are within the same range as those reported here (Table 7).

The influence of element availability in substrate is suggested by the fact that the species with higher K concentrations, and the mean of site values, correspond to the Guánica Ridge and Dwarf forests, in agreement with the results of rock analyses.

Physiological differences may be indicated for the case of *Tetrazygia elaeagnoides*, one of the few Melastomataceae tree species recorded for *mogote* tops in the northern karst belt [17,24]. This species is a strong Al accumulator, a characteristic common among the members of the Melastomataceae, but it is remarkable that it also showed the highest concentrations of Fe and Mn in the whole data set. In addition, the two legumes included in this data set, *P. arboreum* and *L. glaucifolius*, are among the species with highest concentrations of N.

Concentrations of heavy metals (Fe and Mn) in the Liu et al. study [36] were also about two times higher than in the present paper (2 vs. 0.7 mmol/kg). Data on Al in karst vegetation were not found in the available literature. Excluding the Al accumulators sampled in Guajataca site, concentrations reported here are below 10 mmol/kg (Table 4), much lower than values reported by Masunaga et al. [37] for non-accumulator trees in Sumatra rain forest (21.5).

4.2. The Ca/Al Relationship: Al Accumulators in Limestone Forests

In an extensive analysis of tree material in a tropical rain forest in West Sumatra, Masunaga et al. [37,38] were able to identify a wide variation in Al accumulation in tree foliage. They modified the criteria for defining Al-accumulator species using the 1 g/kg leaf concentration boundary (37 mmol/kg), introducing categories of accumulators between 1 and 3 g Al/kg (37–111 mmol/kg) and strong accumulators with >3 g Al/kg (>111 mmol/kg). For non-Al accumulators, Al was positively correlated with Ca ($r = 0.62$, $n = 469$) and Mg ($r = 0.23$). For strong accumulators, the correlation was not significant at the 1% level. For the non-accumulators, Masunaga et al. [37] indicate that the Al/Ca ratio averages 0.04 ± 0.02, whereas for our data set this ratio averaged0.011 ± 0.005, due to the much higher Ca concentration in leaf tissues (509 vs. 367 mmol Ca/kg).

The elemental concentration of leaves from rainforests on Ca-rich soils in West Sumatra is characterized by a large range of variation of most metals, particularly Al, Fe, and Mn, and includes several metal accumulator species [38]. The rainforest leaves of the West Sumatra forest have quite similar concentrations of P, S, K, Ca, and Mg, but much higher concentrations of Al, Fe, and Mn than the karst forest species reported here. These comparisons show that in spite of huge differences in rainfall, the main factor regulating concentration of cations in leaves is the elemental availability from soil, with the exception of metal accumulators. Aluminum accumulators are numerous in the West Sumatra data set [37], whereas only two tree species were detected in the karst forests studied.

Why is there such a clear relationship between Ca and Al concentrations in non-Al accumulators, both in rain and moist and dry karst forests? We suggest that root activity in calcareous substrate includes the secretion of organic acids in the rhizosphere, leading to the dissolution of the carbonatic rocks, thereby releasing Ca and other metals present, as well as P. This is one of the specific mechanisms described to counteract low Fe availability in calcareous soils [14]. Organic acids may act as chelating agents facilitating Fe uptake, but also the uptake of other metallic cations present in the soil solution. In addition to the high Al-Ca correlation, our data set shows that Ca is linearly, but more weakly, related to the concentrations of Mg and Fe ($p = 0.05$), but no relationship was detected with Mn.

4.3. Isotopic Relationships

The significant higher values of $\delta^{13}C$ in both dry karst forest sites were expected. The restrictions in water supply determine long-term reductions in stomatal conductance leading to reduction in ^{13}C discrimination in the photosynthetic process [26].

Variation of N isotopic values under natural conditions can be associated with intrinsic biological characteristics of the species involved or with ecosystems nutrient cycling properties:

1) Occurrence of mycorrhiza may decrease values compared to non-mycorrhizal plants [39,40].
2) Symbiotic N_2 fixation (as in legumes-*Rhizobium* association) results in lower values because this process does not discriminate against the heavier N isotope [41].
3) Conditions inhibiting denitrification processes (water saturated soils) lead to more negative values in vegetation, the opposite effect occurs in drier areas where denitrification operates freely [42,43].
4) High levels of N availability and P limitation lead to higher values [40,44].

There are few studies on mycorrhizal associations in woody vegetation on calcareous substrates in the Caribbean. However, a recent report concluded that seasonal rainfall has a strong influence on the diversity of mycorrhizal fungi but not on infectivity [45]. Therefore, we may assume that mycorrhiza is not responsible for the differences observed. Furthermore, legumes represent only a small fraction of the species analyzed (2 out of 56). Finally, P deficiencies assessed through the N to P ratios were similar for moist and dry sites. The main factor determining lower leaf $\delta^{15}N$ values in the moist north karst species compared to the drier southern karst region seems to be the difference in rainfall. In *mogote* tops, however, humidity conditions may be highly heterogeneous due to low water retention capacity of calcareous substrates, and the random occurrence of rock cracks and crevices where water may be retained.

5. CONCLUSIONS

- Karst vegetation in Puerto Rico appears to be P limited based both on lower P concentrations and high N/P ratios in leaves. The latter are similar to those of rain forests on P poor soils.

- Source of cations originate both from calcareous rocks (Ca and Mg) and allochthonous sediments from volcaniclastic rocks (Al and Fe).
- Calcium is the predominant cation in leaf tissue and its concentration explains more than 70% of the variation in ash content.
- The leaf concentrations of K and Na are larger in the dry karst sites, probably because of their coastal location. Calcium was higher in moist karst sites, probably due the larger availability of soluble Ca in the substrate.
- The tree species of both moist and dry karst forest accumulate small amounts of Al, which is linearly correlated with the total accumulation of Ca in leaf tissue. This correlation may derive from the process of element solubilization in the rhizosphere driven by the secretion of organic acids.
- The $\delta^{13}C$ values of leaf tissue are significantly higher in the dry karst forest, revealing the limitations of water supply.
- The $\delta^{15}N$ values of leaf tissue were, with one exception, significantly lower in moist karst sites. The probable cause is that denitrification tends to predominate in drier environments.

ACKNOWLEDGMENTS

Field work and chemical analyses were conducted using IITF facilities. Alberto Rodríguez (formerly at IITF) provided assistance during field work, and Marcos Caraballo (University of Puerto Rico) helped with the species identification. Edwin López Soto from IITF lab revised the methods section and the whole data set.

REFERENCES

1. Goldich, S.S.; Bergquist, H.R. *Aluminous Lateritic Soil of the Sierra de Bahoruco Area Dominican Republic, W. I.*; Bulletin 953-C; United States Department of the Interior, Geological Survey, United States Government Printing Office: Washington, DC, USA, 1947.
2. Asprey, G.F.; Robbins, R.G. The vegetation of Jamaica. *Ecol. Monogr.* **1953**, *23*, 359–412.
3. Lötschert, W. Die Übereinstimmung von geologischer Unterlage und Vegetation in der Sierra de los Organos (Westkuba). *Ber. Deutsch. Bot. Ges.* **1958**, *71*, 55.
4. Lugo, A.E.; Miranda Castro, L.; Vale, A.; López, T.M.; Hernández Prieto, E.; García Martinó, A.; Puente Rolón, A.R.; Tossas, A.G.; McFarlane, D.A.; Miller, T.; et

al. *Puerto Rican Karst—A Vital Resource*; General Technical Report WO-65; USDA Forest Service: Washington, DC, USA, 2001.

5. Díaz del Olmo, F.; Cámara Artigas, R. *Karst Tropical de Colinas, Tipología y Evolución en el Pliocuaternario en República Dominicana*; Flor, G., Ed.; Actas de la XI Reunión Nacional de Cuaternario, XI Reunión Nacional de Cuaternario: Oviedo, Spain, 2003; pp. 123–129.
6. Day, M. Human interaction with Caribbean karst landscapes: Past, present and future. *Acta Carsologica* **2010**, *39*, 137–146.
7. Kelly, D.L.; Tanner, E.V.J.; Kapos, V.; Dickinson, T.A.; Goodfriend, G.A.; Fairbairn, P. Jamaican Limestone Forests: Floristics, structure and environment of three examples along a rainfall gradient. *J. Trop. Ecol.* **1988**, *4*, 121–156.
8. Borhidi, A. Dry coastal ecosystems of Cuba. In *Ecosystems of the World. Dry Coastal Ecosystems Africa, America, Asia and Oceania*; van der Maarel, E., Ed.; Elsevier: Amsterdam, The Netherlands, 1993; Volume 2, pp. 423–452.
9. Areces-Mallea, A.E.; Weakley, A.S.; Li, X.; Sayre, R.G.; Parrish, J.D.; Tipton, C.V.; Bouche, T. *A Guide to Caribbean Vegetation Types: Preliminary Classification System and Descriptions*; Panagopoulos, N., Ed.; The Nature Conservancy International Headquarters: Washington, DC, USA, 1999; p. 166.
10. Alemán-Gonzalez, W.B. *Karst Map of Puerto Rico*; US Geological Survey: Reston, VA, USA, 2010.
11. Briggs, R.P. The blanket sands of northern Puerto Rico. In Proceedings of the Transactions of the 3rd Caribbean Geology Conference, Kingston, Jamaica, 2–11 April 1962; Jamaica Geological Survey Publication: Kingston, Jamaica, 1966; pp. 60–69.
12. Monroe, W.H. *The Karst Landforms of Puerto Rico*; U.S. Geological Survey Professional Paper 899; Government Printing Office: Washington, DC, USA, 1976; p. 69.
13. Mengel, K.; Kirkby, E.A. *Principles of Plant Nutrition*, 5th ed.; Kluwer Academic Publishers: Dordrecht, The Netherlands, 2001; p. 849.
14. Marschner, H. *Mineral Nutrition of Higher Plants*, 2nd ed.; Academic Press: Cambridge, MA, USA, 1995.
15. Kinzel, H. Influence of limestone, silicates and soil pH on vegetation. In *Physiological Plant Ecology III. Responses to Chemical and Biological Environment*; Lange, O.L., Nobel, P.S., Osmond, C.B., Ziegler, H., Eds.; Encyclopedia of Plant Physiology; Springer: Berlin, Germany, 1983; Volume 12, pp. 201–244.
16. Medina, E.; Cuevas, E.; Molina, S.; Lugo, A.E.; Ramos, O. Structural variability and species diversity of a dwarf Caribbean dry forest. *Caribb. J. Sci.* **2012**, *46*, 203–215.
17. Alvarez Ruiz, M.; Acevedo Rodriguez, P.; Vazquez, M. Quantitative description of the structure and diversity of the vegetation in the Limestone Forest of Rio Abajo, Arecibo-Utuado, Puerto Rico. *Acta Cient.* **1997**, *11*, 21–66.
18. Acevedo-Rodríguez, P.; Axelrod, F.S. Annotated checklist for the tracheophytes of Río Abajo forest reserve, Puerto Rico. *Caribb. J. Sci.* **1999**, *35*, 265–285.
19. Murphy, P.G.; Lugo, A.E. Structure and biomass of a subtropical dry forest in Puerto Rico. *Biotropica* **1986**, *18*, 89–96.
20. Monsegur-Rivera, O.A. Vascular Flora of the Guánica dry Forest, Puerto Rico. Master's Thesis, University of Puerto Rico, Mayagüez Campus, San Juan, Puerto Rico, 2009.
21. Chinea, J.D. The Forest Vegetation of the Limestone Hills of Northern Puerto Rico. Ph.D. Thesis, Cornell University, Ithaca, NY, USA, 1980.
22. Aukema, J.E.; Carlo, T.A.; Collazo, J.A. Landscape assessment of tree communities in the northern karst region of Puerto Rico. *Plant Ecol.* **2007**, *189*, 101–115.
23. Molina Colon, S.; Lugo, A.E. Recovery of a subtropical dry forest after abandonment of different land uses. *Biotropica* **2006**, *38*, 354–364.
24. Axelrod, F.S. *A Systematic Vademecum to the Vascular Plants of Puerto Rico*; Botanical Research Institute of Texas: Fort Worth, TX, USA, 2011.

25. Sánchez, M.J.; Lopez, E.; Lugo, A.E. *Chemical and Physical Analysis of Selected Plants and Soils from Puerto Rico (1981–2000)*; Gen. Tech. Rep. GTR-IITF-45; U.S. Department of Agriculture, Forest Service, International Institute of Tropical Forestry: Rio Piedras, PR, USA, 2015; p. 85.
26. Farquhar, G.D.; Hubick, K.T.; Condon, A.G.; Richards, R.A. Carbon isotope fractionation and plant water use efficiency. In *Stable Isotopes in Ecological Research*; Rundel, P.W., Ehleringer, J.R., Nagy, K.A., Eds.; Ecological Studies; Springer: New York, NY, USA, 1989; Volume 68, pp. 21–40.
27. Handley, L.L.; Raven, J.A. The use of natural abundance of nitrogen isotopes in plant physiology and ecology. *Plant Cell Environ.* **1992**, *15*, 965–985.
28. Viera Martínez, C.A.; Abelleira Martínez, O.J.; Lugo, A.E. Estructura y química del suelo en un bosque de *Castilla elastica* en el carso del norte de Puerto Rico: Resultados de una calicata. *Acta Cient.* **2008**, *22*, 29–35.
29. McGroddy, M.E.; Daufresne, T.; Hedin, L.O. Scaling of C:N:P stoichiometry in forests worldwide: Implications of terrestrial Redfield-type ratios. *Ecology* **2004**, *85*, 2390–2401.
30. Tinoco-Ojanguren, C.; Vázquez-Yánes, C. Especies CAM in la selva húmeda tropical de Los Tuxtlas, Veracruz. *Bol. Soc. Bot. Mex.* **1983**, *45*, 150–153.
31. Ovington, J.D.; Olson, J.S. Biomass and Chemical Content of El Verde Lower Montane Rain Forest Plants. In *A Tropical Rain Forest: A Study of Irradiation and Ecology at El Verde, Puerto Rico*; Odum, H.T., Pigeon, R.F., Eds.; National Technical Information Service: Springfield, VA, USA, 1970; Volume H-2, pp. H53–H75.
32. Townsend, A.R.; Cleveland, C.C.; Asner, G.P.; Bustamante, M.M.C. Controls over foliar N:P ratios in tropical rain forests. *Ecology* **2007**, *88*, 107–118.
33. Hättenschwiler, S.; Aeschlimann, B.; Coûteaux, M.-M.; Roy, J.; Bonal, D. High variation in foliage and leaf litter chemistry among 45 tree species of a neotropical rainforest community. *New Phytol.* **2008**, *179*, 165–175.
34. Rentería, L.Y.; Jaramillo, V.J. Rainfall drives leaf traits and leaf nutrient resorption in a tropical dry forest in Mexico. *Oecologia* **2011**, *165*, 201–211.
35. Rossatto, D.R.; Alvim Carvalho, F.; Haridasan, M. Soil and leaf nutrient content of tree species support deciduous forests on limestone outcrops as a eutrophic ecosystem. *Acta Bot. Bras.* **2015**, *29*, 231–238.
36. Liu, C.; Liu, Y.; Guo, K.; Wang, S.; Yang, Y. Concentrations and resorption patterns of 13 nutrients in different plant functional types in the karst region of south-western China. *Ann. Bot.* **2014**, *113*, 873–885.
37. Masunaga, T.; Kubota, D.; Hotta, M.; Wakatsuki, T. Mineral composition of leaves and bark in aluminum accumulators in a tropical rain forest in Indonesia. *Soil Sci. Plant Nutr.* **1998**, *44*, 347–358.
38. Masunaga, T.; Kubota, D.; Hotta, M.; Wakatsuki, T. Nutritional characteristics of mineral elements in tree species of tropical rain forest, west Sumatra, Indonesia. *Soil Sci. Plant Nutr.* **1997**, *43*, 405–418.
39. Högberg, P. ^{15}N natural abundance in soil-plant systems. *New Phytol.* **1997**, *137*, 179–203.
40. Craine, J.M.; Elmore, A.J.; Aidar, M.P.M.; Bustamante, M.; Dawson, T.E.; Hobbie, E.A.; Kahmen, A.; Mack, M.C.; McLauchlan, K.K.; Michelsen, A.; et al. Global patterns of foliar nitrogen isotopes and their relationships with climate, mycorrhizal fungi, foliar nutrient concentrations, and nitrogen availability. *New Phytol.* **2009**, *183*, 980–992.
41. Shearer, G.; Kohl, D.H. N_2-fixation in field settings: Estimations based on natural ^{15}N abundance. *Aust. J. Plant Physiol.* **1986**, *13*, 699–756.
42. Martinelli, L.A.; Piccolo, M.C.; Townsend, A.R.; Vitousek, P.M.; Cuevas, E.; Mcdowell, W.; Robertson, G.P.; Santos, O.C.; Treseder, K. Nitrogen stable isotopic composition of leaves and soil: Tropical versus temperate forests. *Biogeochemistry* **1999**, *46*, 45–65.

43. Medina, E.; Izaguirre, M.L. N_2-fixation in tropical American savannas evaluated by the natural abundance of ^{15}N in plant tissues and soil organic matter. *Trop. Ecol.* **2004**, *45*, 87–95.
44. Nardoto, G.B.; Ometto, J.P.H.B.; Ehleringer, J.R.; Higuchi, N.; Bustamante, M.M.; Martinelli, L.A. Understanding the influences of spatial patterns on N availability within the Brazilian Amazon forest. *Ecosystems* **2008**, *11*, 1234–1246.
45. Guadarrama, P.; Castillo, S.; Ramos-Zapata, J.A.; Hernández-Cuevas, L.V.; Camargo-Ricalde, S.L. Arbuscular mycorrhizal fungal communities in changing environments: The effects of seasonality and anthropogenic disturbance in a seasonal dry forest. *Pedobiologia* **2014**, *57*, 87–95.

CHAPTER 18

Remote Sensing of Aboveground Biomass in Tropical Secondary Forests: A Review

J. M. Barbosa,[1] E. N. Broadbent,[2] and M. D. Bitencourt[1]

[1]Department of Ecology, Institute of Biosciences, University of São Paulo, 05508-090 São Paulo, SP, Brazil

[2]Sustainability Science Program, Kennedy School of Government, Harvard University, Cambridge, MA 02138, USA

ABSTRACT

Tropical landscapes are, in general, a mosaic of pasture, agriculture, and forest undergoing various stages of succession. Forest succession is comprised of continuous structural changes over time and results in increases in aboveground biomass (AGB). New remote sensing methods, including sensors, image processing, statistical methods, and uncertainty evaluations, are constantly being developed to estimate biophysical forest changes. We review 318 peer-reviewed studies related to the use of remotely sensed AGB estimations in tropical forest succession studies and summarize their geographic distribution, sensors and methods used, and their most frequent ecological inferences. Remotely sensed AGB is broadly used in forest management studies, conservation status evaluations, carbon source and sink investigations, and for studies of the relationships between environmental conditions and forest structure. Uncertainties in AGB estimations were found to be heterogeneous with biases related to sensor type, processing methodology, ground truthing availability, and forest characteristics. Remotely sensed AGB of successional forests is more reliable for the study of spatial patterns of forest succession and over large time scales than that of individual stands. Remote sensing of temporal patterns in biomass requires further study, in particular, as it is critical for understanding forest regrowth at scales useful for regional or global analyses.

1. INTRODUCTION

Secondary and disturbed forests comprise roughly 30–50% of the area covered by tropical forests [1–4]. These forests play important roles as habitats for animal and plant species [5–7] and store significant amount of carbon per unit area [8]. Secondary forests emerge following natural or human disturbances, such as clearing, selective logging, introduction of invasive species, storms, or wild fires. Forest succession is a natural response to these disturbances and occurs at varying rates and in different directions, as indicated through species composition and forest structure, depending on environment conditions [9]. Changes in structure and species composition during forest succession typically result in substantial increases in aboveground biomass [10, 11]. Thereby spatial and temporal changes in aboveground biomass (AGB) can be useful indicator of the velocity and direction of the forest succession and patterns of AGB distribution and change through time can help to understand how forest structure is related to the natural environmental conditions [12, 13] and global carbon cycle [14].

The use of continuous forest metrics obtained using remote sensing (RS) data, for example, aboveground biomass, tree diameter, canopy height, and canopy closure, enables improved characterization of the forest [15] and allows an improved understanding of the ecological drivers behind land cover dynamics [16]. However, challenges exist to estimate biophysical data via available remote sensing technologies, such as optical, radar, and LiDAR data, which form the main tools capable of estimating forest AGB [12, 17–19]. One issue is due to the varying spatial, temporal, radiometric, and spectral resolutions unique to each sensor system, resulting in different advantages and disadvantages to AGB estimation [20, 21].

The use of estimated forest biomass in ecology depends on the cost-benefit relationship that includes image accessibility, availability of image processing techniques, and data confidence. The available literature provides an overall view of RS on forest biomass estimation [20, 21] and evaluates specific sensors and techniques [19]. However, an integrated effectiveness evaluation of how the AGB estimated with RS data can be used to study the progress of the forest succession and the secondary forest characteristics that interfere in the biomass estimation is missing. Therefore, there is a need for a critical review covering the advantages and limitations of the use of remotely sensed AGB in the evaluation of drivers and paths of the forest succession. In this review we discuss how remote sensing of forests biomass helps to identify forest successional patterns. Our specific objectives are (a) to discuss subtle characteristics of secondary forests that challenge the use of remotely sensed AGB estimations in forest succession studies, (b) to summarize the most frequent ecological inferences discussed and methods used in the literature, and (c) to evaluate the implications of methodological and technological issues in the AGB estimation. We first analyze the potential use of remotely sensed AGB estimations as an ecological variable and then discuss the implications for studies of forest succession. We then summarize the ecological issues mentioned in and the methodological characteristics of the selected studies. We conclude with a discussion of the implications of the findings and highlight opportunities for future investigation.

2. DATABASE

Relevant journal articles were found by using keyword searches in the Web of Science and Scopus Databases (accessed in January 2014) and by looking through the reference lists of previous narrative reviews [16, 19, 20]. All searches included the words "tropical forest," "aboveground (or above ground or above-ground) biomass or carbon stock," and "remote sensing" (or "sensed"). Our database selected by the keyword searches consisted of 318 articles and reviews. We found a continuous increase on the number of publications about biomass and carbon stock on tropical forests over the last 29 years (Figure 1). In addition, the higher number of published articles in the last 5–8 years may be related to the increased global interest on climate change. Research institutions from the United States, Brazil, United Kingdom, Germany, and France published most of these articles (Figure 2). The selected articles were published in 101 journals; however, 23% of them were published only in two journals: "Remote Sensing of Environment" and "International Journal of Remote Sensing."

This total database was used to identify the main ecological questions dealt with in the literature. The studies selected by our literature search discussed different ecological problems, including biological invasions, carbon sources and sinks, conservation status, habitat suitability, technical management comparisons, and biomass map elaboration (Table 1). These subjects were discussed in the present review because each is closely related to the forest succession process.

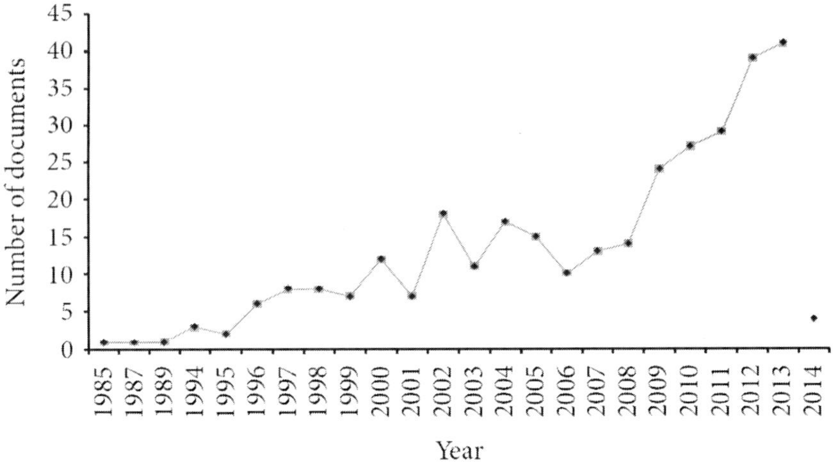

Figure 1: Number of articles published over the last 29 years.

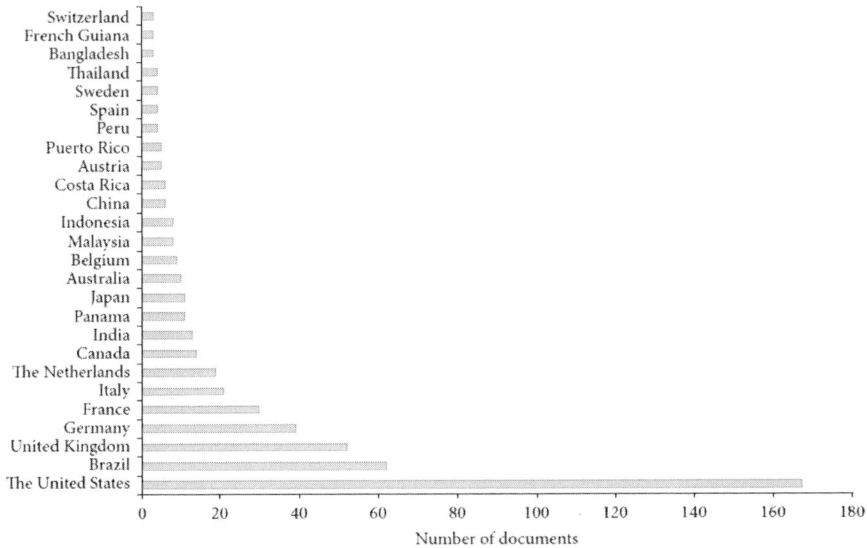

Figure 2: Number of articles published by author's countries. The countries included in the figure have more than two publications, considering our search criteria.

Table 1: Major issues discussed in the reviewed references and the information derived from the aboveground biomass (AGB) estimation by using remote sensing data.

Topic	Information derived	Reference
Forest regrowth	(i) Identifying forest regenerating processes using temporal AGB comparison (ii) Forest structure and AGB changing across forest successional gradient (iii) Secondary forest detection	[28, 29, 37, 38, 88, 104–106]
Human impact on forest biomass	(i) AGB differences due to disturbance regimes (ii) Effect of invasive tree species on the stored AGB and forest structure	[26, 28, 49, 86, 102]
Forest responses to ecological conditions	(i) AGB differences between vegetation physiognomies (ii) Forest structure differences related to topoedaphic gradients, climatic conditions, or past land use	[12, 13, 53, 73, 94, 104]
Carbon sink, source, and storage	(i) Carbon sequestration (ii) Carbon amount before specific human impact (iii) Identifying management action that can improve carbon storage	[12, 26, 29, 30, 102, 106–111]
Modeling evaluation	(i) Testing modeling approach (ii) Image processing techniques (iii) Influence of different spatial scale (iv) Simulation approach	[17, 18, 30, 37, 41, 78, 87, 88, 98, 102, 112–121]
Sensor evaluation	(i) Testing the performance of specific satellite data to estimate biomass and forest structure (ii) Testing the AGB accuracy between successional stages (iii) Testing performance of specific satellite data to temporal comparison	[28, 50, 104, 105, 107, 111–113, 116, 118, 121–131]

We refined our final database using the new following criteria: (a) the search words had to be reported either in the title, abstract, author keywords, or plus keywords; and (b) the aboveground biomass estimation had to be performed using biophysical structure (i.e., length, diameter, and crown size) and remote sensing data. Our final database consisted of 50 articles, which all met the refined criteria and directly or indirectly discuss forest succession patterns and processes. Figure 3 shows the geographic location and spatial resolution of all selected studies. The studies were highly concentrated on Brazilian Amazon, Costa Rica, and Borneo and many of the papers shared the same studied location to collect field data. Approximately, 50% of these studies were done by institutions located outside the studied country.

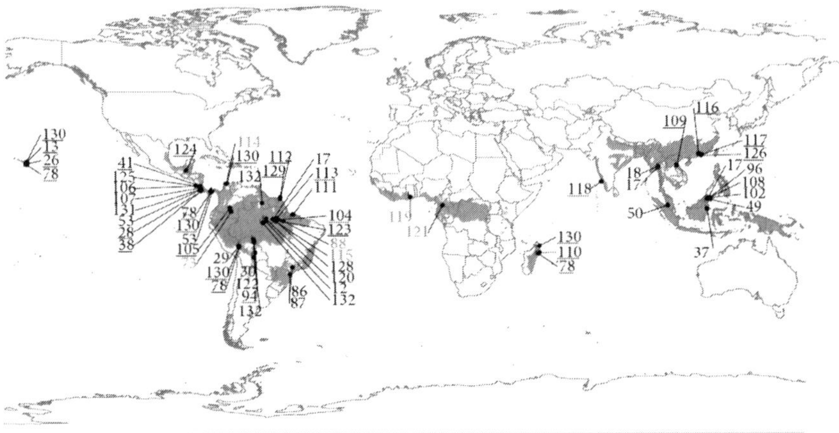

Figure 3: Geographic location of the selected studies. Green area refers to the distribution of the tropical forest biome. Number refers to the article identification on the references (underline black number = spatial resolution ≤10 m; black numbers = spatial resolution >10 m and ≤30 m; grey numbers = spatial resolution >30 m).

3. ESTIMATED ABOVEGROUND BIOMASS AS A VARIABLE IN FOREST SUCCESSION STUDIES

Forest patches regrowth from initial to mature condition by forest succession processes [22]. The opposite way, from mature to initial forest stages, also occurs due to forest disturbances in a direct (i.e., clear cut, selective logging) or indirect way (i.e., edge effect in fragmented landscapes) [23–25]. The progress and regress of forest succession result in continuous modifications in species composition and biophysical features. Successively, there is a replacement of species, local environmental changes, and tree growth or increased mortality. Thereby, many studies investigate how fast the velocity of the forest regeneration is [22], how invasive species change the regrowth tendency [26], what the main environmental features determining the direction of the forest succession are [13], or what the level of the carbon storage in different forest stages is [27].

Historically, these questions have been addressed by ground-based surveys, which are frequently sparsely sampled, inventoried at a small scale, and time consuming [28]. Remote sensing technology adds new possibilities for analysis, allowing obtaining data from inaccessible areas using satellite images, which can reduce biases caused by nonrandom plot locations. Satellite images have been largely used to map forest recovery by forest/nonforest maps or successional stand classifications (initial, advanced stages). However, biomass maps obtained from RS data can be used to identify forest succession patterns because AGB integrates important forest structure information associated with forest regrowth or disturbances, such as tree height, diameter, number of trees by area, and wood density (Figure 4), which can be measured and validated in the field.

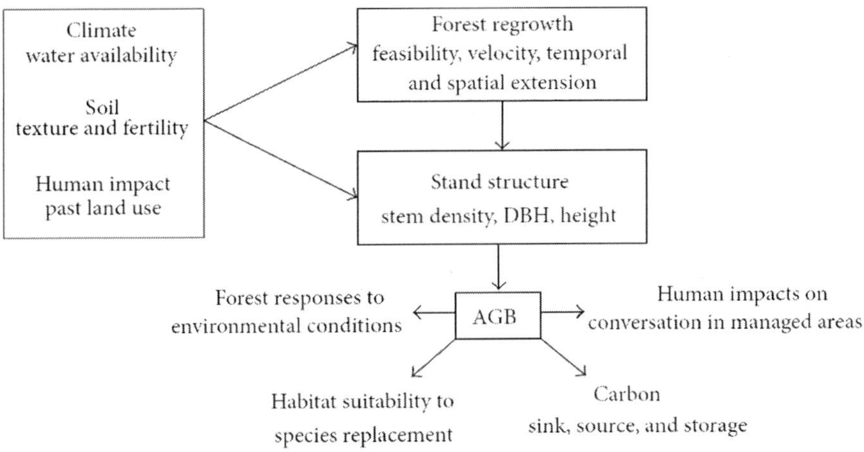

Figure 4: Climate, soil, and human activities determine the forest stand structure and the forest regrowth pattern. The estimated aboveground biomass (AGB) may be used as a variable to answer ecological issues, such as forest responses to environmental conditions, habitat suitability, carbon storage, and conservation in managed areas (adapted from Sarker et al. [132]).

3.1. Dynamics of AGB during Forest Regrowth from Field and RS Measurements

In tropical forests, young fallows show biomass increases ranging from 1 to 15 Mg ha^{-1} yr^{-1} after short-term shifting cultivation [10, 27, 29, 30]. A mean biomass increment of 5.3 Mg ha^{-1} yr^{-1} has been documented in Atlantic rainforest over a 22-year study of undisturbed forest plots [31]. Thus, secondary forests can rapidly grow up to 100 Mg ha^{-1} of AGB or less than 20 Mg ha^{-1} within 20 years after land abandonment. The rate of regrowth depends on the availability of seed propagules, vegetation type, environmental conditions, site productivity, intensity of the past land use, and/or human intervention [32]. In tropical forest regrowth, leaf AGB tends to increase for 20 years and wood AGB increases for 40 years before stabilizing [10]. These time periods seem short when compared to some modeling estimations and field observations that have indicated that the tropical forest needs approximately 80–160 years to reach the mature stage (90% of similar species) [33], although Liebsch et al. [34] found an increase of approximately 57% in species richness over 25 years of forest regrowth in the Brazilian Atlantic Forest. During regrowth, the species richness and the abundance of fauna also increase rapidly and can recover in the first 10–20 years for animals that are not dependent on late-stage plant species [35].

The challenge is to obtain reliable estimates of AGB associated with regrowth using temporal remotely sensed data. Although landscape scale land cover dynamics have been accurately assessed through RS using classification schemes, such as forest/nonforest (e.g., [36]), the study of forest regrowth using continuous variables, such as AGB, can provide new perspectives on regeneration patterns. The ability to track AGB through time enables the assessments of the directional changes in forest dynamics [16]. Although limitations to RS estimation of temporal variation in AGB exist, such as those found by [37] who showed that they overestimated forest regrowth rates as a result of forest fires and a severe El Niño drought before the image dates and uncorrected atmospheric effects; a growing number of studies are developing and implementing methodological techniques to minimize such uncertainties [38–41].

The spatial and temporal evaluation of forest regrowth patterns is also an important issue. Areas belonging to the same region may show different forest area increase as a consequence of differences in human demography and industrialization [42, 43]. Moreover, the losses (by degradation) and accumulations (by forests regrowth) of biomass may change without a change in forest area [44]. The structure and change dynamics of an old-growth tropical forest was studied by Kellner et al. [38]. In 8.5 years, the authors found an equal number of positive and negative transitions of mean canopy height estimated by using LiDAR data, indicating steady-state dynamics of disturbance and regeneration. Therefore, unknown tendencies in forest regrowth patterns associated with AGB changes can appear at different time and spatial ranges of analysis.

Forest regrowth has also been identified in remote sensing studies by using forest age [30, 45] or successional stand classes [46, 47]. However, it is common for one forest age class to exhibit different structural features due to the soil and geographic features or land-use history [11]. In addition, forest age or successional stand information is not always available for some forest areas [45].

3.2. Natural and Anthropogenic Disturbance Impacts on Forest AGB

Changes in the natural course of forest succession may occur due to a large number of disturbances, including wild and intentional fires, selective logging, climate change, and/or the introduction of invasive species. Forest responses to natural disturbances (i.e., storms and fire) are also a feasible use of the estimated AGB data [48]. These disturbance factors may reshape the biophysical structure of the forest or reorder the plant community, resulting in AGB dissimilarity among forest patches with different historical human intervention. Some of the reviewed studies evaluated spatial and temporal changes of tree biomass in logged forests under different management techniques. Tangki and Chappell [49] demonstrated that high-lead harvesting techniques resulted in smaller remaining tree biomasses than logging using tractor skidders. The authors explained that the AGB differences are related to the higher levels of collateral damage caused during high-lead harvesting. In other cases, logged areas exhibited lower mean maximum crown size diameters than old-growth forests [28]. Aerial photographs revealed that tree height/diameter allometric relationships exhibit regression slope differences between logged and old-growth forests [50].

The impacts of tree species invasion on forest structures and AGB are also well studied and particularly in Hawaii, USA. Invasive tree species can rapidly spread through seeds or vegetative propagation and show opportunistic behaviours [51]. This is expected to lead to changes in overall forest AGBs. Asner et al. [26] used airborne LiDAR data to quantify forest structure and found that invasion by the tree Morella faya altered the three-dimensional structure of the rainforests, including the mean tree height, but not the total amount of AGB at the landscape level. The proliferation of tree species, Psidium cattleianum and Ficus rubiginosa, however, corresponded to decreases of 19–38% in tropical forest AGB [13].

3.3. Forest Responses to Environmental Conditions

Environmental conditions and resource availability can also affect primary net production [52] and, consequently, AGB changes throughout forest succession. Differences in the spatial and temporal distributions of precipitation across bioclimatic zones constrain forest AGB in some locations and were found to be an important ecological issue discussed in the reviewed literature. The increase in the number of dry months produced a reduction in forest AGB, showing the influence of net water deficits in the forest structure at regional [53] to continental scales [12]. In a study by Asner et al. [13], using airborne LiDAR data, biomass declined 53–84% in areas with increasing elevation, due to changes in temperature and humidity. On the contrary, Alves et al. [54] found AGB increment with elevation increase in the Brazilian Atlantic coast. All these examples demonstrate how environmental conditions can modulate spatial and temporal patterns of forest AGB.

Another common theme in the literature focuses on spatial distribution of biomass as related to differences among forest types. Biomes or vegetation physiognomies are closely related to environmental and climatic conditions. Wang and Qi [18] reported RS AGB differences between dry dipterocarp forests, mixed deciduous forests, and tropical evergreen forests. These results illustrate the potential use of RS AGB data to distinguish between vegetation types. However, Saatchi et al. [12] found that RS AGB data from Amazon forest was not a useful tool for distinguishing between vegetation types due to the low correlation between forest biomass and vegetation types in this biome. Asner et al. [55] integrated airborne LiDAR with maps of ecosystem types and found significant differences related to forest types and geologic substrate, allowing them to scale up the detailed LiDAR measurements to a large portion of the Southern Peruvian Amazon. Vieira et al. [11] and Saatchi et al. [12], however, found that AGB variations in tropical forests were more due to differences in soil type and land-use history than forest types. Thereby, the relationship between estimated AGB and forest succession patterns needs to be interpreted considering specific characteristics of forest types and the influence of environmental conditions among different scales of analyses.

3.4. Carbon Source, Sink, and Storage

The spatial and temporal change of AGB during the forest succession is controlled by a conjunct of environmental, biological, and human factors, as seen until now. The observed forest AGB can be directly related to tree carbon storage. Dry biomass is expected to contain approximately 50% carbon [8, 56]. The relationship between AGB and carbon opens up a wide range of integrated discussions of the use of forest biomass for carbon monitoring [57, 58]. Brazil, the Democratic Republic of Congo, and Indonesia have the highest carbon stocks and the largest tropical forest areas, indicating that great effort should be expended in the further verification and monitoring of data in these countries [59, 60].

The effect of the increase in atmospheric carbon concentration on climate change is the flagship topic for the study and implementation of environmental carbon sinks as a result of the forest succession. The concept of carbon sinks has been criticized because of estimation biases and the difficulties associated with their quantification and verification using the available technology [61]. The difficulty in measuring changes in biomass and the lack of spatially explicit data are the main cause of variation among carbon estimates at the same area [44]. To overcome these limitations, Hall et al. [19] proposed a new orbital mission capable of estimating forest carbon storage around the world (see Section 4.1.1 for details).

3.5. Habitat Suitability

Forest structural complexity can affect animal species distributions and abundance patterns [62]. Structural attributes and microclimatic conditions change during the forest regrowth process, which can determine habitat suitability for some species [35]. A good relation of habitat heterogeneity and species diversity depends on the taxonomic group under study, how the animal guild studied perceives the vegetation structure, and the spatial resolution of the observation [63]. Biophysical forest characteristics obtained through RS data can generate continuum models related to ecological gradients and can be used to produce habitat suitability models [64].

Gradual changes in RS AGB and forest structure through space and time assessed by optical, radar, or LiDAR data have also been used to investigate the relationship between habitat heterogeneity and species diversity. Kalacska et al. [15] obtained a suitable RS model to estimate Shannon index in a tropical forest. However, the authors noted that remotely sensed data focusing on the canopy surface (i.e., surface reflection measured by Hyperion) might be inappropriate for assessing relationships between subcanopy elements, such as understory species richness. Even though the relationship between species diversity and habitat suitability has been widely documented in field-based research, the empirical knowledge of this relation may be seriously affected by the lack of information over a broad range of environmental conditions [63]. Such problems demonstrate the necessity of large-scale studies related to habitat suitability and the need to understand how RS data interact with field-based variables.

4. BIOPHYSICAL MEASUREMENT AND MODELING

Successful estimation and modeling of AGB over large scales require (a) correct selection and application of remote sensing, (b) being coupled with field data for calibration and validation, and being integrated into (c) an appropriate modeling approach (Figure 5). In this review, we focus on the information required to determine forest successional processes and spatial distribution.

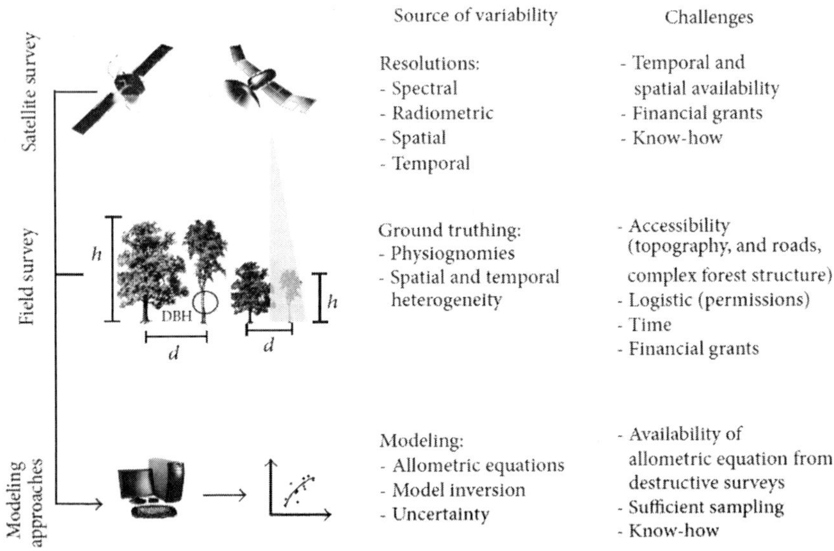

Figure 5: Source of variability and challenges related to each stage of aboveground biomass estimation and modelling.

4.1. Data Sources

4.1.1. Remote Sensing

Selection of an appropriate source of data requires first the identification of the ecological question being asked and identification of the limitations and advantages of each sensor. Overall, the combination or fusion of optical and radar data has the potential to achieve the highest accuracies in the estimated AGB [65, 66]. Optical data can be used to improve the accuracy of radar data [12], for example, by removing leaf backscatter and compensating leaf attenuation to woody structures in each pixel [18]. Radar data (e.g., JERS-1 images) may be especially helpful as the microwaves are independent of cloud cover, a frequent problem in tropical forests. Although the trunk and branch biomass estimations coming from radar are better than the optical data estimations, the response cannot be always the same in terrains with steeper slopes [20].

Other data sources used in the studies evaluated include the Hyperion satellite sensor and aircraft based HYDICE sensor, with high spectral (220 and 210 narrow bands, resp.) and radiometric resolutions. Whereas Landsat has 8-bit resolution with 256 information levels, Hyperion data has 16 bits with 65,536 information levels ranging from data range ~0.4 to 2.5 μm [67]. The greater spectral resolution provided by Hyperion can result in improved sensor capabilities allowing measurements of smaller variations in reflected energy [68, 69], which can be related to forest biomass or phytochemical characteristics of the canopy. The capacity of hyperspectral sensors to differentiate tree phytochemical features enables more detailed estimates of forest structural data in the forest succession studies, partly related to improved mapping of canopy chemistry related to the tree species composition. Some research initiatives have achieved successful measurements of forest structure and species composition by merging LiDAR (tree height) and hyperspectral (phytochemical) data [70, 71], which is an important ongoing focus for studies on the direction and rate of the forest succession.

LiDAR systems can be used to obtain both top of canopy and within canopy structural information as some systems can have effective signal penetration through the forest canopy, resulting in 3D vegetation structure data [53, 72] with spatial resolution varying according to the system design and flight altitude above the study area. LiDAR is now considered to be the state of the art for remote sensing based biomass estimation but has significant disadvantages related to the high cost per area, lack of historical data for temporal analysis, and the unavailability of large-scale datasets making regional or global AGB estimation with airborne LiDAR unfeasible [59, 73]. To mitigate this problem, there is a need for a satellite based platform providing global LiDAR data optimized for AGB estimation with errors less than 10 to 25 Mg C ha^{-1} [19].

The sensors used in the reviewed studies mainly differed in cost and temporal, spatial, spectral, and radiometric resolutions (Table 2). The widespread use of Landsat TM data may be attributed to the low cost of this data. Despite this, the main limitation of the Landsat images is the reflectance saturation at higher biomass values [20]. However, the longer time span of available information (since 1984) is an advantage of Landsat data that can play an important role in temporal forest regrowth analyses.

Multitemporal imagery can be an important source of data for the study of forest regrowth patterns, but the data processing should account for atmospheric corrections and phenological effects. Atmospheric conditions can influence optical RS data through two processes: the absorption and the scatter of solar radiance [74]. The main problem is that images from different dates were acquired under different atmospheric conditions, and preprocessing analyses need to accurately correct for these differences

[75]. The comparison of multiple images from different months to monitor successional changes also needs to consider signal effect resulting from phenological variations in the canopy. Seasonal changes in solar zenith angle, variations in the amount of leaves in the canopy, and the growth of understory vegetation are the main features that can confound remotely sensed signals [76].

Table 2: Remote sensing data used in the studies reviewed.

Sensor	Data used	Spatial resolution	Cost	Number of case studies
LANDSAT/MSS, TM, and ETM+	6 optical bands	30 m	Low	12
USGS/LIDAR	Airborne laser altimeter	1.2–25 m	Very high	8
AIRSAR*	Radar X, C, L, and P band	1.5–10 m	High	5
SAR/JERS-1	7 optical bands and radar L band	18 to 24 m	Low	5
ALOS Palsar/AVNIR-2	Radar L band and 4 optical bands	4.8 m to 9.6 m	Medium	4
IKONOS/IKONOS	4 optical bands	4 m	High	4
SPOT-5/HRS	5 optical bands	2.5 to 5 m	High	3
Terra/MODIS	7 optical bands	250 to 500 m	Low	2
NOAA/AVHRR	5 optical bands	1.1 Km	Low	2
SIR-C	X, C, and L bands	15 to 25 m	Medium	2
GLAS/ICEsat	Laser altimeter satellite	0.75 to 75 m	High	1
Quickbird/Quickbird	4 optical bands	2.4 m	High	1
EO-1/Hyperion	220 optical bands	30 m	Very high	1
NASA/QuikSCAT	Radar L and P bands	12 m	Medium	1
Aerial photographs	RGB true color composite	—	High	1
CHRIS/PROBA	62 optical bands	18 m	Medium	1
VHR Google	RGB true color composite	1 m	Low	1
HYDICE**	210 optical bands	0.75 to 3.75 m	Very high	1
ERS-1	Radar X band	25	Medium	1
TropiSAR	Radar L and P bands	1.2	Very high	1

* AIRSAR: airborne synthetic-aperture radar.
** HYDICE: hyperspectral digital imagery collection experiment.

4.1.2. Ground Truthing

The reviewed studies obtained ground truthing field data from plots or through the point quadrant method. The total sampled area ranged from 1.5 to 272 ha, but in 70% of the studies the area ranged from 1 ha to 9.5 ha. Access difficulties may create restrictions to the sampling of larger areas. Studies with small sampling areas tended to overestimate AGB per hectare, but this tendency is not always significant [77]. Even so, ground truthing of only small areas may result in site-specific models and limit the transferability of the model to other areas [17]. To minimize this condition some studies have used large field databases merging data from different regions [12, 73, 78].

The suitability of the method depends on the scale of the study area and the structural heterogeneity of the forest [79, 80]. The greater the sampling spatial scale, the greater the probability of having good representation of forest structural heterogeneity in the region, indicating that a broad sample area may be required. Moreover, there is a need for a minimum knowledge about the spatial distribution of the forest structure studied [81] and the local influence of the topography and soil on the AGB. Tropical trees tend to exhibit an aggregation pattern [82] and high variability within short spatial distances in forest structure and biomass [83], resulting in a complex mosaic that can be difficult to fully sample. The use of plots greater than 0.1 ha has been shown to be more reliable than smaller plots, which is also related to decreased stochastic small-scale disturbances having disproportionate weighting on total measured structure and AGB [83], maintain greater spatial overlap between ground-reference and sensor data, capture greater forest variability, and reduce edge noise from neighbouring trees [84]. The use of the point quadrant method has the advantage of evaluating forest heterogeneity at less effort and lower cost than plot-based sampling, but this method can underestimate densities and basal areas [85]. The studies that used point quadrant methods also showed high correlation values between field and RS data [18, 86]. Nevertheless, the use of this method needs to be better evaluated as a ground truthing option because such a method has the potential to be used in large-scale studies with reduced fieldwork [87]. Furthermore, a hierarchical sampling scheme of plots, quadrants, and subquadrants can be a good way to include this complexity [88]. Data from the smallest scale (subquadrant) can be treated as random draws from an unobserved quadrant and modeled to obtain uncertainty parameters for the regression of quadrant and satellite data [89].

4.2. Statistical Modeling

4.2.1. Allometric Equations to Estimate Forest Biomass

The studies reviewed used between 1 and 16 different allometric equations per study to estimate AGB. Different allometric relations between biomass and tree diameter or height are set by environmental (e.g., ground slope, climatic conditions, and light level) or biological factors [90] and can explain up to 90–95% of the ground AGB [11, 91]. Testing allometric model accuracies, Vieira et al. [92] applied several allometric equations (developed by different authors) to the same diameter at breast height (DBH) and height and found a difference of 5–7% in AGB estimations, indicating a source of error dependent upon the equation choice. Another source of error in the forest AGB estimation may be the lack of information from nonarboreal sources (such as lianas, palms, tree ferns, bamboo, and epiphytes), which can account for more than 10% of the AGB in Atlantic Forest areas [92].

The choice of the equation used to estimate AGB may depend on the height and DBH range and on the species composition of the forest studied.

Attention to these factors may improve the suitability of the biomass estimation using allometric equations [88, 93]. Another option would be to include wood density in the biomass estimation, providing a more accurate representation of tree biomass than those relying only on DBH and height [94]. When species-level data are not available, a mean wood density by forest type, family, or genus can be used [53].

4.2.2. Remotely Sensed Biomass Estimation by Model Inversion

The main statistical tool used in the studies to relate field and RS data was parametric statistic (90% of the studies). This statistic approach uses continuous biomass data as dependent variable and RS data as independent variable. Modeled biomass values from tropical forests can display nonlinear trends in the residuals, indicating the use of nonlinear models [45]. Li et al. [88] obtained better adjusted -squared and mean squared error values using nonlinear models than linear models and presented a clear framework to develop a nonlinear biomass model for tropical forests. However, about 50% of the studies used linear regression (simple or multiple). Therefore, the choice between linear or nonlinear approaches depends on their statistical requirements. In both cases, linear or nonlinear equations, the resulted equations are frequently used to predict biomass data in areas not surveyed with field work. Generalized linear models (GLM) have also been used to analyze the suitability of AGB models using Akaike Information Criterion and -squared values to evaluate model performances [87, 95]. GLM models allow us to explore combinations of factors, covariates, and interactions among variables through a process called iteratively reweighted least squares. This overcomes the basic problem related to regression models that is the change in the variance as a consequence of the transformation of the data to attain normality [96].

Finally, other predictive models used in the reviewed studies were decision tree and neural networks. Decision trees use hierarchical set of rules to define classes and it does not assume any a priori statistical characteristics for the input data layers [12]. The use of neural networks is an alternative that can play a significant role in obtaining better accuracy [97]. However, neural network models have a substantial subjective component and can be relatively poor for semantically related field and remotely sensed data [98].

4.3. Uncertainty

In the modeling context, uncertainty is the lack of knowledge of the true values of a variable, which can be estimated by the range and likelihood of possible values [57]. The overall uncertainty in AGB estimation using RS is directly related to the relationship between a set of potential errors coming from processing satellite data (Section 4.1.1), field data (Section 4.1.2), statistical modeling (Section 4.2), and the ecological variables studied. Beyond the uncertainty from each variable, the errors can come from the interaction among them, for example, in the conflict of scale between satellite and field data samples. The challenge is to balance how the errors influence the ecological data interpretation and how the errors from each modeling step influence the overall error.

Most studies considered were mainly performed in flat forest areas. In optical images, ground slope can shade areas, resulting in decreased reflectance values compared to flat areas [99] and the generation of spectral differences between forests at the same successional stage [76]. These topographic effects can be minimized by different algorithmic corrections [100], but biases may still be included in analyses [87].

The position errors of the sensor data can also reduce the modeling accuracy in pixel-based modeling [76]. Two common strategies were used to minimize this problem: the correlation of the field data with a 3 × 3 or 5 × 5 pixel window around the ground data [15, 98, 101] and the use of a mean radiance from a forest patch [18, 49]. These approaches can also reduce the confidence level of the model due to the high area disproportion between the field area plot (i.e., 0.09 ha) and the pixel window from, for example, the Landsat data (2.25 ha in a 5 × 5 window).

The root mean square error (RMSE) and determination coefficient values were widely used to compare the accuracies of the AGB estimates. The error estimate had a mean value of 41.2 Mg ha^{-1} (SD = 29.3 Mg ha^{-1}), considering the reviewed studies that present this analysis in the results (17 studies from the 46 analyzed). Commonly, the RMSE is obtained from the comparison of model predictions with independent data. Data can be randomly separated in two groups, one to implement the model and one to check its performance. This process can be repeated many times to obtain randomized RMSE values. Finally, all values are used to build up the overall error probability density function, which can be used to evaluate the magnitude and trend direction of the errors [57].

Li et al. [88] found that the AGB model might show inflated residual variances toward higher biomass values. The authors noted that this can be associated with the application of the same allometric equation to trees with different sizes and forest with different structure complexities. Errors in forest AGB estimation can be random or systematic [83]. Minor AGB errors have been found in forests with low biomass [12, 53, 102].

Despite these inherent modeling errors, biomass estimation using exclusively field data is far from being easy and is free of errors [17]. Studies based only on field information are accurate at the local scale but may fail to take into account the large spatial variability of biomass. AGB estimations using only field data over large areas showed an overall sampling error of approximately ±20% [54, 103]. Moreover, field-based measurements and estimates are time consuming, labor intensive, and difficult to implement in remote areas [20]. Challenges exist in the collection of both RS and field data. However, the integration of these different sources of information is crucial to better AGB estimates in any sites.

5. SYNTHESIS AND FUTURE RESEARCH

In this review we highlight the advantages and disadvantages of estimating forest biomass from remote sensing. Remote sensing of aboveground biomass is critical to understand forest succession and landscape dynamics. Forest biomass relates to unique regrowth patterns, including rates and directions, which are partly determined by natural and anthropogenic disturbances. As species composition changes occur simultaneously to changes in forest structure and related aboveground biomass, study approaches limited to forest structure may have high errors in correctly identifying forest successional stages. This problem may be alleviated through fusion of sensors capable of deriving detailed structural (e.g., radar and LiDAR) and phytochemical (e.g., hyperspectral images) features of the forest canopy.

Access to historical ground truthing information remains one of the main difficulties in multitemporal comparisons of AGB. Most of the selected studies indicated forest regrowth as a potential subject to be analyzed by AGB estimation but do not address the topic. Further interpretation of tropical forest ecological processes, as can be derived from remotely sensed forest structure, is required [13]. Nevertheless, the potential for the use of remotely sensed biomass estimations in forest succession studies in tropical forests remains high. More detailed analyses on the spatial distribution of AGB are merited for improved forest management, conservation status evaluation, carbon source and sink investigations, and better understanding implications of environmental conditions, including climatic change, on forest structure and composition.

Our review indicates that accuracy and precision assessment are the most important challenges related. The uncertainty estimates need to be robust and transparent, due to the current importance for decision makers [58]. In addition, the evaluation of the studies reviewed here indicates that errors in AGB estimations can in some cases be the same as annual AGB changes by regrowth process, making AGB comparisons on a short time scale unfeasible using imprecise or biased data. However, long time scale lag comparisons are still possible if attention is paid to spectral variations between satellite images related to atmospheric conditions and satellite signal degradation over time. Hall et al. [19] highlighted the importance of a new space mission with a technology able to obtain biophysical forest estimates with lower errors in short time scale comparisons.

A reliable AGB estimation depends on the sensor type, processing strategy and forest characteristics. In focusing attention only on the RS data process, we may fail to identify critical ecological patterns. Our results suggest that sampling choices depend on intrinsic forest features and sources of noise related, for example, to topographic characteristics. In addition, we discussed some positive and negative features of each sensor for obtaining AGB estimations. The choice of data source depends on trade-offs between advantages and disadvantages of methodological and technological features.

When making a modeling choice, attention must be paid to potential sources of bias in the biomass regression models, including (a) sample size, (b) accuracy of the allometric equations, (c) image calibration, (d) the methodology used for temporal adjustments between the sampled biomass data and satellite observations, (e) the multicollinearity of the polynomial model [88], (f) the fact that 30–50% of the forest biomass is hidden by emergent trees [94], (g) the increased number of lianas in the intermediate successional stage [15], (h) topography, and (i) the past land-use context in the study of forest regrowth dynamics.

ACKNOWLEDGMENTS

The authors express their gratitude to E. Sebastián González, Paulo I. de Knegt Lopez de Prado, G. Kuntschik, J. A. Quintanilha, and L. R. Tambosi for commenting on earlier versions of this paper.

REFERENCES

1. F. Achard, H. D. Eva, H.-J. Stibig et al., "Determination of deforestation rates of the world's humid tropical forests," Science, vol. 297, no. 5583, pp. 999–1002, 2002.
2. Grainger, "Difficulties in tracking the long-term global trend in tropical forest area," Proceedings of the National Academy of Sciences of the United States of America, vol. 105, no. 2, pp. 818–823, 2008.
3. G. P. Asner, T. K. Rudel, T. M. Aide, R. Defries, and R. Emerson, "A contemporary assessment of change in humid tropical forests," Conservation Biology, vol. 23, no. 6, pp. 1386–1395, 2009.
4. J. Blaser, A. Sarre, D. Poore, and S. Johnson, Status of Tropical Forest Management, Technical Series No. 38, International Tropical Timber Organization, Yokohama, Japan, 2011, Edited by ITTO.

5. S. J. Wright, "Tropical forests in a changing environment," Trends in Ecology & Evolution, vol. 20, no. 10, pp. 553–560, 2005.
6. R. L. Chazdon, C. A. Peres, D. Dent et al., "The potential for species conservation in tropical secondary forests," Conservation Biology, vol. 23, no. 6, pp. 1406–1417, 2009.
7. D. H. Dent and S. J. Wright, "The future of tropical species in secondary forests: a quantitative review," Biological Conservation, vol. 142, no. 12, pp. 2833–2843, 2009.
8. R. Martin and S. C. Thomas, "A reassessment of carbon content in tropical trees," PLoS ONE, vol. 6, no. 8, Article ID e23533, 2011.
9. E. Lugo, "The emerging era of novel tropical forests," Biotropica, vol. 41, no. 5, pp. 589–591, 2009.
10. S. Brown and A. E. Lugo, "Tropical secondary forests," Journal of Tropical Ecology, vol. 6, no. 1, pp. 1–32, 1990.
11. C. G. Vieira, A. S. de Almeida, E. A. Davidson, T. A. Stone, C. J. Reis de Carvalho, and J. B. Guerrero, "Classifying successional forests using Landsat spectral properties and ecological characteristics in eastern Amazônia," Remote Sensing of Environment, vol. 87, no. 4, pp. 470–481, 2003.
12. S. S. Saatchi, R. A. Houghton, R. C. Dos Santos Alvalá, J. V. Soares, and Y. Yu, "Distribution of aboveground live biomass in the Amazon basin," Global Change Biology, vol. 13, no. 4, pp. 816–837, 2007.
13. G. P. Asner, R. F. Hughes, T. A. Varga, D. E. Knapp, and T. Kennedy-Bowdoin, "Environmental and biotic controls over aboveground biomass throughout a tropical rain forest," Ecosystems, vol. 12, no. 2, pp. 261–278, 2009.
14. Grace, "Understanding and managing the global carbon cycle," Journal of Ecology, vol. 92, no. 2, pp. 189–202, 2004.
15. M. Kalacska, G. A. Sanchez-Azofeifa, B. Rivard, T. Caelli, H. P. White, and J. C. Calvo-Alvarado, "Ecological fingerprinting of ecosystem succession: estimating secondary tropical dry forest structure and diversity using imaging spectroscopy," Remote Sensing of Environment, vol. 108, no. 1, pp. 82–96, 2007.
16. Q. Chambers, G. P. Asner, D. C. Morton et al., "Regional ecosystem structure and function: ecological insights from remote sensing of tropical forests," Trends in Ecology & Evolution, vol. 22, no. 8, pp. 414–423, 2007.
17. G. M. Foody, D. S. Boyd, and M. E. J. Cutler, "Predictive relations of tropical forest biomass from Landsat TM data and their transferability between regions," Remote Sensing of Environment, vol. 85, no. 4, pp. 463–474, 2003.
18. Wang and J. Qi, "Biophysical estimation in tropical forests using JERS-1 SAR and VNIR imagery. II. Aboveground woody biomass," International Journal of Remote Sensing, vol. 29, no. 23, pp. 6827–6849, 2008.
19. F. G. Hall, K. Bergen, J. B. Blair et al., "Characterizing 3D vegetation structure from space: mission requirements," Remote Sensing of Environment, vol. 115, no. 11, pp. 2753–2775, 2011.
20. Lu, "The potential and challenge of remote sensing-based biomass estimation," International Journal of Remote Sensing, vol. 27, no. 7, pp. 1297–1328, 2006.
21. Eisfelder, C. Kuenzer, and S. Dech, "Derivation of biomass information for semi-arid areas using remote-sensing data," International Journal of Remote Sensing, vol. 33, no. 9, pp. 2937–2984, 2012.
22. R. L. Chazdon, S. G. Letcher, M. van Breugel, M. Martínez-Ramos, F. Bongers, and B. Finegan, "Rates of change in tree communities of secondary Neotropical forests following major disturbances," Philosophical Transactions of the Royal Society B: Biological Sciences, vol. 362, no. 1478, pp. 273–289, 2007.
23. J. Groeneveld, L. F. Alves, L. C. Bernacci et al., "The impact of fragmentation and density regulation on forest succession in the Atlantic rain forest," Ecological Modelling, vol. 220, no. 19, pp. 2450–2459, 2009.
24. B. A. Santos, C. A. Peres, M. A. Oliveira, A. Grillo, C. P. Alves-Costa, and M. Tabarelli, "Drastic erosion in functional attributes of tree assemblages in Atlantic forest

fragments of northeastern Brazil," Biological Conservation, vol. 141, no. 1, pp. 249–260, 2008.
25. Tabarelli, A. V. Lopes, and C. A. Peres, "Edge-effects drive tropical forest fragments towards an early-successional system," Biotropica, vol. 40, no. 6, pp. 657–661, 2008.
26. G. P. Asner, R. E. Martin, D. E. Knapp, and T. Kennedy-Bowdoin, "Effects of Morella faya tree invasion on aboveground carbon storage in Hawaii," Biological Invasions, vol. 12, no. 3, pp. 477–494, 2010.
27. C. Uhl, R. Buschbacher, and E. A. S. Serrao, "Abandoned pastures in eastern Amazonia. I. Patterns of plant succession," Journal of Ecology, vol. 76, no. 3, pp. 663–681, 1988.
28. B. Clark, J. M. Read, M. L. Clark, A. M. Cruz, M. F. Dotti, and D. A. Clark, "Application of 1-m and 4-m resolution satellite data to ecological studies of tropical rain forests," Ecological Applications, vol. 14, no. 1, pp. 61–74, 2004.
29. Naughton-Treves, "Deforestation and carbon emissions at tropical frontiers: a case study from the Peruvian Amazon," World Development, vol. 32, no. 1, pp. 173–190, 2004.
30. H. Helmer, M. A. Lefsky, and D. A. Roberts, "Biomass accumulation rates of Amazonian secondary forest and biomass of old-growth forests from Landsat time series and the Geoscience Laser Altimeter System," Journal of Applied Remote Sensing, vol. 3, no. 1, Article ID 033505, pp. 1–31, 2009.
31. S. G. Rolim, R. M. Jesus, H. E. M. Nascimento, H. T. Z. do Couto, and J. Q. Chambers, "Biomass change in an Atlantic tropical moist forest: the ENSO effect in permanent sample plots over a 22-year period," Oecologia, vol. 142, no. 2, pp. 238–246, 2005.
32. E. Swanson, J. F. Franklin, R. L. Beschta et al., "The forgotten stage of forest succession: early-successional ecosystems on forest sites," Frontiers in Ecology and the Environment, vol. 9, no. 2, pp. 117–125, 2011.
33. M. R. Guariguata and R. Ostertag, "Neotropical secondary forest succession: changes in structural and functional characteristics," Forest Ecology and Management, vol. 148, no. 1–3, pp. 185–206, 2001.
34. D. Liebsch, M. C. M. Marques, and R. Goldenberg, "How long does the Atlantic Rain Forest take to recover after a disturbance? Changes in species composition and ecological features during secondary succession," Biological Conservation, vol. 141, no. 6, pp. 1717–1725, 2008.
35. M. E. Bowen, C. A. McAlpine, A. P. N. House, and G. C. Smith, "Regrowth forests on abandoned agricultural land: a review of their habitat values for recovering forest fauna," Biological Conservation, vol. 140, no. 3-4, pp. 273–296, 2007.
36. M. A. Zambrano, E. N. Broadbent, M. Schmink, S. G. Perz, and G. P. Asner, "Deforestation drivers in Southwest Amazonia: comparing smallholder farmers in Iñapari, Peru, and Assis Brasil, Brazil," Conservation & Society, vol. 8, no. 3, pp. 157–170, 2010.
37. Wijaya, V. Liesenberg, and R. Gloaguen, "Retrieval of forest attributes in complex successional forests of Central Indonesia: modeling and estimation of bitemporal data," Forest Ecology and Management, vol. 259, no. 12, pp. 2315–2326, 2010.
38. J. R. Kellner, D. B. Clark, and S. P. Hubbell, "Pervasive canopy dynamics produce short-term stability in a tropical rain forest landscape," Ecology Letters, vol. 12, no. 2, pp. 155–164, 2009.
39. T. Hudak, E. K. Strand, L. A. Vierling et al., "Quantifying aboveground forest carbon pools and fluxes from repeat LiDAR surveys," Remote Sensing of Environment, vol. 123, pp. 25–40, 2012.
40. M. Ryan, T. Hill, E. Woollen et al., "Quantifying small-scale deforestation and forest degradation in African woodlands using radar imagery," Global Change Biology, vol. 18, no. 1, pp. 243–257, 2012.
41. S. Antonarakis, S. S. Saatchi, R. L. Chazdon, and P. R. Moorcroft, "Using LiDAR and Radar measurements to constrain predictions of forest ecosystem structure and function," Ecological Applications, vol. 21, no. 4, pp. 1120–1137, 2011.

42. S. R. Baptista and T. K. Rudel, "A re-emerging Atlantic forest? Urbanization, industrialization and the forest transition in Santa Catarina, southern Brazil," Environmental Conservation, vol. 33, no. 3, pp. 195–202, 2006.
43. S. R. Baptista, "Metropolization and forest recovery in southern Brazil: a multiscale analysis of the Florianópolis city-region, Santa Catarina state, 1970 to 2005," Ecology and Society, vol. 13, no. 2, article 5, 2008.
44. R. A. Houghton, "Aboveground forest biomass and the global carbon balance," Global Change Biology, vol. 11, no. 6, pp. 945–958, 2005.
45. W. Liu, C. Song, T. A. Schroeder, and W. B. Cohen, "Predicting forest successional stages using multitemporal Landsat imagery with forest inventory and analysis data," International Journal of Remote Sensing, vol. 29, no. 13, pp. 3855–3872, 2008.
46. T. Neeff, R. M. Lucas, J. R. D. Santos, E. S. Brondizio, and C. C. Freitas, "Area and age of secondary forests in Brazilian Amazonia 1978–2002: an empirical estimate," Ecosystems, vol. 9, no. 4, pp. 609–623, 2006.
47. H. Sirén and E. S. Brondizio, "Detecting subtle land use change in tropical forests," Applied Geography, vol. 29, no. 2, pp. 201–211, 2009.
48. K. A. Dolan, G. C. Hurtt, J. Q. Chambers, R. O. Dubayah, S. Frolking, and J. G. Masek, "Using ICESat's Geoscience Laser Altimeter System (GLAS) to assess large-scale forest disturbance caused by hurricane Katrina," Remote Sensing of Environment, vol. 115, no. 1, pp. 86–96, 2011.
49. Tangki and N. A. Chappell, "Biomass variation across selectively logged forest within a 225-km^{2}region of Borneo and its prediction by Landsat TM," Forest Ecology and Management, vol. 256, no. 11, pp. 1960–1970, 2008.
50. T. Okuda, M. Suzuki, S. Numata et al., "Estimation of aboveground biomass in logged and primary lowland rainforests using 3-D photogrammetric analysis," Forest Ecology and Management, vol. 203, no. 1–3, pp. 63–75, 2004.
51. E. Busby, P. Vitousek, and R. Dirzo, "Prevalence of tree regeneration by sprouting and seeding along a rainfall gradient in Hawai'i," Biotropica, vol. 42, no. 1, pp. 80–86, 2010.
52. D. A. Clark, S. Brown, D. W. Kicklighter et al., "Net primary production in tropical forests: an evaluation and synthesis of existing field data," Ecological Applications, vol. 11, no. 2, pp. 371–384, 2001.
53. J. B. Drake, R. G. Knox, R. O. Dubayah et al., "Above-ground biomass estimation in closed canopy Neotropical forests using lidar remote sensing: factors affecting the generality of relationships," Global Ecology & Biogeography, vol. 12, no. 2, pp. 147–159, 2003.
54. L. F. Alves, S. A. Vieira, M. A. Scaranello et al., "Forest structure and live aboveground biomass variation along an elevational gradient of tropical Atlantic moist forest (Brazil)," Forest Ecology and Management, vol. 260, no. 5, pp. 679–691, 2010.
55. P. Asner, G. V. N. Powell, J. Mascaro et al., "High-resolution forest carbon stocks and emissions in the Amazon," Proceedings of the National Academy of Sciences of the United States of America, vol. 107, no. 38, pp. 16738–16742, 2010.
56. S. Brown and A. E. Lugo, "The storage and production of organic matter in tropical forests and their role in the global carbon cycle," Biotropica, vol. 14, no. 3, pp. 161–187, 1982.
57. IPCC, 2006 IPCC Guidelines for National Greenhouse Gas Inventories, National Greenhouse Gas Inventories Programme, Kanagawa, Japan, 2006, Edited by H.S. Eggleston, L. Buendia, K. Miwa, T. Ngara and K. Tanabe.
58. IPCC, "Expert meeting on national forest GHC inventories," in National Forest GHC Inventories—A Stock Taking, H. S. Eggleston, N. Srivastava, K. Tanabe, and J. Baasansuren, Eds., IGES, Hayama, Japan, 2010.
59. S. S. Saatchi, N. L. Harris, S. Brown et al., "Benchmark map of forest carbon stocks in tropical regions across three continents," Proceedings of the National Academy of Sciences of the United States of America, vol. 108, no. 24, pp. 9899–9904, 2011.
60. FAO, State of the World's Forests, Food and Agriculture Organization, Rome, Italy, 2011.

61. Freedman, G. Stinson, and P. Lacoul, "Carbon credits and the conservation of natural areas," Environmental Reviews, vol. 17, pp. 1–19, 2009.
62. M. Heindl and H. Winkler, "Vertical lek placement of forest-dwelling manakin species (Aves, Pipridae) is associated with vertical gradients of ambient light," Biological Journal of the Linnean Society, vol. 80, no. 4, pp. 647–658, 2003.
63. Tews, U. Brose, V. Grimm et al., "Animal species diversity driven by habitat heterogeneity/diversity: the importance of keystone structures," Journal of Biogeography, vol. 31, no. 1, pp. 79–92, 2004.
64. Fischer and D. L. Lindenmayer, "Beyond fragmentation: the continuum model for fauna research and conservation in human-modified landscapes," Oikos, vol. 112, no. 2, pp. 473–480, 2006.
65. T. M. Kuplich, "Classifying regenerating forest stages in Amazônia using remotely sensed images and a neural network," Forest Ecology and Management, vol. 234, no. 1–3, pp. 1–9, 2006.
66. S. G. Zolkos, S. J. Goetz, and R. Dubayah, "A meta-analysis of terrestrial aboveground biomass estimation using lidar remote sensing," Remote Sensing of Environment, vol. 128, pp. 289–298, 2013.
67. Pearlman, S. Carman, C. Segal, P. Jarecke, P. Clancy, and W. Browne, "Overview of the hyperion imaging spectrometer for the NASA EO-1 mission," in Proceedings of the IEEE International Geoscience and Remote Sensing Symposium (IGARSS '01), vol. 7, pp. 3036–3038, Sydney, Australia, July 2001.
68. J. Legleiter, W. A. Marcus, and R. L. Lawrence, "Effects of sensor resolution on mapping in-stream habitats," Photogrammetric Engineering & Remote Sensing, vol. 68, no. 8, pp. 801–807, 2002.
69. S. Thenkabail, E. A. Enclona, M. S. Ashton, C. Legg, and M. J. de Dieu, "Hyperion, IKONOS, ALI, and ETM+ sensors in the study of African rainforests," Remote Sensing of Environment, vol. 90, no. 1, pp. 23–43, 2004.
70. G. P. Asner, D. E. Knapp, J. Boardman et al., "Carnegie Airborne Observatory-2: increasing science data dimensionality via high-fidelity multi-sensor fusion," Remote Sensing of Environment, vol. 124, pp. 454–465, 2012.
71. J. B. Féret and G. P. Asner, "Semi-supervised methods to identify individual crowns of lowland tropical canopy species using imaging spectroscopy and LiDAR," Remote Sensing, vol. 4, no. 8, pp. 2457–2476, 2012.
72. L. Imhoff, "Theoretical analysis of the effect of forest structure on synthetic aperture radar backscatter and the remote sensing of biomass," IEEE Transactions on Geoscience and Remote Sensing, vol. 33, no. 2, pp. 341–352, 1995.
73. J. A. Anaya, E. Chuvieco, and A. Palacios-Orueta, "Aboveground biomass assessment in Colombia: a remote sensing approach," Forest Ecology and Management, vol. 257, no. 4, pp. 1237–1246, 2009.
74. F. Vermote, D. Tanré, J. L. Deuzé, M. Herman, and J.-J. Morcrette, "Second simulation of the satellite signal in the solar spectrum, 6s: an overview," IEEE Transactions on Geoscience and Remote Sensing, vol. 35, no. 3, pp. 675–686, 1997.
75. Latorre, O. A. C. Júnior, A. P. F. Carvalho, and Y. E. Shimabukuro, "Correção atmosférica: conceito e fundamentos," Espaço & Geografia, vol. 5, pp. 153–178, 2002.
76. Song and C. E. Woodcock, "Monitoring forest succession with multitemporal landsat images: factors of uncertainty," IEEE Transactions on Geoscience and Remote Sensing, vol. 41, no. 11, pp. 2557–2567, 2003.
77. S. Brown and A. E. Lugo, "Aboveground biomass estimates for tropical moist forests of the Brazilian Amazon," Interciência, vol. 17, pp. 8–18, 1992.
78. G. P. Asner, J. Mascaro, H. C. Muller-Landau et al., "A universal airborne LiDAR approach for tropical forest carbon mapping," Oecologia, vol. 168, no. 4, pp. 1147–1160, 2012. View at Publisher
79. Luckman, J. Baker, T. M. Kuplich, F. Y. Corina da Costa, and C. F. Alejandro, "A study of the relationship between radar backscatter and regenerating tropical forest

biomass for spaceborne SAR instruments," Remote Sensing of Environment, vol. 60, no. 1, pp. 1–13, 1997.

80. A. Houghton, K. T. Lawrence, J. L. Hackler, and S. Brown, "The spatial distribution of forest biomass in the Brazilian Amazon: a comparison of estimates," Global Change Biology, vol. 7, no. 7, pp. 731–746, 2001.
81. Tomppo, H. Olsson, G. Ståhl, M. Nilsson, O. Hagner, and M. Katila, "Combining national forest inventory field plots and remote sensing data for forest databases," Remote Sensing of Environment, vol. 112, no. 5, pp. 1982–1999, 2008.
82. P. Hubbell, "Tree dispersion, abundance, and diversity in a tropical dry forest," Science, vol. 203, no. 4387, pp. 1299–1309, 1979.
83. Saatchi, M. Marlier, R. L. Chazdon, D. B. Clark, and A. E. Russell, "Impact of spatial variability of tropical forest structure on radar estimation of aboveground biomass," Remote Sensing of Environment, vol. 115, no. 11, pp. 2836–2849, 2011.
84. W. Frazer, S. Magnussen, M. A. Wulder, and K. O. Niemann, "Simulated impact of sample plot size and co-registration error on the accuracy and uncertainty of LiDAR-derived estimates of forest stand biomass," Remote Sensing of Environment, vol. 115, no. 2, pp. 636–649, 2011.
85. A. Lindsey, J. D. Barton Jr., and S. R. Miles, "Field efficiencies of forest sampling methods," Ecology, vol. 39, no. 3, pp. 428–444, 1958.
86. D. Bitencourt, H. N. de Mesquita Jr., G. Kuntschik, H. R. da Rocha, and P. A. Furley, "Cerrado vegetation study using optical and radar remote sensing: two Brazilian case studies," Canadian Journal of Remote Sensing, vol. 33, no. 6, pp. 468–480, 2007.
87. J. M. Barbosa, I. Melendez-Pastor, J. Navarro-Pedreño, and M. D. Bitencourt, "Remotely sensed biomass over steep slopes: an evaluation among successional stands of the Atlantic Forest, Brazil," ISPRS Journal of Photogrammetry and Remote Sensing, vol. 88, pp. 91–100, 2014.
88. Li, P. Mausel, E. Brondizio, and D. Deardorff, "A framework for creating and validating a non-linear spectrum-biomass model to estimate the secondary succession biomass in moist tropical forests," ISPRS Journal of Photogrammetry and Remote Sensing, vol. 65, no. 2, pp. 241–254, 2010.
89. M. Wilson, J. A. Silander, A. Gelfand, and J. H. Glenn, "Scaling up: linking field data and remote sensing with a hierarchical model," International Journal of Geographical Information Science, vol. 25, no. 3, pp. 509–521, 2011.
90. A. King, S. J. Davies, S. Tan, and N. S. M. Noor, "The role of wood density and stem support costs in the growth and mortality of tropical trees," Journal of Ecology, vol. 94, no. 3, pp. 670–680, 2006.
91. M. D. Burger and W. B. C. Delitti, "Allometric models for estimating the phytomass of a secondary Atlantic Forest area of southeastern Brazil," Biota Neotropica, vol. 8, no. 4, pp. 131–136, 2008.
92. A. Vieira, L. F. Alves, M. Aidar et al., "Estimation of biomass and carbon stocks: the case of the Atlantic Forest," Biota Neotropica, vol. 8, no. 2, pp. 21–29, 2008.
93. W. Nelson, R. Mesquita, J. L. G. Pereira, S. G. A. de Souza, G. T. Batista, and L. B. Couto, "Allometric regressions for improved estimate of secondary forest biomass in the central Amazon," Forest Ecology and Management, vol. 117, no. 1–3, pp. 149–167, 1999.
94. N. Broadbent, G. P. Asner, M. Peña-Claros, M. Palace, and M. Soriano, "Spatial partitioning of biomass and diversity in a lowland Bolivian forest: linking field and remote sensing measurements," Forest Ecology and Management, vol. 255, no. 7, pp. 2602–2616, 2008.
95. Kajisa, T. Murakami, N. Mizoue, N. Top, and S. Yoshida, "Object-based forest biomass estimation using Landsat ETM+ in Kampong Thom Province, Cambodia," Journal of Forest Research, vol. 14, no. 4, pp. 203–211, 2009.
96. M. Bolker, Ecological models and data in R, Princeton University Press, Princeton, NJ, USA, 2008.

97. J. C. Ingram, T. P. Dawson, and R. J. Whittaker, "Mapping tropical forest structure in southeastern Madagascar using remote sensing and artificial neural networks," Remote Sensing of Environment, vol. 94, no. 4, pp. 491–507, 2005.
98. M. Foody, M. E. Cutler, J. McMorrow et al., "Mapping the biomass of Bornean tropical rain forest from remotely sensed data," Global Ecology & Biogeography, vol. 10, no. 4, pp. 379–387, 2001.
99. D. Riaño, E. Chuvieco, J. Salas, and I. Aguado, "Assessment of different topographic corrections in Landsat-TM data for mapping vegetation types (2003)," IEEE Transactions on Geoscience and Remote Sensing, vol. 41, no. 5, pp. 1056–1061, 2003.
100. Y. Gao and W. Zhang, "A simple empirical topographic correction method for ETM+ imagery," International Journal of Remote Sensing, vol. 30, no. 9, pp. 2259–2275, 2009.
101. M. A. Lefsky, W. B. Cohen, S. A. Acker, G. G. Parker, T. A. Spies, and D. Harding, "Lidar remote sensing of the canopy structure and biophysical properties of Douglas-Fir Western Hemlock Forests," Remote Sensing of Environment, vol. 70, no. 3, pp. 339–361, 1999.
102. Köhler and A. Huth, "Towards ground-truthing of spaceborne estimates of above-ground life biomass and leaf area index in tropical rain forests," Biogeosciences, vol. 7, no. 8, pp. 2531–2543, 2010.
103. J. Chave, C. Andalo, S. Brown et al., "Tree allometry and improved estimation of carbon stocks and balance in tropical forests," Oecologia, vol. 145, no. 1, pp. 87–99, 2005.
104. J. R. Santos, C. C. Freitas, L. S. Araujo et al., "Airborne P-band SAR applied to the aboveground biomass studies in the Brazilian tropical rainforest," Remote Sensing of Environment, vol. 87, no. 4, pp. 482–493, 2003.
105. D. H. Hoekman and M. J. Quiñones, "Land cover type and biomass classification using AirSAR data for evaluation of monitoring scenarios in the Colombian Amazon," IEEE Transactions on Geoscience and Remote Sensing, vol. 38, no. 2 I, pp. 685–696, 2000.
106. J. P. Arroyo-Mora, G. A. Sánchez-Azofeifa, M. E. R. Kalacska, B. Rivard, J. C. Calvo-Alvarado, and D. H. Janzen, "Secondary forest detection in a Neotropical dry forest landscape using Landsat 7 ETM+ and IKONOS imagery," Biotropica, vol. 37, no. 4, pp. 497–507, 2005.
107. D. A. Coomes, R. B. Allen, N. A. Scott, C. Goulding, and P. Beets, "Designing systems to monitor carbon stocks in forests and shrublands," Forest Ecology and Management, vol. 164, no. 1–3, pp. 89–108, 2002.
108. C. Hurtt, R. Dubayah, J. Drake et al., "Beyond potential vegetation: combining lidar data and a height-structured model for carbon studies," Ecological Applications, vol. 14, no. 3, pp. 873–883, 2004
109. C. Morel, S. S. Saatchi, Y. Malhi et al., "Estimating aboveground biomass in forest and oil palm plantation in Sabah, Malaysian Borneo using ALOS PALSAR data," Forest Ecology and Management, vol. 262, no. 9, pp. 1786–1798, 2011.
110. S. Descloux, V. Chanudet, H. Poilvé, and A. Grégoire, "Co-assessment of biomass and soil organic carbon stocks in a future reservoir area located in Southeast Asia," Environmental Monitoring and Assessment, vol. 173, no. 1–4, pp. 723–741, 2011.
111. J. B. Drake, R. O. Dubayah, D. B. Clark et al., "Estimation of tropical forest structural characteristics, using large-footprint lidar," Remote Sensing of Environment, vol. 79, no. 2-3, pp. 305–319, 2002.
112. S. Eckert, H. R. Ratsimba, L. O. Rakotondrasoa, L. G. Rajoelison, and A. Ehrensperger, "Deforestation and forest degradation monitoring and assessment of biomass and carbon stock of lowland rainforest in the Analanjirofo region, Madagascar," Forest Ecology and Management, vol. 262, no. 11, pp. 1996–2007, 2011.
113. D. Lu, P. Mausel, E. Brondízio, and E. Moran, "Relationships between forest stand parameters and Landsat TM spectral responses in the Brazilian Amazon Basin," Forest Ecology and Management, vol. 198, no. 1–3, pp. 149–167, 2004.

114. Proisy, P. Couteron, and F. Fromard, "Predicting and mapping mangrove biomass from canopy grain analysis using Fourier-based textural ordination of IKONOS images," Remote Sensing of Environment, vol. 109, no. 3, pp. 379–392, 2007.
115. da Silva Narvaes, A. de Queiroz da Silva, and J. R. dos Santos, "Evaluation of the interaction between SAR L-band signal and structural parameters of forest cover," in Proceedings of the IEEE International Geoscience and Remote Sensing Symposium (IGARSS '07), pp. 1607–1610, Barcelona, Spain, June 2007.
116. M. Simard, V. H. Rivera-Monroy, J. E. Mancera-Pineda, E. Castañeda-Moya, and R. R. Twilley, "A systematic method for 3D mapping of mangrove forests based on Shuttle Radar Topography Mission elevation data, ICEsat/GLAS waveforms and field data: application to Ciénaga Grande de Santa Marta, Colombia," Remote Sensing of Environment, vol. 112, no. 5, pp. 2131–2144, 2008.
117. F. J. Ponzoni, L. S. Galvao, V. Liesenberg, and J. R. Santos, "Impact of multi-angular CHRIS/PROBA data on their empirical relationships with tropical forest biomass," International Journal of Remote Sensing, vol. 31, no. 19, pp. 5257–5273, 2010.
118. E. Nichol and M. L. R. Sarker, "Improved biomass estimation using the texture parameters of two high-resolution optical sensors," IEEE Transactions on Geoscience and Remote Sensing, vol. 49, no. 3, pp. 930–948, 2011.
119. L. R. Sarker and J. E. Nichol, "Improved forest biomass estimates using ALOS AVNIR-2 texture indices," Remote Sensing of Environment, vol. 115, no. 4, pp. 968–977, 2011.
120. Luckman, J. Baker, T. M. Kuplich, F. Y. Corina da Costa, and C. F. Alejandro, "A study of the relationship between radar backscatter and regenerating tropical forest biomass for spaceborne SAR instruments," Remote Sensing of Environment, vol. 60, no. 1, pp. 1–13, 1997.
121. H. T. Minh, T. le Toan, F. Rocca, S. Tebaldini, M. M. D'Alessandro, and L. Villard, "Relating P-band synthetic aperture radar tomography to tropical forest biomass," IEEE Transactions on Geoscience and Remote Sensing, vol. 52, no. 2, pp. 967–979, 2014.
122. Ploton, R. Pélissier, C. Proisy et al., "Assessing aboveground tropical forest biomass using Google Earth canopy images," Ecological Applications, vol. 22, no. 3, pp. 993–1003, 2012.
123. M. Foody and P. J. Curran, "Estimation of tropical forest extent and regenerative stage using remotely sensed data," Journal of Biogeography, vol. 21, no. 3, pp. 223–244, 1994.
124. G. M. Foody, R. M. Green, R. M. Lucas, P. J. Curran, M. Honzak, and I. do Amaral, "Observations on the relationship between SIR-C radar backscatter and the biomass of regenerating tropical forests," International Journal of Remote Sensing, vol. 18, no. 3, pp. 687–694, 1997.
125. S. Boyd, G. M. Foody, and P. J. Curran, "The relationship between the biomass of Cameroonian tropical forests and radiation reflected in middle infrared wavelengths (3.0–5.0 μm)," International Journal of Remote Sensing, vol. 20, no. 5, pp. 1017–1023, 1999.
126. A. Salas, M. J. Ducey, E. Rignot, and D. Skole, "Assessment of JERS-1 SAR for monitoring secondary vegetation in Amazonia: I. Spatial and temporal variability in backscatter across a chrono-sequence of secondary vegetation stands in Rondonia," International Journal of Remote Sensing, vol. 23, no. 7, pp. 1357–1379, 2002.
127. T. Neeff, L. V. Dutra, J. R. dos Santos, C. da Costa Freitas, and L. S. Araujo, "Tropical forest measurement by interferometric height modeling and P-band radar backscatter," Forest Science, vol. 51, no. 6, pp. 585–594, 2005.
128. M. A. Castillo-Santiago, M. Ricker, and B. H. J. de Jong, "Estimation of tropical forest structure from SPOT-5 satellite images," International Journal of Remote Sensing, vol. 31, no. 10, pp. 2767–2782, 2010.

129. M. L. Clark, D. A. Roberts, J. J. Ewel, and D. B. Clark, "Estimation of tropical rain forest aboveground biomass with small-footprint LiDAR and hyperspectral sensors," Remote Sensing of Environment, vol. 115, no. 11, pp. 2931–2942, 2011.

130. C. Baraloto, S. Rabaud, Q. Molto et al., "Disentangling stand and environmental correlates of aboveground biomass in Amazonian forests," Global Change Biology, vol. 17, no. 8, pp. 2677–2688, 2011.

131. G. P. Asner and J. Mascaro, "Mapping tropical forest carbon: calibrating plot estimates to a simple LiDAR metric," Remote Sensing of Environment, vol. 140, pp. 614–624, 2014.

132. M. L. R. Sarker, J. Nichol, B. Ahmad, I. Busu, and A. A. Rahman, "Potential of texture measurements of two-date dual polarization PALSAR data for the improvement of forest biomass estimation," ISPRS Journal of Photogrammetry and Remote Sensing, vol. 69, pp. 146–166, 2012.

Index